**DRILL HALL LIBRARY
MEDWAY**

Nuclear Analytical Techniques for Metallomics and Metalloproteomics

Nuclear Analytical Techniques for Metallomics and Metalloproteomics

Edited by

Chunying Chen
National Center for Nanoscience and Technology, Beijing 100190, P. R. China

Zhifang Chai and Yuxi Gao
Key Lab of Nuclear Analytical Techniques, Institute of High Energy Physics, Chinese Academy of Sciences, Beijing 100049, P. R. China

RSCPublishing

ISBN: 978-1-84755-901-2

A catalogue record for this book is available from the British Library

© Royal Society of Chemistry 2010

All rights reserved

Apart from fair dealing for the purposes of research for non-commercial purposes or for private study, criticism or review, as permitted under the Copyright, Designs and Patents Act 1988 and the Copyright and Related Rights Regulations 2003, this publication may not be reproduced, stored or transmitted, in any form or by any means, without the prior permission in writing of The Royal Society of Chemistry or the copyright owner, or in the case of reproduction in accordance with the terms of licences issued by the Copyright Licensing Agency in the UK, or in accordance with the terms of the licences issued by the appropriate Reproduction Rights Organization outside the UK. Enquiries concerning reproduction outside the terms stated here should be sent to The Royal Society of Chemistry at the address printed on this page.

The RSC is not responsible for individual opinions expressed in this work

Published by The Royal Society of Chemistry,
Thomas Graham House, Science Park, Milton Road,
Cambridge CB4 0WF, UK

Registered Charity Number 207890

For further information see our website at www.rsc.org

Foreword

The systematic acquisition of information relevant to genomes, gene transcripts, and proteins and their global and comprehensive analysis (genomics and proteomics), has been a fundamental contribution of analytical chemistry to the progress of research in life sciences. The functions of many proteins, referred to as metalloproteins, depends critically on their interaction with a metal, usually copper, iron, molybdenum, or zinc. There are proteins which are expressed as a defense mechanism of an organism against heavy metal stress, and others which serve within an organism as transporters of essential nutrient ions, contaminants and metal probes. The role of the pool of metal-binding metabolism products of enzymatic or biochemical reactions is critical in mechanisms by which a metal is sensed, stored or incorporated as a co-factor in a cell and by which an organism must regulate a proper assimilation and incorporation of trace metals. Therefore, the chemistry of a cell needs to be characterized not only by its characteristic genome in the nucleus and protein content, a proteome, but also by the distribution of the metals and metalloids among the different species and cell compartments, the metallome.

The systematic study of interactions and functional connections of metal ions and their species with genes, proteins, metabolites, and other biomolecules within organisms and ecosystems has given birth to the field of metallomics as a complement to genomics and proteomics. Metallomics is a transdisciplinary research area with an impact on geochemistry, clinical biology and pharmacology, plant and animal physiology, and nutrition. Its ultimate goal is to provide a global and systematic understanding of the metal uptake, trafficking, role and excretion in biological systems. The most important area of metallomics, metalloproteomics, deals with the detection, identification, and distribution of the entirety of metalloproteins (metalloproteome), their structural and functional characterisation and the description of the metal-coordination environments.

The progress in metallomics and metalloproteomics is critically dependent on advances in methods for *in silico* (bioinformatics), *in vitro* and *in vivo* analysis bringing information on the identity, concentration, and localisation of elements and their species. Nuclear analytical techniques are a basic instrumental toolbox enabling and facilitating the acquisition of the relevant analytical information. They include, in the broader sense, not only neutron activation, scattering and diffraction and Mössbauer spectroscopy, but also X-ray fluorescence, particle-induced X-ray emission (PIXE) and absorption (EXAFS and XANES) spectrometry, and, last but not least, elemental mass spectrometry with isotopic resolution (ICP MS) and are often accompanied by high-resolution separation techniques for the species resolution.

It is my great pleasure to introduce the first monographic book which addresses the key aspects of applications of nuclear analytical techniques to metallomics and metalloproteomics studies. The book is a collection of chapters written by eminent Chinese experts addressing the different analytical techniques and their applications. A prominent place is devoted to X-ray techniques and their aspects for multielement analysis with local resolution, structural analysis of metalloproteins and oxidation state and metal coordination environment in metalloproteins and metal-containing metabolites. Multielemental detection in chromatography and gel electrophoresis is discussed in the context of NAA (molecular NAA) pioneered for bio-inorganic speciation studies in the editor's laboratory. The discussion of the state-of-the-art and potential contribution of ICP MS for isotopic analysis of metal-containing biomolecules completes the panorama of instrumental techniques. Two comprehensive chapters are dedicated to particular application areas of interaction of iron and metallodrugs with biomolecules. The picture of recent research on metals in biology is completed by issues related to the characterisation of metal-based nanomaterials, their interactions with biological molecules and toxicology. The coverage of all the topics is, on one hand, sufficiently didactic to be useful for post-graduate students and newcomers to the field of analytical metalloproteomics and metallomics, and, on the other hand, sufficiently detailed to present some unknown facets of these emerging areas to scientists already involved in this fascinating research.

Ryszard Lobinski
Research Director at the CNRS
Laboratory of Analytical Bioinorganic and Environmental Chemistry, Pau

Preface

The intent of this book is to provide readers with a comprehensive view of application of advanced nuclear and relevant analytical techniques for a new scientific frontier "metallomics and metalloproteomics", which are becoming the hot topics in bioanalytical chemistry and life sciences.

This book is comprised of 11 chapters which are written by experts from disciplines as diverse as analytical chemistry, nuclear analytical chemistry, environmental science, molecular biology and medicinal chemistry in order to identify potential "hot spots" of metallomics and metalloproteomics. Various nuclear analytical techniques, such as neutron activation analysis, X-ray fluorescence, isotope tracer, Mössbauer spectrometry, X-ray absorption spectrometry, and neutron scattering and diffraction, are powerful tools to study a number of the basic issues in metallomics and metalloproteomics. The hybrid techniques with HPLC and other separation techniques have been involved as the coupling of a powerful selective separation with a sensitive and element-specific detection. Nuclear analytical techniques provide useful information about both chemical species and structural characterization of metalloproteins and metals in biological systems. Therefore, the scientific fundamentals of new approaches, like isotopic techniques combined with ICP-MS/ESI-MS/MS, the synchrotron radiation-based techniques, X-ray absorption spectroscopy, X-ray diffraction and neutron scattering, as well as their various applications with a focus on some selected elements, *e.g.* mercury, selenium, chromium, arsenic, iron, and metal-based medicines are critically reviewed, which can help understand their impacts on human health. Thus, it will be of particular interest to researchers, scientists, engineers, post-graduate students and newcomers working in the fields of environmental and industrial chemistry, biochemistry, nutrition, toxicology and medicine.

It is evident that this is a multidisciplinary research field. The metallomics and metalloproteomics studies will not only be beneficial to professional scientists and graduate students in the above fields, but also be hot in medical fields to find the reason how diseases happen. Thus, besides providing useful information of related methods and techniques, this book will be beneficial to diagnosis of the certain diseases and help develop the new drugs based on the targeted metalloproteins and genes.

It is foreseeable that for chemical speciation and -omics studies nuclear analytical techniques with their advanced characteristics of high sensitivity, good accuracy, high space and time resolution, and suitability for on-line and *in situ* measurement will benefit from the tremendous development of science and technology in the new century. The advancement of nuclear analytical techniques, including the improvement of the state of the art of HTXAS, as well as the development of new generation of spallation neutron source will greatly facilitate the metallomics and metalloproteomics studies in the near future. The major limitation of nuclear analytical techniques is relatively difficult access to large facilities and occupational protection to irradiation along with the running costs and radioactivity hazards.

As a key lab of Chinese Academy of Sciences, most of the important nuclear analytical techniques are available and also open to external users from various disciplines. In addition, the synchrotron radiation facility in our institute provides useful beamlines for metallomics and metalloproteomics-related researches. Therefore, based on these convenient conditions, we have developed related methodology and application over the past 20 years. In this book, we outline our main achievements and current progress in the metallomics and metalloproteomics on several important elements (Hg, Se, As, REEs, *etc.*) and nanomaterials made within the last 10 years.

We have carried out extensive investigations from bulk analysis and speciation studies of trace elements to the new emerging filed of metallomics and metalloproteomics in biological and environmental sciences by nuclear and related analytical techniques. We express our special appreciation to the Natural Science Foundation of China (NSFC) for the continuously financial supports from the 1980s, such as NSFC major projects: Molecular toxicology of several typical environmental pollutants by advanced nuclear analytical techniques (*e.g.* 2004–2007, No. 10490180 and 1998–2003, No. 19935020) and major projects approved by Chinese Academy of Sciences, such as "Application of Advanced Nuclear Analytical Techniques in the Environmental Sciences" in the years 2001–2004. The authors also thank to a research project titled "Tracking Metals in Cells by Metallomics: Insights into the Metal-associated Pathophysiological Processes" under the NSFC/RGC Joint Research Scheme from the RGC in Hong Kong and NSFC in the mainland, respectively, which started in 2009 (No. 20931160430). The authors acknowledge the financial support from the Ministry of Science and Technology of China (MOST) in recent years. We initiate and extend this research field of metallomic and metalloproteomics to a fast developing research area as nanoscience and nanotechnology under the support by the National Key Basic Research Program of China (973 Program) as titled "Health and Safety Impacts of Nanotechnology Exploring Solutions" from MOST. The systematic study of metallic nanomaterials, nanometallomics, is therefore proposed for the first time.

Preface

We would like to thank all contributing authors for their hard work and great effort they have dedicated to this book. Dr Ryszard Lobinski is highly indebted for his promise to write the Foreword. We would like to express our sincere appreciation to Dr. Merlin Fox, Commissioning Editor and Rosalind Shattock, Commissioning Administrator from the Royal Society of Chemistry. Without their kind help and support, the publication of this book would be truly impossible.

Chunying Chen and Zhifang Chai
Beijing, China

Contents

List of Abbrevations	xix
About the Editors	xxv
List of Contributors	xxvii

Chapter 1	Introduction			1
	Ying Qu, Yu-Feng Li, Ru Bai, Chunying Chen and Zhifang Chai			
	1.1	Background		1
	1.2	Metallomics and Metalloproteomics		2
		1.2.1	Trace Elements, Chemical Species and Speciation Analysis	2
		1.2.2	Metallomics	6
		1.2.3	Metalloproteomics	11
		1.2.4	Relationships Among Metallomics and Other "-omics"	15
	1.3	Nuclear Analytical Techniques		15
		1.3.1	General Analytical Approaches in Metallomics and Metalloproteomics	15
		1.3.2	Specific Nuclear Analytical Techniques	17
	1.4	Applications of Advanced Nuclear Analytical Techniques for Metallomics and Metalloproteomics		20
		1.4.1	Nuclear Analytical Techniques for Multielemental Quantification	20
		1.4.2	Nuclear Analytical Techniques for Metallome and Metalloproteome Distribution	21
		1.4.3	Nuclear Analytical Techniques for the Structural Analysis of Metallomes and Metalloproteomes	23

Nuclear Analytical Techniques for Metallomics and Metalloproteomics
Edited by Chunying Chen, Zhifang Chai and Yuxi Gao
© Royal Society of Chemistry 2010
Published by the Royal Society of Chemistry, www.rsc.org

1.5	Purposes of the Book and Layout of the Chapters	29
1.6	Outlook	36
References		37

Chapter 2 Neutron Activation Analysis — 44
Zhiyong Zhang

2.1	Introduction	44
	2.1.1 Activation and Analysis	45
	2.1.2 Sensitivities	47
	2.1.3 Strengths and Limitations	42
2.2	Development of NAA for Metallomics and Metalloproteomics	48
	2.2.1 Differential Centrifugation–NAA	48
	2.2.2 Liquid Chromatography–NAA	50
	2.2.3 Electrophoresis–NAA	53
2.3	Conclusion	59
References		59

Chapter 3 X-ray Fluorescence — 62
Yuxi Gao

3.1	Background	62
3.2	The Physics of X-ray Fluorescence	63
3.3	XRF Facilities	64
	3.3.1 Primary Radiation Source	65
	3.3.2 Optics	67
	3.3.3 Detectors	68
	3.3.4 Electronics	69
	3.3.5 Wavelength Dispersive XRF	70
3.4	Analytical Procedures with EDXRF	71
	3.4.1 Sample Preparation	71
	3.4.2 Element Measuring	71
	3.4.3 Data Processing	72
	3.4.4 Sensitivity, Limit of detection, and Precision	73
3.5	Other XRF Techniques	75
	3.5.1 Micro-XRF	75
	3.5.2 Total Reflection XRF	75
	3.5.3 Grazing-exit XRF	77
3.6	Applications of EDXRF in Metallomics and Metalloproteomics	77
	3.6.1 XRF Methods Used for Elemental Analysis in Protein Fractions after Biochemical Separation	79
	3.6.2 Electrophoresis and Sample Preparation for SRXRF Measurement	85

		3.6.3	XRF as an On-line Detector of Capillary Electrophoresis and Other Separation Techniques	88
	3.7	Outlook and Challenges		90
	References			91

Chapter 4 Isotopic Techniques Combined with ICP-MS and ESI-MS 95
Meng Wang, Weiyue Feng and Zhifang Chai

	4.1	Inductively Coupled Plasma–Mass Spectrometry		95
		4.1.1	Introduction	95
		4.1.2	ICP as a High-temperature Ionization Source	96
		4.1.3	Mass Analyzers for ICP-MS	97
		4.1.4	ICP-MS Coupled Techniques	98
	4.2	ESI-MS		99
		4.2.1	Introduction	99
		4.2.2	Electrospray and Related Ionization Techniques	101
	4.3	Isotopic Tracer Techniques Combined with ICP-MS in the Study of Metallomics		103
		4.3.1	Introduction	103
		4.3.2	Examples of Applications	105
	4.4	Isotope Dilution Analysis in the Quantitative Study of Proteins		107
		4.4.1	Introduction	107
		4.4.2	Species-specific Method	108
		4.4.3	Species-unspecific Method	110
	4.5	Isotope Tagging and Labeling Techniques for Protein Quantification		115
		4.5.1	Introduction	115
		4.5.2	Chemical Labeling Methods for Protein Quantification	116
		4.5.3	Metabolic Labeling for Protein Quantification	120
		4.5.4	Hetero-elements used as Elemental Tags	121
	4.6	Conclusions		122
	References			122

Chapter 5 Mössbauer Spectroscopy 128
Yang Qiu and Chunying Chen

	5.1	Introduction and Fundamentals		128
		5.1.1	The Mössbauer Effect	128
		5.1.2	Hyperfine Structure	132

	5.2	Equipment and Experiments	137
		5.2.1 Equipment for a Mössbauer Spectrometer	137
		5.2.2. The Mössbauer Spectrum	140
	5.3	Applications for Chemical Speciation and Metalloproteins	143
		5.3.1 Sample Preparation	144
		5.3.2 Some Examples of Mössbauer Spectroscopy Applied to Metalloprotein Studies	144
		5.3.3 Mössbauer Spectroscopy for the Study of Elemental Speciation	153
	5.4	Other Techniques for Hyperfine Interaction Studies	154
		5.4.1 Nuclear Resonance Spectroscopy	154
		5.4.2 Electron Paramagnetic Resonance	155
		5.4.3 Resonance Raman Spectroscopy	157
	5.5	Limitations and Conclusions	159
	References		159
Chapter 6	**X-ray Absorption Spectroscopy**		**163**
	Yu-Feng Li and Chunying Chen		
	6.1	Introduction	163
	6.2	Basics of X-Ray Absorption Spectroscopy	164
		6.2.1 X-ray Absorption and Fluorescence	164
		6.2.2 Samples and Sample Preparation	168
		6.2.3 XAS Measurement	168
		6.2.4 Data Analysis	170
	6.3	Application of XAS in Metallomics and Metalloproteomics	172
		6.3.1 Fingerprints Studies and Quantitative Speciation by XANES	172
		6.3.2 Fingerprints and Structural Information by EXAFS	177
		6.3.3 Micro-XAS in Metallomics and Metalloproteomics	179
	6.4	Combination of XAS and other Techniques in Metallomics and Metalloproteomics	185
		6.4.1 Combination of XAS with Separation Techniques in Metallomics and Metalloproteomics	186
		6.4.2 Combination of XAS with XRF in Metallomics and Metalloproteomics	186
		6.4.3 Combination of XAS with Protein Crystallography in Metallomics and Metalloproteomics	188

	6.4.4	Combination of XAS with Computational Chemistry in Metallomics and Metalloproteomics	191
	6.4.5	Combination of XAS with Neutron Scattering in Metallomics and Metalloproteomics	192
	6.4.6	Combination of XAS with Circular Dichroism in Metallomics and Metalloproteomics	194
	6.4.7	Combination of XAS with Nuclear Magnetic Resonance in Metallomics and Metalloproteomics	196
	6.4.8	Combination of XAS with Raman Spectroscopy in Metallomics and Metalloproteomics	197
	6.4.9	Combination of XAS with Electron Spin Resonance in Metallomics and Metalloproteomics	199
6.5	Conclusions and Outlook	201
References		206

Chapter 7 Protein Crystallography for Metalloproteins 212
Zengqiang Gao, Haifeng Hou and Yuhui Dong

7.1	Introduction	212
7.2	Structure Determination by Protein Crystallography	214
7.3	Structure Determination Using the Multi-wavelength Anomalous Dispersion Method	225
	7.3.1 Theoretical Background	225
	7.3.2 Experimental Strategies	226
7.4	Crystal Structure Made Clear: Structure and Function of SmdCD	229
References		235

Chapter 8 Applications of Nuclear Analytical Techniques for Iron-omics Studies 239
Guangjun Nie, Motao Zhu and Bo Ning

8.1	Chemistry of Iron	239
8.2	Physiology of Iron	240
8.3	Cellular and Systemic Iron Metabolism Regulation	240
	8.3.1 Cellular Iron Uptake	240
	8.3.2 Iron Storage	241
	8.3.3 Coordination of Iron Uptake and Storage	242
	8.3.4 Iron Hormone Regulates Systemic Iron Homeostasis	243
8.4	Mitochondrial Iron Metabolism	245

8.5	Molecular Mechanism of Impaired Iron Homeostasis in Neurodegenerative Disorders		247
	8.5.1	Iron Dysregulation and Neurodegenerative Diseases	247
	8.5.2	Iron Metabolism in the Central Nervous System	248
	8.5.3	Participation of Iron in Neurodegenerative Diseases	248
8.6	Examples of Nuclear Analytical Techniques in Iron Metabolism Studies		250
	8.6.1	Synchronous Radiation-based Analytical Techniques	250
	8.6.2	Particle-induced X-ray Emission	251
	8.6.3	Neutron Activation Analysis	255
	8.6.4	Radioactive and Enriched Stable Isotope-based Techniques	256
	8.6.5	Mössbauer Spectroscopy	257
	8.6.6	Speciation Analysis by Pre-separation Procedures in Combination with Nuclear Analytical Techniques	259
References			260

Chapter 9 Nuclear-based Metallomics in Metal-based Drugs 265
Ruiguang Ge, Ivan K. Chu and Hongzhe Sun

9.1	Introduction		265
9.2	Cellular Distribution and Metabolism of Metallodrugs		267
	9.2.1	Hydrolysis of Platinum Compounds	267
	9.2.2	Cellular Localization of Metallodrugs	271
	9.2.3	Pharmacokinetics of Metallodrugs	276
9.3	Metallodrug–Biomolecule Interactions		278
	9.3.1	Platinated-DNA Adducts	279
	9.3.2	Metallodrug–Protein Interactions	282
	9.3.3	Platinated DNA–Protein Interactions	288
9.4	Conclusions and Perspectives		290
References			291

Chapter 10 Application of Integrated Techniques for Micro- and Nano-imaging Towards the Study of Metallomics and Metalloproteomics in Biological Systems 299
Lili Zhang and Chunying Chen

10.1	Introduction	299
10.2	X-ray Fluorescence	300

	10.2.1	Environmental Science	302
	10.2.2	Life Science	309
10.3	Particle Induced X-ray Emission		320
	10.3.1	Plants	322
	10.3.2	Animals	325
	10.3.3	Cellular Imaging	326
10.4	Mass Spectrometry Imaging		328
	10.4.1	Matrix-assisted Laser Desorption Ionization Mass Spectrometry	329
	10.4.2	Secondary Ion Mass Spectrometry	329
	10.4.3	Laser Ablation Inductively Coupled Plasma Mass Spectrometry	330
	10.4.4	Near-field LA-ICP-MS: A Novel Elemental Analytical Technique for Nano-imaging	333
10.5	Tomography		334
10.6	Conclusions		337
References			337

Chapter 11 Nuclear-based Metallomics in Metallic Nanomaterials: Nanometallomics — 342
Yu-Feng Li, Liming Wang, Lili Zhang and Chunying Chen

11.1	Introduction		342
11.2	Nanometallomics and its Study Area		344
11.3	Nuclear Analytical Techniques for Characterization of Metallic Nanomaterials		344
	11.3.1	Size Characterization	345
	11.3.2	Oxidation State Analysis	346
	11.3.3	Electronic Configuration and Coordination Geometry	348
11.4	Quantification and Distribution of Metallic Nanomaterials		352
	11.4.1	Neutron Activation Analysis for Quantification	352
	11.4.2	ICP-MS for Quantification	356
	11.4.3	Distribution of Metallic Nanomaterials in Biological Systems	357
11.5	Structural Analysis for the Bio-nano Interaction		372
11.6.	Conclusion and Outlook		379
References			381

Subject Index — 385

List of Abbreviations

AAS	Atomic absorption spectrometry
AD	Alzheimer's disease
ADMET	Absorption, distribution, metabolism, excretion, and toxicity
AEY	Auger electron yield
AMT	Alpha-methyltyrosine
AMU	Atomic mass unit
API	Atmospheric pressure ionization
APS	Advanced photon source
ANL	Argonne National Laboratory
BBB	Blood–brain barrier
BSA	Bovine serum albumin
CaM	Calmodulin
CBDCA	Cyclobutanedicarboxylate
CC	Collision cell
CD	Circular dichroism
CDA	Cytidine deaminase
cDTPA	Cyclic anhydride diethylenetriaminepentaacetic
CE	Capillary electrophoresis
cICAT	Cleavable ICAT
CID	Collision-induced dissociation
CNS	Central nervous system
CNTs	Carbon nanotubes
CODH	Carbon monoxide dehydrogenase
Cp	Ceruloplasmin
CPO	Chloroperoxidase
Cryo-EM	Cryo-electron microscope
CW ENDOR	Continuous wave electron nuclear double resonance
dCMP	Deoxycytidine-5'-monophosphate
dCTP	Deoxycytidine-5'-triphosphate
DFT	Density functional theory
DIHEN	Direct-injection high-efficiency nebulizer
DL	Detection limit

DLS	Dynamic light scattering
DMA	Dimethylarsinic acid
DMSA	Meso-2,3-dimercaptosuccinic acid
DMSF	Dimethylsulfoxide
DOTA	1,4,7,10-tetraazacyclododecane-N,N',N'',N'''-tetraacetic acid
DRC	Dynamic reaction cell
DTPA	Diethylenetriamine-N,N,N,N-pentaacetic acid
dTTP	Deoxythymidine-5'-triphosphate
DGNAA	Delayed gamma-ray neutron activation analysis
dUMP	Deoxyuridine-5'-monophosphate
dUTP	Deoxyuridine-5'-triphosphate
eALAS	Erythroid-specific 5-aminolevulinic acid synthase
EAP	Experimental analytical precision
ECAT	Element-coded affinity tag
ED	Energy-dispersive
EDM	Energy dispersive monochromator
EDXRF	Energy-dispersive X-ray fluorescence
EFG	Electric field gradient
EIMS	Electron impact mass spectrometry
ENPs	Engineered nanoparticles
EPR	Electron paramagnetic resonance
ESI	Electrospray ionization
ESI-MS	Electrospray ionization–mass spectrometry
ESR	Electron spin resonance
ESRF	European Synchrotron Radiation Facility
EVISA	European Virtual Institute of Speciation Analysis
EXAFS	Extended X-ray absorption fine structure
FDH_H	Formate dehydrogenase H
Fe-S	Iron–sulfur
FeMoco	Iron–molybdenum cofactor
FP	Fundamental parameters
FUR	Ferric uptake regulator
FY	Fluorescence yield
GC	Gas chromatograph
GC-MS	Gas chromatography/mass spectrometry
GD	Guanine deaminase
GE	Gel electrophoresis
GE-XRF	Grazing-exit XRF
Hb	Hemoglobin
HCC	Hepatocellular carcinoma
HEWL	Hen egg white lysozyme
HiPIP	High-potential iron–sulfur protein
HMG	High mobility group
HPLC	High-performance liquid chromatography
HRP	Horseradish peroxidase

List of Abbreviations

HO-1	Heme oxygenase 1
HOs	Heme oxygenases
HSA	Human serum albumin
HT	High-throughput
HTXAS	High-throughput X-ray absorption spectroscopy
ICAT	Isotope-coded affinity tags
ICP-IDMS	Inductively coupled plasma–isotope dilution mass spectrometry
ICP-MS	Inductively coupled plasma mass spectrometry
ICP-OES	Inductively coupled plasma–optical emission spectroscopy
ICPL	Isotope-coded protein labeling
ID	Instilled dose
IDA	Isotope dilution analysis
IE	Ion exchange
IEF	Isoelectric focusing
IEF-PAGE	Isoelectrofocusing PAGE
IMAC	Immobilized metal affinity chromatography
INAA	Instrumental neutron activation analysis
IPGs	Immobilised pH gradients
IREs	Iron response elements
IRPs	Iron regulatory proteins
IUPAC	International Union of Pure and Applied Chemistry
JAERI	Japan Atomic Energy Research Institute
KED	Kinetic energy discrimination
LA	Laser ablation
LA-ICP-MS	Laser ablation inductively coupled plasma mass spectrometry
LAXS	Large-angle X-ray scattering
LC	Liquid chromatography
LC-XANES	Linear combination X-ray absorption near edge structure
LD	Limits of detection
LIP	Labile iron pool
MAD	Multi-wavelength anomalous dispersion
MALDI-MS	Matrix-assisted laser desorption/ionization mass spectrometry
MALDI-TOF	Matrix-assisted laser desorption ionization-time of flight
MALDI-TOF-MS	Matrix-assisted laser desorption/ionization time of flight mass spectrometer
MALDI	Matrix-assisted laser desorption ionization
MCA	Multi-channel analyzer
MCR	Methyl-coenzyme M reductase
MD	Molecular dynamics
MECT	Metal element chelate tag

MeCAT	Metal-coded affinity tag
micro-XANES	Microscopic X-ray absorption near-edge structure spectroscopy
micro-XRF	Microscopic X-ray fluorescence analysis
MIMOS	Miniaturized Mössbauer spectrometer
MM	Molecular mechanics
MMA	Monomethyl arsonic acid
Mn-AMPP	Manganese(II)-activated aminopeptidase P
MoNAA	Molecular neutron activation analysis
MPTP	1-methyl-4-phenyl-1,2,3,6-tetrahydropyridine
MS	Mass spectrometry
MSI	Mass spectrometry imaging
MR	Molecular replacement
MS	Mass spectrometer
MT	Metallothionein
MtFt	Mitochondrial ferritin
MV-NIS	Measles virus vector expressing the *NIS* gene
MWCNTs	Multiwall carbon nanotubes
NAA	Neutron activation analysis
NATs	Nuclear analytical techniques
NCEs	New chemical entities
NEXAFS	Near-edge X-ray absorption fine structure
NFT	Neurofibrillary tangles
Nic-NHS	Nicotinoyloxy succinimide
NiO	Nickel oxide
NM	Neuromelanin
NmDMSA	Meso-2,3-dimercaptosuccinic acid coated nanomaghemites
NMR	Nuclear magnetic resonance
NO	Nitric oxide
NS	Neutron scattering
NTBI	Non-Tf-bound iron
ON	Olfactory nerve
OV	Olfactory ventricle
ORF	Open reading frames
PAD	Pixel array detector
PAGE	Polyacrylamide gel electrophoresis
PD	Parkinson's disease
PDB	Protein data bank
PDF	Peptide deformylase
PET	Positron emission tomography
Pf	*Pyrococcus furiosus*
PGNAA	Prompt gamma-ray neutron activation analysis
PIXE	Particle induced X-ray emission
PIXE	Proton induced X-ray emission
PSs	Polysaccharides

List of Abbreviations

PX	Protein crystallography
PIXE	Proton induced X-ray emission
PN	Perineuronal net
Pt-Cu	Platinum–copper
QM	Quantum mechanics
Rb	Rubredoxin
RBP	Riboflavin binding protein
REEs	Rare-earth elements
RNAA	Radiochemical neutron activation analysis
ROS	Reactive oxygen species
RP	Reversed-phase
RR	Resonance Raman
RSC	Royal Society of Chemistry
RSD	Reproducibility with size-dependent
SANS	Small angle neutron scattering
SAXS	Small angle X-ray scattering
SCA	Single-channel analyzer
SDS-PAGE	Sodium dodecyl sulfate–polyacrylamide gel electrophoresis
SeCys	Selenocysteine
SE	Size exclusion
SEC	Size-exclusion chromatography
SEM	Scanning electron microscopy
SILAC	Stable isotope labeling by amino acids in cell
SIMS	Secondary ion mass spectroscopy
Sm-dCD	*Streptococcus mutans* dCD
SN	Substantia nigra
SNAA	Speciation neutron activation analysis
SNP	Sodium nitroprusside
SOD	Superoxide dismutase
SPECT	Single-photon emission computed tomography
SR	Synchrotron radiation
SR-µXRF	Synchrotron radiation X-ray fluorescence with microbeam
SRIXE	Synchrotron radiation-induced X-ray emission
SR-TXRF	Synchrotron radiation total reflection X-ray fluorescence
SRXRF	Synchrotron radiation X-ray fluorescence
TADs	TRNA-specific adenosine deaminases
TBP	TATA binding protein
TEM	Transmission electron microscopy
TEY	Total electron yield
Tf	Transferrin
Tf-Fe	Tf-bound iron
TfR	Tf receptor
TfR2	Tf receptor 2

TIMS	Thermal ionization mass spectrometry
TOF	Time-of-flight
TPEN	N,N,N',N'-tetrakis-(2-pyridylmethyl)-ethylenediamine
TXRF	Total reflection XRF
UTRs	Untranslated regions
WD	Wavelength-dispersive
WDXRF	Wavelength-dispersive X-ray fluorescence
WT	Wild-type
XANES	X-ray absorption near edge structure
XAS	X-ray absorption spectroscopy
XE	X-ray emission
XLSA/A	XLSA with ataxia
XRD	X-ray diffraction
XRF	X-ray fluorescence
XRFM	X-ray fluorescence microtomography
WFA	Wisteria floribunda agglutinin
2D	Two-dimensional
3D	Three-dimensional
2-DE	Two-dimensional gel electrophoresis
6-OHDA	6-hydroxydopamine
μ-PIXE	Micro-PIXE
μ-SRXRF	Microbeam synchrotron radiation X-ray fluorescence

About the Editors

Professor Chunying Chen received her Bachelor's degree in Chemistry (1991) and obtained her PhD degree (1996) in Biomedical engineering from Huazhong University of Science and Technology of China. She held postdoctoral positions at the Key Laboratory of Nuclear Analytical Techniques, Institute of High Energy Physics of Chinese Academy of Sciences (1996–1998) and at the Medical Nobel Institute for Biochemistry of Karolinska Institute, Sweden (2001–2002). Dr. Chen currently is a principal investigator at CAS Key Laboratory for Biological Effects of Nanomaterials and Nanosafety in National Center for Nanoscience and Technology of China. She has authored or co-authored over 70 peer-reviewed papers or book chapters and three patents. She is a member of the International Union of Pure and Applied Chemistry (IUPAC), and Chinese Society of Toxicology.

Her current research interests include the development of analytical approaches for metallomics and metalloproteomics such as isotopic labeling and synchrotron radiation based techniques that allow the detection of interaction of metal and molecular species; the potential toxicity of nanoparticles used for nanotechnology applications; the therapy for malignant tumor using nanoparticles for their immunomodulatory effects, drug delivery and tumor targeting; extensive *in vitro* and *in vivo* studies in cellular uptake and intracellular trafficking of nanoparticles and tissue and cellular targeting for cancer treatment; the improvement of HIV vaccine treatment by novel nanotechnology using nanomaterials as potential non-viral vectors; public health impact of long-term exposure of metals in susceptible population strata.

Professor Zhifang Chai is a radiochemist working at the Multidisciplinary Initiative Centre, Institute of High Energy Physics, The Chinese Academy of Sciences. He graduated from Fudan University, China, in 1964. As a fellow of the Alexander von Humboldt Foundation, Germany, he worked at Cologne University from 1980 to 1982 in the field of nuclear technology and its applications. Later, he worked in France, the USA, the Netherlands and Japan. He has long been involved in the methodology of nuclear analytical techniques and their multidisciplinary applications, especially in the study of the chemical speciation of trace elements in environmental and biological systems. He has authored or co-authored over 332 papers in peer-reviewed journals, 6 Chinese books and 3 English books. He is a titular member of Analytical Chemistry Division, International Union for Pure and Applied Chemistry, Fellow of Royal Society of Chemistry, UK, and many domestic and international scientific societies. He is a member of editorial board or advisory committee of 10 international and national journals. In 2005 he was awarded the George von Hevesy Award – the premier international award of excellence to honour outstanding achievements in radioanalytical and nuclear chemistry. In 2007 he was elected as a member (Academician) of the Chinese Academy of Sciences. His present interest is to develop novel nuclear analytical methods for the study of metallomics.

Yuxi Gao is an associate professor at Institute of High Energy Physics, Chinese Academy of Sciences. He obtained his PhD degree in environmental sciences in 2000 from Research Center for Eco-Environmental Sciences, Chinese Academy of Sciences. His current research focuses on the methodology of metallomics and metalloproteomics based on the nuclear analytical techniques; the applications of metallomics and metalloproteomics tourniquets on the environmental and biomedical research; the homeostasis of trace elements and their regulatory mechanism; the structure, function and structure-function relationship of important metalloproteins.

List of Contributors

Ru Bai
CAS Key Laboratory for Biological Effects of Nanomaterials & Nanosafety, National Center for Nanoscience and Technology of China & Institute of High Energy Physics, Chinese Academy of Sciences, Beijing 100190, P. R. China
Email: bair@nanoctr.cn

Zhifang Chai
CAS Key Laboratory of Nuclear Analytical Techniques, Institute of High Energy Physics, Chinese Academy of Sciences, Beijing 100049, P. R. China
Email: chaizf@ihep.ac.cn

Chunying Chen[*]
[1]CAS Key Laboratory for Biological Effects of Nanomaterials & Nanosafety, National Center for Nanoscience and Technology of China & Institute of High Energy Physics, Chinese Academy of Sciences, Beijing 100190, China; and [2]CAS Key Laboratory of Nuclear Analytical Techniques, Institute of High Energy Physics, Chinese Academy of Sciences, Beijing 100049, P. R. China
Email: chenchy@nanoctr.cn; Tel: +86-10-82545560; Fax: +86-10-62656765

Ivan K. Chu
Department of Chemistry and Open Laboratory of Chemical Biology, The University of Hong Kong, Hong Kong, P. R. China

Yuhui Dong[*]
Beijing Synchrotron Radiation Facility, Institute of High Energy Physics, Chinese Academy of Sciences, Beijing 100049, P. R. China
Email: dongyh@ihep.ac.cn; Tel: +86-10-88233090; Fax: +86-10-88233201

Weiyue Feng[*]
CAS Key Laboratory of Nuclear Analytical Techniques, Institute of High Energy Physics, Chinese Academy of Sciences, Beijing 100049, P. R. China
Email: fengwy@ihep.ac.cn; Tel: +86-10-88233209; Fax: +86-10-88235294

[*]corresponding author

Yuxi Gao*
CAS Key Laboratory of Nuclear Analytical Techniques, Institute of High Energy Physics, Chinese Academy of Sciences, Beijing 100049, P. R. China
Email: gaoyx@ihep.ac.cn; Tel: +86-10-88233212; Fax: +86-10-88235294

Zengqiang Gao
Beijing Synchrotron Radiation Facility, Institute of High Energy Physics, Chinese Academy of Sciences, Beijing 100049, P. R. China
Email: gaozq@ihep.ac.cn

Ruiguang Ge
[1]Department of Chemistry and Open Laboratory of Chemical Biology, The University of Hong Kong, Hong Kong, China; and [2]The Laboratory of Integrative Biology, College of Life Sciences, Sun Yat-Sen University, Guangzhou 510006, P. R. China.

Haifeng HOU
Beijing Synchrotron Radiation Facility, Institute of High Energy Physics, Chinese Academy of Sciences, Beijing 100049, China
Email: houhf@ihep.ac.cn

Yufeng Li
CAS Key Laboratory of Nuclear Analytical Techniques, Institute of High Energy Physics, Chinese Academy of Sciences, Beijing 100049, China
Email: liyf@ihep.ac.cn

Guangjun Nie*
CAS Key Laboratory for Biological Effects of Nanomaterials & Nanosafety, National Center for Nanoscience and Technology of China & Institute of High Energy Physics, Chinese Academy of Sciences, Beijing 100190, P. R. China
Email: niegj@nanoctr.cn; Tel: +86-10-82545529; Fax: +86-1062656765

Bo Ning
CAS Key Laboratory for Biological Effects of Nanomaterials & Nanosafety, National Center for Nanoscience and Technology of China & Institute of High Energy Physics, Chinese Academy of Sciences, Beijing 100190, P. R. China
Email: ningb@nanoctr.cn

Ying Qu
CAS Key Laboratory for Biological Effects of Nanomaterials & Nanosafety, National Center for Nanoscience and Technology of China & Institute of High Energy Physics, Chinese Academy of Sciences, Beijing 100190, P. R. China
Email: quy@nanoctr.cn

---

*corresponding author

List of Contributors

Hongzhe Sun[*]
Department of Chemistry and Open Laboratory of Chemical Biology, The University of Hong Kong, Hong Kong, P. R. China
E-mail: hsun@hkucc.hku.hk; Tel: +852-28598974; Fax: +852-28571586

Liming Wang
CAS Key Laboratory for Biological Effects of Nanomaterials & Nanosafety, National Center for Nanoscience and Technology of China & Institute of High Energy Physics, Chinese Academy of Sciences, Beijing 100190, P. R. China
Email: wanglm@nanoctr.cn

Meng Wang
CAS Key Laboratory of Nuclear Analytical Techniques, Institute of High Energy Physics, Chinese Academy of Sciences, Beijing 100049, P. R. China
Email: wangmeng@ihep.ac.cn

Lili Zhang
CAS Key Laboratory for Biological Effects of Nanomaterials & Nanosafety, National Center for Nanoscience and Technology of China & Institute of High Energy Physics, Chinese Academy of Sciences, Beijing 100190
Email: zhangll@nanoctr.cn

Zhiyong Zhang[*]
CAS Key Laboratory of Nuclear Analytical Techniques, Institute of High Energy Physics, Chinese Academy of Sciences, Beijing 100049, P. R. China
Email: zhangzhy@ihep.ac.cn; Tel: +86-10-88233215; Fax: +86-10-88235294

Motao Zhu
CAS Key Laboratory of Nuclear Analytical Techniques, Institute of High Energy Physics, Chinese Academy of Sciences, Beijing 100049, P. R. China
Email: zhumt@ihep.ac.cn

---

[*]corresponding author

CHAPTER 1
Introduction

YING QU,[a] YU-FENG LI,[b] RU BAI,[a] CHUNYING CHEN[a,b,*] ZHIFANG CHAI[b]

[a] CAS Key Laboratory for Biological Effects of Nanomaterials & Nanosafety, National Center for Nanoscience and Technology of China & Institute of High Energy Physics, Chinese Academy of Sciences, Beijing 100190, China; [b] CAS Key Laboratory of Nuclear Analytical Techniques, Institute of High Energy Physics, Chinese Academy of Sciences, Beijing 100049, China

1.1 Background

The terms "-ome" and "-omics" have been widely adopted by scientists. The "-omics" informally refers to the studies in biology, while the related "-omes" addresses the objects of study in such fields. The suffix "-ome" is thought to derive from the Latin prefix "omni-", meaning total or complete. Thus, "-omes" are intended to be a comprehensive description of all of the relevant components, both known and unknown, in a particular biomolecular subset.[1] "-Omes" can provide an easy short-hand to encapsulate a field; for example, a proteome refers to the protein complement of an entire organism, tissue type, or cell, and its associated field "proteomics" is clearly recognizable as relating to the study field of proteins on a large scale. The term "-omics" represents the rigorous study of various collections of molecules, biological processes, or physiological functions and structures as systems.

Compared to the well known genomics and proteomics, metallomics and metalloproteomics are relatively new fields. They are receiving great attention in the investigation of trace elements in biology and expected to develop as an interdisciplinary science complementary to genomics and proteomics. In the

---

*Corresponding author: Email: chenchy@nanoctr.cn; Tel: +86-10-82545560; Fax: +86-10-62656765.

Nuclear Analytical Techniques for Metallomics and Metalloproteomics
Edited by Chunying Chen, Zhifang Chai and Yuxi Gao
© Royal Society of Chemistry 2010
Published by the Royal Society of Chemistry, www.rsc.org

first chapter of this book, the history and definition of metallomics and metalloproteomics will be introduced. In addition, the current application of nuclear or nuclear-related analytical techniques for metallomics and metalloproteomics will be overviewed.

1.2 Metallomics and Metalloproteomics

1.2.1 Trace Elements, Chemical Species and Speciation Analysis

The term "trace elements" dates back to the early 20th century, in recognition of the fact that many elements occurred at such low concentrations that their presence could only just be detected.[2] In analytical chemistry, a trace element is an element in a sample that has an average concentration of less than 100 parts per million atoms, or less than $100\,\mu g\,g^{-1}$. In biochemistry, a trace element is a chemical element that is needed in minute quantities for the proper growth, development, and physiology of the organism and it is also referred to as a micronutrient.[3] So far, 117 elements in total have been observed, of which 92 occur naturally on Earth. However, living organisms are composed of about 26 elements, and only six of those 26 make up practically all of the weight of most living things. The other 20 elements essential for life are present in very small amounts, some of them are in such tiny amounts that they are correspondingly called "trace elements". Besides their phenotypic and phylogenetic characteristics, only 11 elements appear to be approximately constant and predominant in all biological systems, which are called major elements. In the human body, these constitute 99.9% of the total number of atoms present, but just four of them (C, O, H, and N) correspond to 99% of the total and the other seven elements (Na, K, Ca, Mg, P, S, and Cl) represent only about 0.9%.[4]

Trace elements play an important role in the functioning of life.[5] Essential trace elements, acting as catalytic or structural components of larger molecules, have specific functions and are indispensable for life. In addition to the long-known deficiencies of iron and iodine, signs of deficiency for chromium, copper, zinc, and selenium have been identified in free-living populations. It is considered that marginal or severe trace element imbalances can be risk factors for several diseases of public health importance. However, the cause and effect relationships will depend on a more complete understanding of basic mechanisms of action, and more importantly, on better analytical procedures and functional tests to determine marginal trace element status in human.[6]

The biological effect of an element is not only dependent on the total concentration, but also highly related to its chemical forms present in biological systems, e.g. the oxidation state, the nature of the ligands or even the molecular structure.[7] Dramatic examples are chromium, tin and mercury, to name just a few. Cr(VI) ions are considered far more toxic than Cr(III). Although the inorganic forms of tin and mercury are less toxic or even do not show toxic

Introduction 3

properties, the alkylated forms are highly toxic. Dialkylmercury derivatives are considered extremely toxic, while mercuric selenide has a relatively low toxicity and accumulates as an apparently benign detoxification product in marine animals,[8] and methylmercury cysteine proves to be much less toxic than methylmercury chloride in a zebra fish larvae model system.[9] Therefore, to produce qualitative and quantitative information on chemical compounds that affect the quality of life, chemical forms of specific element should be considered carefully.

Nowadays, there is increasing awareness of the importance of the chemical form in which an element is present in biological systems. More and more chemical speciation information on a given element is demanded in most fields of research. In fact, many environmental, toxicological, pharmacological, nutritional, and biological issues today require reliable information on the actual chemical species present, rather than total element concentrations as usually provided by atomic techniques in routine laboratories. In recent years, trace element speciation has become a worldwide trend in current analytical chemistry. Often these different chemical forms of a particular element or its compounds are referred to as "species".

Species and speciation, words borrowed from the biological sciences originally, have been adopted by those in analytical chemistry and accepted in such diverse fields as toxicology, clinical chemistry, geochemistry, and environmental chemistry, expressing the idea that the specific chemical forms of an element should be considered individually.[2] The concept of "speciation" dates back to 1954 when Goldberg introduced it to improve the understanding of the biogeochemical cycling of trace elements in seawater.[10] Since then, this development has been growing exponentially to the point that research on trace element analysis is being conducted. Today, research is almost exclusively focused on trace element species.

Although the concept of speciation is now widely appreciated in many fields, these terms have been used in a number of different ways and created confusion. In a report on the 1984 workshop "The Importance of Chemical Speciation in Environmental Processes", Bernhard et al. pointed out that the usage of speciation varies among different fields ranging from evolutionary changes to distinctions based on chemical state.[5] The term has been used in no fewer than four different ways including the reaction specificity, transformation of species, the distribution of species, or the analytical activity to determine the concentrations of species. They concluded that authors should either avoid the term altogether or clearly define it.

In an attempt to end the confusion that existed regarding the usage of the term speciation, the International Union of Pure and Applied Chemistry (IUPAC), (three IUPAC Divisions represented by the Commission on Microchemical Techniques and Trace Analysis, the Commission on Fundamental Environmental Chemistry, and the Commission on Toxicology) provided IUPAC Recommendations in 2000 in an attempt to define what is a chemical species, what is speciation and what is speciation analysis.[11] It was agreed that straightforward, standard terminology is important for interdisciplinary communication and also for communication to non-scientists, such as legislators and consumer groups. In their paper, IUPAC provided a clear definition of chemical speciation, distinguishing it from fractionation, and where necessary suggested less ambiguous alternative expressions to those in use.

The IUPAC has defined elemental speciation in chemistry as follows:[11]

1. *Chemical species*. Chemical elements: a specific form of an element defined with regard to isotopic composition, electronic or oxidation state, and/or complex or molecular structure.
2. *Speciation analysis*. Analytical chemistry: analytical activities of identifying and/or measuring the quantities of one or more individual chemical species in a sample.
3. *Speciation of an element*; *speciation*. The distribution of an element amongst defined chemical species in a system.

Normally, the toxicity of metal ions is described by measuring median lethal dose (LD_{50}) values or median lethal concentration (LC_{50}). However, some authors reported toxicities using IC_{50} (median inhibitory concentration) and EC_{50} (median effective concentration) terms. LD_{50} values of some forms of metal ions are listed in Table 1.1 (The values are taken from the book *Instrumental Methods in Metal Ion Speciation*).[12]

Thus, it is clear that the metal-associated ligands can profoundly affect their solubility, absorption, bioavailability and the effects of exposure, and then the subsequent toxicity *in vivo*.

Speciation is essential and in urgent need today. The measurement of the total concentration of a particular element cannot provide information about the actual physicochemical form of the element. It is required for the understanding of its toxicity, biotransformation and so forth. Also, new developments in analytical instrumentation and methodology now often allow us to identify and measure the species present in a particular system. Elemental speciation analysis has now become a part of everyday life for environmental and analytical chemists; at the same time, it is also important to industry, academia, and governmental and legislative bodies.[12] Therefore, the number of research papers is increasing continuously every year, especially in "metal speciation". The growth of the speciation of the elements over a 30–year period is shown in Figure 1.1 (Information obtained from the ISI website of science: http://apps.isiknowledge.com/CitationReport).

More than six thousand (6132) papers with an emphasis on "metal speciation" have been published during the past 30 years (1979–2008), with the total number of citations approaching a hundred thousand (96 769). Along with the development of advanced nuclear analytical techniques, there has been an obvious boom in the number of papers written since the 1990s (information obtained from ISI website of science http://apps.isiknowledge.com/CitationReport). Growth in the number of research papers on "metal speciation" over a 30-year period (1979–2008) and the number of times they have been cited are shown in Figures 1.2 and 1.3.

Table 1.1 LD_{50} values of some different forms of metal ions.[12]

Chemical species	Forms of the compounds	Animal	LD_{50} $(mg\,kg^{-1})$
Arsenicals	Potassium arsenite	Rat	14
	Calcium arsenate	Rat	20
	Monomethylarsonate (MMA)	Rat	700–1800
	Dimethylarsinate (DMA)	Rat	700–2600
	Arsenobetaine	Rat	>100 000
Lead compounds	Pb-arsenite	Rat	100
	Pb-naphthinate	Rat	5100
	Pb-nitrate	Guinea pig	1300
	Pb-chloride	Guinea pig	2000
	Pb-fluoride	Guinea pig	4000
	Pb-oleate	Guinea pig	8000
Mercury compounds	HgCl	Rat	210
	HgI	Mouse	110
	$HgNO_3$	Mouse	388
	HgO	Rat	8
	$MeHg^+$	–	10
	EtHg	–	40
Selenium compounds	Na-selenite(IV)	Rat	3.5 (MLD)
	Na-selenate(VI)	Rat	5.7 (MLD)
	DL-Selenocysteine or methionine	Rat	4 (MLD)
	Dimethylselenide	Rat	1600
	Trimethyl-Se-Cl	Rat	50
	Se(IV) and Se(VI)	Rabbit	2
	Selenium disulfide	Mouse	48
Organotin compounds	Bis(tributyltin) oxide	Rat	150–234
	Trimethyltin hydroxide	Rat	540
	Triphenyltin gydroxide	Rat	125
	Trimethyltin acetate	Rat	9.1
	Triethyltin acetate	Rat	4
	Tributyltin acetate	Rat	>4000
Antimony compounds	Sb(III) or Sb(V)	Rat	100
	Sb_2O_3	Rat	3250
	Sb_2S_3	Rat	1000
	Sb_2O_5	Rat	4000
	Sb_2S_5	Rat	1599

For more comprehensive descriptions about speciation, readers are referred to the *Handbook of Elemental Speciation: Techniques and Methodology* by Cornelis et al.,[2] *Instrumental Methods in Metal Ion Speciation* by Ali and Aboul-Enein,[12] and reviews by Tack and Verloo,[13] and Caruso and Montes-Bayon,[14] respectively.

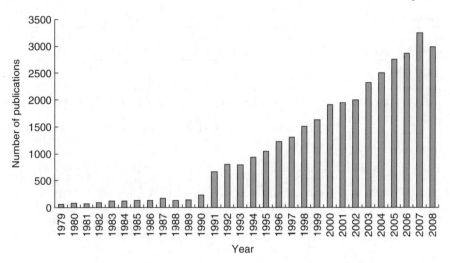

Figure 1.1 Growth in the number of research papers in "speciation" over a 30-year period (1979–2008).

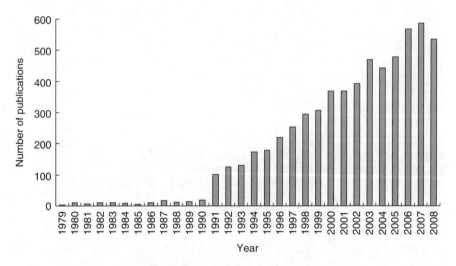

Figure 1.2 Growth in the number of research papers in "metal speciation" over a 30-year period (1979–2008).

1.2.2 Metallomics

1.2.2.1 History and Development

In 2001, Williams used the term "metallome" to refer to an element's distribution, equilibrium concentrations of free metal ions or as the free element content in a cellular compartment, cell, or organism.[15] He mentioned that

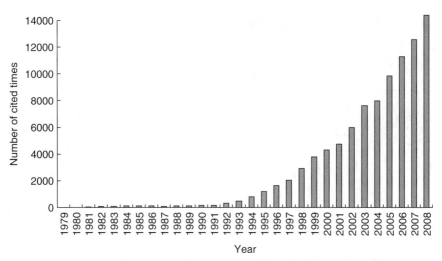

Figure 1.3 Growth in the cited times of research papers in "metal speciation" over a 30-year period (1979–2008).

...*the variety of paths which individual elements follow in any organ adds to the specific character of the organisms. Clearly the paths have evolved to create an elemental distribution which we shall call the metallome, to parallel the nomenclature of protein distribution, the proteome.*

In the 2002 international symposium on bio-trace elements held in Japan, Haraguchi proposed the concept and term of "metallomics" as a new scientific field to integrate the research fields related to biometals. The concept was further elucidated in his successive publication, in which metallomics was defined as bio-trace element science, and metalloproteins, metalloenzymes and other metal-containing biomolecules were defined as "metallomes" in a similar manner to genomes in genomics as well as proteomes in proteomics.[16] Subsequently, the term "metallomics" has been used as the name for the study of metallomes. Szpunar defined metallomics as "... comprehensive analysis of the entirety of metal and metalloid species within a cell or tissue type".[17]

Although it is a rather new concept, metallomics has received great attention. Indeed, many of the papers in recent-year issues have bearings on metallomics, and the subject has been the topic of a number of reviews.[14,15,18–36] The *Journal of Analytical and Atomic Spectrometry*, based at the Royal Society of Chemistry, published two special issues on metallomics in 2004[7] and 2007[18] (see Figure 1.4 for the cover of these two special issues), and a new metallomics and biological elemental speciation web page (<http://www.rsc.org/Publishing/ Journals /JA/ News/ biological_content.asp>) has been set up. In 2009, RSC published a new journal, *Metallomics* (see Figure 1.5 for the journal cover), which covers the research fields related to biometals, and is expected to be the core publication for the emerging metallomics community as they strive to fully understand the role of metals in biological, environmental, and clinical systems. This timely new

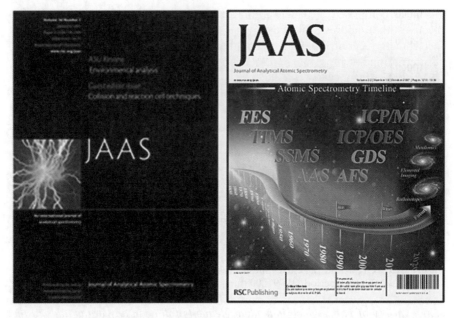

Figure 1.4 Special issues on metallomics of *Journal of Analytical and Atomic Spectrometry (JAAS)*.

Figure 1.5 Selected journal covers of *Metallomics*.

Introduction 9

journal is comprised of papers dealing with techniques, applications, and perspectives that address this emerging field of study and show how atomic spectrometry is contributing to the understanding of biological systems.

International symposiums on metallomics have been held in Nagoya, Japan (ISM 2007 <http://www.ism2007.org>), and Cincinnati, USA (ISM 2009 <http://www.uc.edu /plasmachem/iswm/index.htm>). The symposia brought scientists together from different areas to realize a greater understanding of the role of metals and metal compounds in many biological, chemical, environmental and clinical systems. These meetings included invited presentations by world leaders in the multidisciplinary areas of system biology, bioinorganic chemistry, instrument development, metallomics methods and approaches, environmental and clinical sciences, as well as contributed papers covering the entirety of metallomics. Also, a Metallomics Center of the Americas under the leadership of Professor Joseph A. Caruso, and in partnership with Agilent Technologies (<http://www.che.uc.edu/metallomics/>) has been established at the University of Cincinnati, which is the first large-scale metallomics research center in the world.

IUPAC is currently working on a project "Metal-focussed -omics: guidelines for terminology and critical evaluation of analytical approaches", and the objectives are (1) the definition of terms related to analytical chemistry of interactions of metals with biomolecules in environmental, nutrition and life-sciences, such as metallome, metalloproteome, and the corresponding -omics; and (2) a critical evaluation of analytical techniques suitable for metallomics, and the validity and pertinence of data obtained.

The project targets the speciation analysis community organized around the European Virtual Institute of Speciation Analysis (EVISA, http://www.speciation.net), structural genomic consortia, clinical biochemistry, medicine and health sciences communities (characterization of metal-related diseases and related areas, heteroatom-containing species as new clinical biomarkers), nutrition and metabolic sciences (molecular targets of metal binding for essential nutrients and toxic metals), and environmental toxicology (toxic metals in life sciences and their environmental effects). It should be of interest to regulatory bodies answering the question on what valid information can be obtained in quantitative and routine ways in the metal-related -omics areas.

1.2.2.2 *Definition of Metallomics and Metallomes*

Metallomics is the study of metallomes, interactions and functional connections of metal ions and their species with genes, proteins, metabolites, and other biomolecules within organisms and ecosystems, where a metallome is the entirety of metal and metalloid species present in a cell, cell compartment, tissue or organism, defined according to their identity, quantity, and localization.[19]

A metallomics study is expected to imply:

1. A focus on (essential, beneficial or toxic) metals (e.g. Cu, Zn, Fe, Mn, Mo, Ni, Ca, Cd, Pb, Hg) or metalloids (e.g. As, Se, Sb) in a biological

context. The extension of the term to biologically important non-metals, such as sulfur or phosphorus, is discouraged.
2. A correlation of the element concentration blueprint or element speciation with the genome. This correlation may be statistical (an enrichment of an element coincides with the presence of a particular gene), structural (sequence of a metalloprotein is traceable to a gene) or functional (the presence of a bioligand is the result of a gene-encoded mechanism).
3. A systematic, comprehensive or global approach. An identification of a metal species, however important, without specifying its significance and contribution to the system is not metallomics.

The definition of metallome and metallomics should be seen in the context of speciation of an element (as mentioned above which was defined by IUPAC as the distribution of an element amongst defined chemical species in a system),[11] but distinguished by consideration of the global role of all metals/metalloids in a system. In speciation analysis, analytical activities of identifying and/or measuring the quantities of one or more individual chemical species in a sample, narrowed to metallobiomolecules, can be referred to as metallomics. In other words, metallomics can be considered as a subset (referring to cellular biochemistry) of speciation analysis understood as the identification and/or quantification of elemental species.

The analysis of a metallome (metallomics) will inform us of (1) how an element (metal or metalloid) is distributed among the cellular compartments of a given cell type; (2) its coordination environment, in which a biomolecule is incorporated or by which bioligand it is complexed; and (3) the concentrations of the individual metal species that are present.[17,20]

The different metallospecies in a biological environment which are the analytical targets for metallomics studies have been organized into several classes, and are shown schematically in Figure 1.6.

As shown in Figure 1.6, endogenous metal(loid)-containing biomolecules can be divided mainly into three parts: (1) non-proteinaceous biomacromolecules, (2) metabolites and (3) proteins. Non-proteinaceous biomacromolecules, including DNA, RNA, carbohydrates and lipids, can form complexes with metals non-covalently. Metabolites and proteins can combine with metals both covalently and non-covalently. Metabolites form coordination bonds with metal(loid)s (e.g. organic acids or peptides such as phytochelatins) and also bind with metals covalently (e.g. organoarsenic compounds, organoselenium compounds, arsenolipids and drug metabolipids), both of which constitute a metallo-metabolome. Proteins form complexes with metal(loid)s coordinatively, which could be considered as a metalloproteome. The most important classes of complexes of interest include metal stress proteins, enzymes, transport proteins, and metal sensing proteins. Proteins can also be bound with metals covalently and form a heteroatom-containing proteome together with a metalloproteome. There are also "free" metal ions, inorganic complexes and metal(loid)oxo-anions in biological samples which are defined as ionomes.[21] Because of the important roles in the occurrence of the physiological functions

Introduction

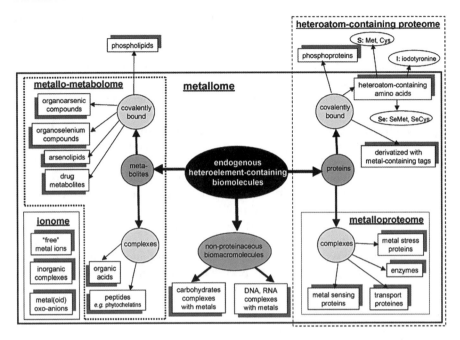

Figure 1.6 Metallome: the different classes of metal species in a biological environment: systematic representation of endogenous metal(loid)-containing biomolecules (adapted from Mounicou et al.[19] and Lobinski et al.[20]). © 2009 The Royal Society of Chemistry. © 2005 Wiley Periodicals, Inc.

in the biological systems, those metallic ions such as alkali and alkaline earth metal ions, which exist mostly as free ions in biological fluids, should also be included in metallomes.[16] All the metallo-metabolome, metalloproteome, non-proteinaceous biomacromolecules and ionome should fall into the category of metallomes.

1.2.3 Metalloproteomics

1.2.3.1 Metalloproteins

Metalloproteins are one of the most diverse classes of proteins that contain metal atoms. The function of these metalloproteins critically depends on the specific interactions between the proteins and the binding metals, such as Cu, Fe, Zn, or Mo. The intrinsic metal atoms in protein structures provide catalytic, regulatory, and structural roles critical to protein function, ranging from electron transfer, substrate binding and activation, transport and storage processes, to regulation of enzymatic activity and gene expression.[22,23] In addition, intrinsic transition metal atoms can be used for crystallographic phasing in favorable cases based on the anomalous signal from the metal atom. Searching the Protein Data Bank (PDB) shows that almost one-fourth of the entries contain a metal atom coordinated to a protein, with Zn being most abundant; Fe, Mg, and Ca are also frequently observed.[24] Overall, about 14% of the non-redundant PDB entries (at 40% identity cutoff) contain at least one of the transition metals (<0.3 nm distance to any protein atom), with a frequency of occurrence of Zn > Fe ≫ all others. Thus, the ability to rapidly and easily identify transition metal content should provide metalloprotein annotations for 10–15% of all proteins.[24]

As mentioned above, Zn is the most abundant transition metal in cells, and plays a vital role in the functionalities of more than 300 enzymes, in the stabilization of DNA, and in gene expression. Other trace metals, such as Se, W, and Mo, are essential in human health and also important in the environment. For example, Se is usually incorporated into antioxidant enzymes as selenocysteine (SeCys), the redox active site of SeCys-containing enzymes, and plays a key role in host oxidative defense. It is predicted that there are a total of 25 SeCys-containing proteins in humans. The most widely known SeCys-containing enzymes are thioredoxin reductase and glutathione peroxidase. Both are ubiquitous and found in bacteria, plants, and mammals, including humans.[25]

When work as the biological catalysts to regulate the biological reactions and physiological functions in biological cells and organs, metalloproteins are called "metalloenzymes" (IUPAC definition: An enzyme that, in the active state, contains one or more metal ions which are essential for its biological function.).[26] Some typical metalloenzymes and metalloproteins are summarized in Table 1.2.

As shown in Table 1.2, metalloenzymes contain specific numbers of metal ions at the active sites in specific proteins. The presence of metal ions allows metalloenzymes to perform functions as biocatalysts for specific enzymatic reactions including gene (DNA, RNA) synthesis, metabolism, antioxidation and so forth.[16,27]

1.2.3.2 Metalloproteomics Studies

It is well known that proteomics analysis aims at high-throughput identification and quantification of the proteins present in the sample. The typical proteomics strategies profit from the knowledge of the studied organism's genome, and consequently the set of the putatively present proteins, and include:

1. *Bottom-up proteomics* based on two-dimensional (2D) gel electrophoresis followed by the tryptic digestion of spots and analysis by mass spectroscopy (MS).[28]
2. *Shotgun proteomics*.[29] The sample is digested with trypsin and the resulting peptides are separated by 2D nanoHPLC and identified with ESMS/MS. The comparison of the data set with a set of proteins potentially synthesized by a given organisms allows the identification of the proteins present.

Introduction 13

Table 1.2 Some typical metalloenzymes (and metalloproteins) and their biological functions.

Metalloenzyme	Molecular weight (kDa)	Number of metal atoms	Biological function
Transferrin	66–68	2 Fe	Transportation of iron
Ferritin	473	1 Fe	Storage of iron
Catalase	225	4 Fe	Decomposition of H_2O_2
Nitrogenase	200–220	24 Fe, 2 Mo	Nitrogen fixation
Cytochrome P-450	50	1 Fe	Metabolisms of steroids and drugs
Carbonic anhydrase	30	1 Zn	Catalyst of H_2CO_3 equilibrium ($H_2CO_3 \leftrightarrow CO_2 + H_2O$)
Caboxypeptidase	34	1 Zn	Hydrolysis of peptide bonds at carboxyl terminal
Alcohol dehydrogenase	150	4 Zn	Dehydrogenation of alcohol ($C_2H_5OH \rightarrow CH_3CHO + H_2$)
Alkaline phosphatase	89	3.5 Zn	Hydrolysis of phosphate esters
DNA polymerase	109	2 Zn	DNA synthesis
RNA polymerase	370	2 Zn	RNA synthesis
Ceruloplasmin	151	6 Cu	Scavenging oxygen radicals; involved in acute-phase reaction of inflammation
Plastocyanin	134	1 Cu	Electron transfer
Cytochrome oxidase	400	2 Cu	Electron transfer
Calmodulin	16	4 Ca	Calcium-binding protein; regulates a number of cellular functions
Superoxide dismutase	32.5	1 Cu, 1 Zn	Catalyses the dismutation of superoxide anion and hydrogen peroxide
Arginase	120	2 Mn	Converts L-arginine into L-ornithine and urea
Selenoprotein P	61	10–12 Se	Not clear (extracellular antioxidant; protects cell membranes against peroxynitrite)
Gluthathionperoxidase	76–92	1 Se	Decomposition of H_2O_2 and organic superoxides
Urease	480	10 Ni	Transformation of urease to ammonia
Metallothionein	3.5–14	Cd, Zn, Hg, Cu, As, Ag etc.	Stores excess metals, metal binding and regulating oxidative stress

3. *Top-down proteomics.*[30] However, the typical analytical approaches to proteomics usually ignore the existence of metal complexes with proteins. The information on the metal–protein interactions is either lost during ionization (e.g. MALDI), on the procedures of sample preparation (because of denaturation), or simply not acquired because of the inadequate ionization efficiency, and, consequently, insufficient sensitivity.[19]

Metalloproteomics is a new subject focusing on the distributions and compositions of all metalloproteins in a proteome (metalloproteome), their structural and functional characterization and their structural metal binding moieties.[7] The specificity of metalloproteomics studies demands the need for a description of the metal-binding sites, metal stoichiometry, and metal-dependent structure or conformation changes as well as the identification and quantification of the metalloproteins. The metalloproteome can be considered as not only a subset of the metallome, but also a very important subset of the proteome.

The study of metalloproteomics would be helpful to the understanding of the biological effects of trace elements and of the pathogenesis of metal-associated diseases. Furthermore, it would supply a basis for the design of inorganic drugs for chemotherapy. Also, metalloproteomics would provide beneficial insight into the biosynthesis process of metalloproteins, including a series of post-translational events such as metal transport, trafficking, metallocenter assembly, cluster exchange, the effects of environmental changes on these processes, and the structure–function relationship of metalloproteins.

Based on a standard proteomics strategy, the workflow for metalloproteomics used in the Laboratory of Nuclear Analytical Techniques of Institute of High Energy Physics, China, is summarized in Figure 1.7. The workflow involves the extraction of proteins from biological samples (organs, tissues with various physiopathologies, cells cultured in various media), separation of the proteins, detection of metalloproteins by elemental analysis methods, measurement of the metalloprotein sequence, and structure. The chemical speciation analysis for specific identification of bioactive metalloproteomes in the biological systems and characterization of their structures are the most important analytical paths to metalloproteomics studies.[31]

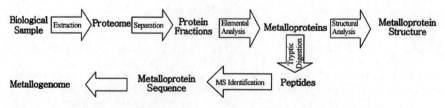

Figure 1.7 Roadmap for metalloproteomics studies used in the Laboratory of Nuclear Analytical Techniques of Institute of High Energy Physics, China.[31] © 2007 The Royal Society of Chemistry.

Introduction

1.2.4 Relationships Among Metallomics and Other "-omics"

Now, over 200 different -omics have been created and used. Listed below are some -omes and -omics which are related to metallomics (Table 1.3). For more -omics terms and concepts, the website "-omes and -omics glossary & taxonomy" (http://www.genomicglossaries.com/content/omes.asp) can be consulted.

DNA is known to be copied to DNA (DNA replication), DNA information can be copied into mRNA, (transcription), and proteins can be synthesized using the information in mRNA as a template (translation).[40] Genomics deals with scientific studies on the genetic information of DNAs and RNAs encoded as the sequences of nucleic acid. Such DNAs and RNAs are generally called genomes in genomics, and play essential roles in protein syntheses. The transcriptome could be defined as a complement of mRNAs transcribed from a cell's genome.[41] Proteins are distributed inside and outside the cell, and they work as enzymes for the synthesis and metabolism of various biological substances inside the cell. The proteome thus can be defined as the protein complement of an entire organism, tissue type, or cell.[32,33] The systematic studies on carbohydrates or lipids in organism are glycomics[42] or lipidomics,[43] respectively. Analysis of the total metabolite pool ("metabollome") offers a means of revealing novel aspects of cellular metabolism and global regulation.[44] The metabollome can be defined as the quantitative complement of all the low molecular weight molecules present in cells in a particular physiological or developmental state. Considering the global role of all metals/metalloids amongst defined chemical species in a system, metallomics is an emerging field addressing the role, uptake, transport, and storage of trace metals essential for life.[19] Metallomics overlaps with genomics, proteomics, metabollomics, glycomics, and lipidomics to form metallogenomics,[45] metalloproteomics, metallometabollmics,[19] metalloglycomics,[46] and even metallolipidomics. All these studies involve the quantitative and simultaneous measurement of the elemental composition of living organisms and changes in this composition in response to physiological stimuli, developmental state, and genetic modifications, and are called ionomics.[38] Ionomics extends the scope of metallomics to include biologically significant non-metals.[38] Elementomics goes further to study all the elements of interest and elemental species, as well as their interactions, transformations, and functions in biological systems. Elementomics may thus be used for the full complement of studied elements and element moieties (free and bound).[36,37] An illustration of the relationships among metallomics and other -omics is shown in Figure 1.8.

1.3 Nuclear Analytical Techniques

1.3.1 General Analytical Approaches in Metallomics and Metalloproteomics

To identify the biometals and their functions in biological systems, it is desirable to integrate all the related but independently established scientific fields

Table 1.3 Various "-omes" and "-omics", and their concepts.

Term	Concept	Reference
Genome	The set of genes of a given organism	32,33
Genomics	The study of the genome of an organism	32,33
Proteome	The entire protein complement of a given genome, that is, the entirety of the proteins that are expressed by the genome	32,33
Proteomics	Study of the proteome of an organism. The term is most commonly associated with the use of MS to identify proteins expressed in a given cell type or tissue under a given set of conditions	32,33
Metallome	The entirety of metal and metalloid species present in a cell or tissue type; their identity, quantity and localization	19
Metallomics	The research field dealing with the study of metallomes and their correlations with genomes and proteomes	19
Metalloproteome	The entirety of metal complexes with proteins in a sample. Note that this term was originally used in a narrower sense and concerned the proteins with enzymatic functions only	31,34
Metalloproteomics	The study focusing on the distributions and compositions of all metalloproteins in a proteome, their structural and functional characterization and their structural metal binding moieties	31,34
Metabolome	The set of metabolites produced as a result of reactions catalyzed by certain proteins (enzymes)	19
Metabolomics	The study of metabolite profiles in biological samples, particularly urine, saliva, and blood plasma; scientists are interested in all, rather than some, of the metabolites in a given sample	35
Elementomics	The study of elements of interest and element species, and their interactions, transformations, and functions in biological systems and elementome may thus be used for the full complement of studied elements and element moieties (free and bound)	36,37
Ionome	Mineral nutrient and trace element composition of an organism and represents the inorganic component of cellular and organismal systems	38
Ionomics	The study of the ionome, involves the quantitative and simultaneous measurement of the elemental composition of living organisms and changes in this composition in response to physiological stimuli, developmental state, and genetic modifications	38
Hetero-atom tagged proteomics	Study of the proteome in which analytical information is acquired by elemental MS either owing to the presence of heteroatom (S, Se, P, I) in a protein or introduced via tagging (derivatization)	17,39

Introduction 17

Table 1.3 (*Continued*)

Term	Concept	Reference
Phosphoproteome	The part of the proteome modified by phorphorylation (a post-translational modification)	17
Selenoproteome	The part of the proteome incorporating selenoamino acids, selenomethionine and selenocysteine. The use of the term is often limited to proteins with genetically encoded selenocysteine only	17

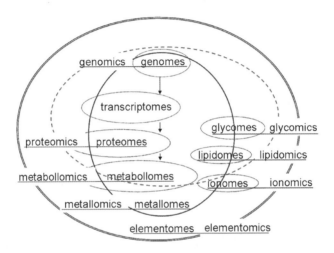

Figure 1.8 A schematic model of the biological system, showing the relationships among metallomics with other -omics.

together. The ten main research subjects in metallomics as well as the corresponding analytical techniques needed to be explored are listed in Table 1.4.

Further, the dedicated analytical approaches to characterize the pool of metal-containing species in living organisms, the metallome, and of their interactions with the genome, transcriptome, proteome and metabolome were grouped into *in vivo* detection, localization, identification and quantification, *in vitro* functional analysis and *in silico* prediction using bioinformatics, which can be illustrated as in Figure 1.9.

Among all these analytical approaches, nuclear analytical techniques with their unique features, like high sensitivity, good accuracy and precision, non-destructiveness, no and/or reduced matrix effect and so on deserve their application in this emerging field.

1.3.2 Specific Nuclear Analytical Techniques

Nuclear analytical techniques deal with nuclear excitations, electron inner shell excitations, nuclear reactions, and/or radioactive decay. Mass, spin and

Table 1.4 Research subjects in metallomics and analytical techniques required in metallomics research.[16] © 2004 The Royal Society of Chemistry

Research subject	Analytical technique
1. Distributions of the elements in biological fluids, cells, organs, *etc.*	Ultratrace analysis, all-elements analysis, one atom detection, one molecule detection
2. Chemical speciation of the elements in biological samples and systems	Hyphenated methods (LC-ICP-MS, GC-ICP-MS, MALDI-MS, ES-MS)
3. Structural analysis of metallomes (metal-binding molecules)	X-ray diffraction analysis, EXAFS
4. Elucidation of reaction mechanisms of metallomes using model complexes (bioinorganic chemistry)	NMR, XPS, laser–Raman spectroscopy, DNA sequencer, amino acids sequencer, time–resolution and spatial–resolution fluorescence detection
5. Identification of metalloproteins and metalloenzymes	LC-ES-MS, LC-MALDI-MS, LC-ICP-MS
6. Metabolisms of biological molecules and metals (metabollomes, metabolites)	LC, GC, LC-MS, GC-MS, ES-MS, API-MS[a], biosensors
7. Medical diagnosis of health and disease related to trace metals on a multielement basis	ICP-AES, ICP-MS, graphite-furnace AAS, autoanalyser, spectrophotometry
8. Design of inorganic drugs for chemotherapy	LC-MS, LC-ICP-MS, stable isotope tracers
9. Chemical evolution of the living systems and organisms on the earth	Isotope ratio measurement (chronological techniques), DNA sequencer
10. Other metal-assisted function biosciences in medicine, environmental science, food science, agriculture, toxicology, biogeochemistry, *etc.*	*In situ* analysis, immunoassay, bio-assay, food analysis, clinical analysis

[a]Atmospheric pressure chemical ionization mass spectrometry.

magnetic moment, excited states and related parameters, and probability of nuclear reactions are nuclear parameters that serve as basis for nuclear analytical techniques. When dealing with a radioactive isotope, the properties of a half-life and the types and energies of the emitted radiation are also involved. In principle, nuclear techniques are based on properties of the nucleus itself, compared to non-nuclear techniques which use properties of the atom as a whole, primarily governed by properties of the electrons arranged in shells. However, fundamentally as well as practically no sharp borderline can be drawn between nuclear and non-nuclear techniques. For example, mass spectrometry deals with ionized atoms, and rarely with the bare nucleus; however, the signal is determined by the mass differences of the nucleus. Therefore, in this book mass spectrometry will be considered as nuclear technique. Some nuclear analytical techniques are not only based on nuclear properties, but on a combination of nuclear and electronic properties, either within a single technique, or within a hyphenation of two techniques. For example, in Mössbauer spectrometry, the nuclear signal is fine-tuned by the electron energy levels, also giving chemical information.

Introduction 19

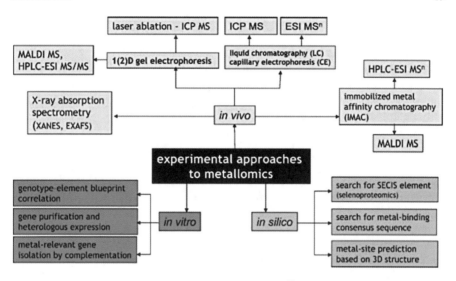

Figure 1.9 Experimental approaches to metallomics.[19] © 2009 The Royal Society of Chemistry.

Some analytical characteristics of nuclear analytical techniques are as follows.[47]

1.3.2.1 Isotopic Analysis Rather than Elemental Analysis

In non-nuclear techniques, isotopes of the same element generally cannot be distinguished, while in nuclear techniques, specific isotopes are measured instead of elements. Therefore, direct quantitative information on the associated elements can be obtained since poly-isotopic elements have constant isotope ratios. Further, isotopes of a given element may be discriminated; therefore, analytical information may be obtained by using elements enriched in respect to a particular stable isotope or labelled with a radioisotope, e.g. in isotope dilution analysis. In addition to analytical information, isotope studies may also yield kinetic and mechanistic information.

1.3.2.2 No or a Limited Effect of Electronic and Molecular Structure

In many non-nuclear analytical techniques, the signal relevant for detection depends on the chemical state of the measurand, and consequently the occurrence of a measurand in more than one chemical and/or physical state may act as interference, leading to erroneous results. Since the nucleus is rather insensitive for such effects, nuclear analytical techniques are, in principle, not sensitive either. However, by linking to different separation techniques, both selectivity and detectability and sensitivity can be assured.

In some nuclear analytical techniques there are specific interactions (coupling) between the energy levels of electrons and nuclei. Although such interactions are rather weak, they may occasionally provide interesting possibilities to give information on electronic and molecular structures. This is the case for analysis via the Mössbauer effect and via NMR. However, it should be noted that only a part of the nucleus is suited for NMR, and the Mössbauer effect can only be applied to a rather small number of nuclei.[48]

1.3.2.3 Penetrating Character of Nuclear Radiation

In most nuclear analytical techniques, the excitation and de-excitation signals penetrate through matter, thereby enabling non-invasive measurements without disturbing the processes to be studied. This is the case where radio waves and a magnetic field are used (NMR), neutrons (neutron scattering/diffraction, NAA), and gamma radiation (activation analysis, Mössbauer spectroscopy, radiotracer investigations, and radioisotope dilution analysis).

1.4 Applications of Advanced Nuclear Analytical Techniques for Metallomics and Metalloproteomics

1.4.1 Nuclear Analytical Techniques for Multielemental Quantification

Inductively coupled plasma mass spectrometry (ICP-MS) is ideal for multielemental quantification. ICP-MS is extremely sensitive, due to the efficient ionization from plasma coupled with the sensitive detection of the mass spectrometer. At its best, the technique can achieve a detection limit of parts per trillion.[49] ICP-MS can detect most elements in biological systems, but sulfur, phosphorus and halogens are not efficiently ionized by the ICP owing to their high ionization energies. Further, a number of polyatomic interferences also hinder the detection of S, P, and transition elements like Fe and V using ICP-MS.[17] The problem of polyatomic interferences can be solved by using either a high resolution (sector field double focusing) mass spectrometer or, in the more commonly available quadrupole mass analyzer, the collision/reaction cell technique.[50,51] Besides the polyatomic interference effect, a memory effect of Hg, U, Os, Hf and Pt, of example, is another important issue in precise detection using ICP-MS. To solve this problem, complexing agents can be added to prevent adherence of the elements to the walls of the spray chamber and the transfer tubing of the sample introduction system.[52]

Neutron activation analysis (NAA) is also a multielemental quantification technique which can simultaneously measure more than 30 elements in a sample. The detection limits of NAA range from 10^{-6} to 10^{-13} g g^{-1} dependent upon the irradiation parameters, measurement conditions, and nuclear parameters of the elements of interest. It has been much used in many areas of fundamental and applied research as well as industrial applications. The theme work of NAA is trace element "fingerprinting". It has been used to study nutritional bio-availability and absorption of essential trace elements in the human using enriched stable isotopes.[53] One of the principal advantages of NAA is that it is nearly free of any matrix interference effects as the vast majority of samples are completely transparent to both the probe (the neutron) and the analytical signal (the gamma ray). Moreover, because NAA can most often be applied instrumentally (no need for sample digestion or dissolution), there is little, if any, opportunity for reagent or laboratory contamination.[54] More detailed information about NAA and the application of NAA in metallomics and metalloproteomics can be found in Chapter 2 (Neutron Activation Analysis) in this book.

In general, ICP-MS and NAA are excellent techniques for multielemental quantification. NAA can achieve non-destructive and *in situ* multielemental analysis for solid samples while ICP-MS needs laser ablation for this purpose.

After obtaining the concentrations of different elements in samples, correlation studies using mathematical methods may lead to further understanding on the roles and their cross-talk of different elements. For example, positive correlation between Se and Hg contents has been found in fish, marine mammals and birds, and animals and workers exposed to inorganic mercury which leads to the study of Hg and Se antagonism or synergism mechanism.[55–58]

1.4.2 Nuclear Analytical Techniques for Metallome and Metalloproteome Distribution

Besides total concentration of multielements, their spatial distribution in samples is also very important in understanding their bioavailability, trophic transfer, and environmental risk. A number of complementary analytical techniques exist for the mapping of elemental distributions in biological tissues including SRXRF (synchrotron radiation X-ray fluorescence) with microbeam (SR-μXRF), microscopic EDX (energy-dispersive X-ray fluorescence), microscopic WDX (wavelength-dispersive X-ray fluorescence), microscopic PIXE (particle-induced X-ray emission), laser ablation ICP-MS, microscopic SIMS (secondary ion mass spectrometry).

SRXRF with microbeam (SR-μXRF) has been used widely in elemental distribution studies. For example, Gao *et al.*[59] mapped the biodistribution of elements in a model organism, *Caenorhabditis elegans*, after exposure to copper nanoparticles with microbeam synchrotron radiation X-ray fluorescence and found the exposure to copper nanoparticles can result in an obvious elevation of Cu and K levels, and a change of bio-distribution of Cu in nematodes. By regulating with a slit or focusing system, such as a Kirkpatrick–Baez mirror system, refractive lenses or with a Fresnel zone plate,[60,61] the beam size can be made at the micrometer or even nanometer level and high spatial resolution is obtained. Carmona *et al.*[62] applied a 90 nm XRF probe for trace metal mapping of single dopaminergic cells and iron is distributed in a granular form into dopamine vesicles, and thin neurite-like processes produced by differentiated cells accumulate copper, zinc, and, to a minor extent, lead.

SRXRF tomography can also be used to perform three-dimensional elemental analysis by measuring a series of projected distributions under various angles which are then back-projected using appropriate mathematical algorithms.[63,64] Since this method involves rotation of the sample over 180° or 360° relative to the primary beam, it is limited to the investigation of relatively small objects. The spatial resolutions for XRF tomography are situated at the 1–2 μm level while, routinely, a resolution of 5 μm is employed.[65] More detailed information and applications of XRF can be found in Chapter 3 (X-ray Fluorescence) in this book.

Other XRF-based techniques like EDX have been coupled to microscopes such as a transmission electron microscope (TEM) or scanning electron microscope (SEM) to map elemental in roots and leaves of *Arabidopsis thaliana*, and mouse liver tissue pyramidal neurons, for example.[66,67] Although both SEM-EDX and TEM-EDX can provide very good spatial resolution at about 10 nm, the detection limits are at about g kg^{-1} level, which may hold back their application in the detection of trace elements in biological samples.[68]

PIXE with a proton microprobe has also been developed[69,70] and has been applied to elemental distribution in plant and animal tissues, and human blood cells and tumors.[71] Further, 3D micro-PIXE also has been developed to perform depth analysis recently and has a spatial resolution of 4 μm by using characteristic titanium K X rays (4.558 keV) produced by 3 MeV protons with beam spot size of ∼1 μm.[72]

With the microprobe technique, LA-ICP-MS has became one of more powerful tools in the quantitative analysis of major and minor elements *in situ* owing to the very high sensitivity of ICP-MS and direct laser sampling to obtain more information from small samples in a micro area. LA-ICP-MS is well established for elemental mapping in the geological sciences[73,74] and has been applied to biological tissues such as plant leaves, snake tails, tree rings, rat brains, pig liver, human brains, and human teeth.[75–77] The spatial resolution of LA-ICP-MS is about 10 μm with a detection limit of below the mg kg^{-1} level.[68] Accurate quantitative analysis by LA-ICP-MS is a challenge because the lack of suitable certified solid standards. Various calibration procedures have been used and the most accurate approach is to have true matrix matching of standards and samples. Commercially available certified reference materials have also been used for calibration.[78,79]

SIMS operates on the principle that bombardment of a material with a beam of ions with high energy (1–30 keV) results in the ejection or sputtering of atoms from the material. A small percentage of these ejected atoms leave as either positively or negatively charged ions, which are referred to as secondary ions. The collection of these sputtered secondary ions and their analysis by mass-to-charge spectrometry gives information on the composition of the sample, with the elements present identified through their atomic mass values. Counting the number of secondary ions collected can also give quantitative data on the sample's composition.[80] SIMS can be used for practically all elements of the periodic table (only the noble gases are difficult to measure because they do not ionize easily) with a detection limit at the ng kg^{-1} level and lateral resolution at

10 µm.[81] Since ions of different mass are measured separately, SIMS is ideally suited for the study of isotopic compositions of small samples.[82] SIMS works by analyzing material removed from the sample by sputtering, and is therefore a locally destructive technique, which may not be a suitable technique for live biological samples, although it has been used in soybean root, grape seeds, and animal tissues and cells.[83–85] During a SIMS measurement, the sample is slowly sputtered (eroded) away and depth profiles can be obtained, which is the 3D measurements.

Taken together, SR-µXRF, PIXE, LA-ICP-MS and SIMS are all very good techniques for multielement imaging and have a spatial resolution of at least µm and a detection limit of at least mg kg^{-1}. Although TEM-EDX and SEM-EDX have much better spatial resolution, the sensitivity is much lower for trace elements than the sensitivity of SR-µXRF, PIXE, LA-ICP-MS and SIMS. When compared with LA-ICP-MS and SIMS, the beam-time limitation for SR-µXRF and PIXE may hinder their wide application.

Besides the above-mentioned techniques for metallome and metalloproteome distribution analysis, isotopic tracing is mainly used for *in vivo* or *in vitro* studies of the absorption, distribution, transportation, storage, retention, metabolism, excretion, and toxicity of trace elements and other materials. Combined with chromatographic and electrophoretic techniques, they can be used for chemical speciation analysis of trace elements in biological samples. The methods have the advantage of high sensitivity (10^{-14} to 10^{-18} g), good accuracy, and convenient operation. Owing to the great development of modern analytical techniques, especially mass spectrometry, the use of enriched stable tracers has rapidly increased recently. The use of enriched stable-isotope tracers is very similar to the use of radioisotopes and there is more general acceptance by scientists who are not familiar with nuclear techniques.

Isotopic techniques will continuously play an important role in the emerging field of metallomics or metalloproteomics. In future, the application of isotopic tracers will probably increase and contribute significantly to the quest for new knowledge. Nevertheless, radioactive experiments need well-trained professionals in a strictly protective environment. In addition, the isotopic effect and radiation effect have to be considered. For more details about this method and the application of these techniques, readers are encouraged to read Chapter 4 (Isotopic Techniques Combined with ICP-MS and ESI-MS) in this book.

1.4.3 Nuclear Analytical Techniques for the Structural Analysis of Metallomes and Metalloproteomes

Techniques such as ray-based techniques and nuclear magnetic resonance (NMR) can be used for the structural characterization of metallomes and metalloproteome.[86] Ray-based techniques can characterize the structure at the atomic level. Rays that can be used for structural analysis include X-rays, gamma rays, or neutron beams.

In the X-ray-based techniques, X-ray crystallography is the most powerful tool for the determination of macromolecular 3D structures at a resolution of 0.15–2 nm but the requirement for a single crystal will greatly limit its application to numerous biological samples. The application of X-ray crystallography for metalloproteomics is illustrated in Chapter 7 (Protein Crystallography for Metalloproteins) in this book.

X-ray absorption spectroscopy (XAS), especially extended X-ray absorption fine structure (EXAFS), may provide an alternative tool for determining the local structure around certain atoms at a resolution of 10^{-4} to 10^{-3} nm without the requirement for crystalline samples.[87] For example, Hg in human hair and blood samples from long-term mercury-exposed populations has been studied using EXAFS and structural information such as bond distances and coordination numbers of Hg were obtained.[88,89] Further, EXAFS can provide a refinement of the structure determined from X-ray crystallography since EXAFS has higher spatial resolution than X-ray crystallography especially in local structures.[90] More detailed information about XAS and its application in metallomics and metalloproteomics study can be found in Chapter 6 (X-ray Absorption Spectroscopy) in this book.

Mössbauer spectroscopy is also a kind of local structure characterization tool but is based on gamma-ray absorption or emission rather than X-ray absorption. The major limitation for the application of Mössbauer spectroscopy in biological samples is that only a few isotopes can be used. The most commonly used isotope is ^{57}Fe, which has been used for the study of ferritins in biological samples, but the low content and natural abundance of Fe may hinder its application.[48] More detailed information can be found in Chapter 5 (Mössbauer Spectroscopy).

Techniques based on the use of neutron beams, such as single-crystal neutron diffraction spectroscopy (SCND), can provide complementary structural information at about 20 nm to that obtained from X-ray-based techniques since light atoms (e.g. hydrogen) can easily be detected even in the presence of heavier ones.[91] Both small-angle X-ray scattering (SAXS) and small-angle neutron scattering (SANS) are structural characterization tools for solid and fluid biological samples at relatively a lower resolution (approx. 100 nm) than single-crystal X-ray or neutron diffraction spectroscopy but again do not need crystalline samples.[92]

Nuclear magnetic resonance (NMR) spectroscopy is a potentially powerful alternative to X-ray crystallography for the determination of macromolecular 3D structures at the same resolution of 0.15–2 nm.[93] NMR has the advantage over crystallographic techniques in that experiments are performed in aqueous solution as opposed to a crystal lattice. Additionally, NMR can be used to determine protein secondary structure content empirically.[94] The major difference in the structures of single crystals determined through NMR or X-ray crystallography is that groups (10–50) of structures, each satisfying the experimental constraints equally well, will be obtained for a one-unit cell when NMR is used, while only one, or, at most, a few structures will be determined when X-ray crystallography is used. Therefore, for NMR, the potential

Introduction

problem is either that the entire ensemble of structures are evaluated or that a mean conformation is produced and then evaluated.[95] Mean structures from ill-defined portions of the polypeptide chain will have non-standard geometries and may cause problems in analysis.

All the techniques used for structure characterization are generally low-throughput considering the number of samples treated at one time and the time

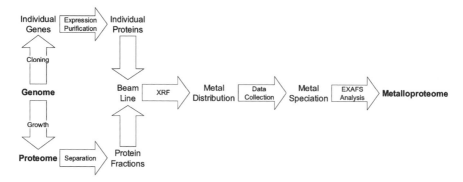

Figure 1.10 Schematic flowchart for the two alternate workflows for high-throughput X-ray absorption spectroscopy (HT-XAS). Synchrotron radiation analysis consists of mapping the metal distribution using X-ray fluorescence (XRF), using XANES for metal speciation and using EXAFS for metal-site structural analysis of the metalloproteome.[96] © 2005 International Union of Crystallography.

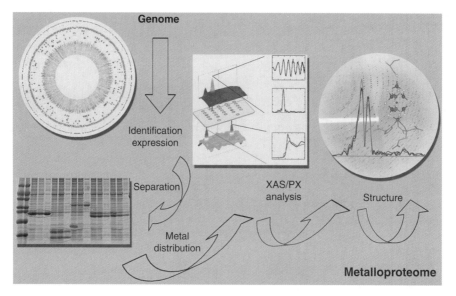

Figure 1.11 The most important steps of the metallogenomics pipeline.[45] © 2005 International Union of Crystallography.

Table 1.5 Features of the main nuclear analytical techniques for chemical element imaging, quantification, and speciation in metallmoics and metalloproteomics studies.

Techniques	NAA	XRF	EDX	PIXE	ICP-MS	XAS	SIMS	Möss.Sp.	NMR	XRD
Related techniques	MAA	SR-μXRF				XANES, EXAFS, SEXAFS				XPD
Exiting radiation	Thermal neutrons	XRF X-ray tube, XRF MP synchrotron X-rays only	Electron 2–50 keV	0.3–4 MeV protons	Plasma	Synchrotron and some rotating anode X-ray sources	Ions, atoms for FABMS	Gamma rays	RF photons	X-ray tube or synchrotron X-rays
Particle emission	Alpha (^{4}He^{2+}), electrons, neutrons and X-ray photons	Characteristic X-rays	Characteristic X-ray photons	Characteristic X-ray photons	Ions	Transmitted or characteristic X-rays	Secondary ions	Gamma rays	RF photons	Diffracted X-rays
Isotopes detection	Yes	No	No	No	Yes	No	Yes	No	No	No
Depth information	None	None	0.5–5 μm, depends on matrix	None	None	Yes, SEXAFS	3–10 nm	None	None	None
Spatial resolution	None	50 nm to 10 μm synchrotron X-rays	0.5–5 μm, depends on matrix	2 μm to 20 mm	None	50 nm X-rays potential with synchrotron X-rays	0.02 μm best, 1–5 μm typical	None	None	50 nm–10 μm synchrotron X-rays, none X-ray tube
Detection limit	0.001–0.1 ppm	1–10 ppm (~ mg kg^{-1})	0.10% (1000 ppm?)	0.1–10 ppm	0.00001–10 ppm	~500 ppm	0.1–10 ppm	1–1000 ppm	10–100 000 ppm, major constituents	1–5% mixtures (XPD)
Elements detected	H–U	Li–U	Na–U (C–U)*	Na–U	Li–U	Li–U	H–U	Isotopes with Mössbauer transitions, approximately 44 elements	Non ferromagnetic elements with spin 1/2 nuclear isotopes (approximately 80)	Li–U
Depth profiling	No	No	In cross-sections	No	No	No	Yes to 2 μm	No	No	No
Chemical information	No	No	No	No	Little	Yes, oxidation state, species, coordination number, some structure	Some bonding	Yes, structure, bonding	Yes, structure and bonding	Yes, species identity and structure
Quantitative analysis	Yes, 2–10%	Yes, high precision	Yes, relative error	Yes, 5%	Yes, high precision with isotope dilution	No	Yes, to 25% relative error	Semi-quantitative	Yes, 1–10%	Yes, to 5–10%

Introduction

	(1)	(2)	(3)	(4)	(5)	(6)	(7)	(8)	(9)	(10)
Imaging	No	Yes	No	Yes	No	No	Yes	No	No	No
Dimension	—	1D to 3D	1D, 2D	1D to 3D	1D, 2D	—	1D to 3D	—	—	—
Mapping	No	Yes	Yes	Yes	No	No	Yes	No	No	No
Line traces	No	Yes	Yes	Yes	No	No	Yes	No	No	No
Molecules detected	No	No	No	No	No	Some	Fragments	Yes, chemical shifts	Yes, chemical shifts	Yes
Crystallography information	No	No	No	No	No	No	No	No	Yes, crude	Yes
Materials	Solids and liquids	Solids and nonvolatile liquids	Any solid	Any solid	Solution	Solids and nonvolatile liquids	Any solid, some liquids	Solid	Liquid and solid	Crystalline solids and polymers
Applications and information obtained	Bulk element analysis	Elemental composition	Elemental analysis of bulk or near surface regions	Elemental composition, location of adsorbed species	Elemental, isotopic composition	Coordination number, oxidation state, bonding of amorphous phases and ID	Elemental, isotopic and molecular composition	Site locations, bonding, structure	Bulk organic ID, bonding and valency, structure	Structure detection and ID of major species in mixtures
Advantages	Most sensitive, little sample prep needed, nondestructive	Many elements, non-destructive, low detection limits, in situ, maps, imaging	Good energy resolution, fast, multiple elements concurrently	Lower background and detection limits over SEM–EDX	Low detection limits for a wide variety of elements, highly sensitive, capable of the quantitative detection	In situ, non-destructive, low detection limits, species determination	Very sensitive, microbeams, imaging, mapping, isotopes	Nondestructive, in situ	No standards needed, water okay, nondestructive	In situ, large library of patterns
Disadvantages	Only elements, resolution 10 μm	High detection limits, element interferences	Special proton accelerator, only elemental information	No molecular characterization, interference of Ar and O adducts	Rare, not many spectra available, poor with mixtures	Destructive, vacuum needed, poor precision	Limited number of elements, need radio isotopes necessary, poor data analysis	Low sensitivity, not great for solids, limited number of elements	nondestructive, Insensitive, only crystalline materials	Disadvantages
Availability	Very rare	XRF common, XRF MP only at synchrotron facilities	Moderately common	Rare	Common	Rare, at synchrotron facilities	Relatively uncommon	Very uncommon	Uncommon	Common
Matrix effects	Less	Yes	Yes	Yes	Yes, severe	Yes	Yes	Yes	Yes	Yes
References	36,69,97,98	69,97–103	68,69,97,104	36,72,97,104	69,105	36,69,96,106,107	36,69,108	31,36,69,109,98,110	36,69,111,112	69,113,114

used for one sample. However, attempts have been made to develop high-throughput techniques like high-throughput XAS (HT-XAS) by automating an array of small samples, rapid data collection of multiple low-volume low-concentration samples; also, data reduction and analysis has been proposed.[96] The schematic flowchart of a high-throughput X-ray absorption spectroscopy (HT-XAS) procedure proposed by Ascone et al.[45] and Scott et al.[96] is shown in Figure 1.10. A composite image to show the most important steps, in detail, of the metallogenomics pipeline is shown in Figure 1.11.[45] As Figures 1.10 and 1.11 show, the study of metalloproteins could be performed in the frame as a metallogenomics programme. Shi and Chance[25] and Ascone et al.[45] highlight the most important steps of metallogenomics: (1) identification of metalloproteins in genomic databases, (2) biochemical and molecular methods to produce proteins incorporating metal ions, (3) metalloprotein purification, and (4) structural methods appropriate to the study of the metalloproteins and in particular metal site(s) as shown in Figure 1.11.

The data of metal-site structures provided by the above procedure could be used to build a database of metal-binding sequence motifs. Such a database would be an important component for a future sequence-to-structure predictor. It is noticeable that biological information about relative expression levels, and post-translational processing and modification, is lost in this pipeline. To obtain more information about the biological relevance of metal sites, one can express or extract the entire proteome (from microorganisms grown under specific conditions, or the protein complement of cells of a given organ or tissue at a specific stage of development, for example), separate proteins and then characterize the individual metalloprotein by XAS. In this case, the expression level of metalloproteins and the protein separating technique are two problems that must be taken into account.

There is a wide range of analytical methods available with which the chemical composition of a sample can be determined. They differ in the covered range of concentration and of elements, in information depth, in accuracy etc. These characteristics can vary for each technique from one instrument to another. They have been estimated for standard instruments and considered the analysis of inorganic elements such as transition metals, within a biological matrix such as an isolated cell or a tissue section. But only a few of them are able to selectively analyse small sample areas. Table 1.5 provides a summary of the variety of analytical methods currently available for chemical element imaging, quantification, and speciation in metallmoics and metalloproteomics studies. Properties such as the elemental range covered, together with the depth of the analyzed sample volume are summarized for every method. The relation between the detectable concentration range and the spatial resolution is displayed in Figure 1.12, which shows that micro-XRF has already covered a relatively large concentration range together with an acceptable range of spatial resolution. Hence, it can be expected that this method will find a wide range of applications in the near future.

Figure 1.12 shows some of the analytical methods and their spatial resolution and concentration range subsequently (adapted from Kanngießer et al.[115]).

Introduction

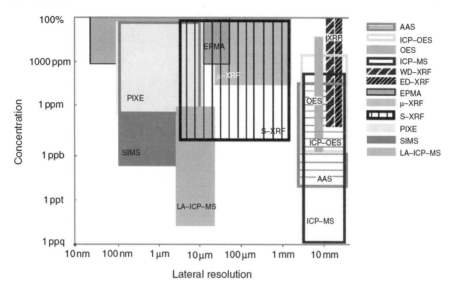

Figure 1.12 Analytical methods and their spatial resolution and concentration range.[115] © 2006 Springer-Verlag Berlin Heidelberg.

Herein, reports on simultaneous multi-elemental speciation in biological samples are summarized in Table 1.6. More examples focusing on specific techniques are included in the subsequent chapters, respectively.

In conclusion, some of the analytical techniques that are discussed above for metallomics studies, especially for high throughput analysis are summarized in Figure 1.13.

1.5 Purposes of the Book and Layout of the Chapters

This book is written by experts from disciplines as diverse as analytical chemistry, nuclear chemistry, environmental science, molecular biology, and medicinal chemistry in order to identify potential "hot spots" of metallomics and metalloproteomics. The scientific fundamentals of new approaches, like isotopic techniques combined with ICP-MS/ESI-MS/MS, the synchrotron radiation-based techniques, X-ray absorption spectroscopy, X-ray diffraction, and neutron scattering, as well as their various applications, with a focus on mercury, selenium, chromium, arsenic, iron and metal-based medicines are critically reviewed, which can help to understand their impacts on human health. The book will be of particular interest to researchers in the fields of environmental and industrial chemistry, biochemistry, nutrition, toxicology, and medicine. Basically, the book has two aims. The first deals with the educational point of view. Chapters 2 to 7 provide the basic concept of each of the selected nuclear analytical techniques and should be understandable by Master and PhD students in chemistry, physics, biology and nanotechnology. The

Table 1.6 Selected papers studying simultaneous multielemental speciation in biological samples using multiple state-of-the art techniques.

Element	Sample (matrix)	Method[a]	Main results	Detection limit	Ref.
As	Human blood, serum	HPLC-ICP-MS	Five As species were separated within 5 min from human blood serum by ion pair HPLC-ICP-MS. 20 μL sample was injected on a C18 column coated with phosphatidylcholine. The mobile phase was composed of 5 mM citrate buffer (pH 4.0)–5 mM SDS–5 mM TMAH–0.2 mM 3-[(3-cholamidopropyl)-dimethylammonio-1-propane sulfonate. The precision of the method, based upon analysis of 15 μg L^{-1} As, was better than 5.9% for all species	As(V): 3.1 ng g^{-1}; As(III): 2.7 ng g^{-1}; MMA: 4.5 ng g^{-1}; DMA: 2.5 ng g^{-1}; AB: 2.5 ng g^{-1}	116
As	Urine, food samples	MS; ICP; LC	Separation of As species was achieved within 3 min by ion-pair liquid chromatography on a commercial monolithic silica column using 2.5 mM tetrabutylammonium bromide, 10 mM phosphate buffer (pH 5.6) and 1.0% (v/v) MeOH as the mobile phase. The precision of the method, based upon analysis of 15 μg L^{-1} As, was better than 5.9% for all species.	As(III): 0.107 μg L^{-1}; As(V): 0.101 μg L^{-1}; MMA: 0.121 μg L^{-1}; DMA: 0.120 μg L^{-1}; AB: 0.084 μg L^{-1}	117
Cd	Bone phantoms	XRF	The comparison of three XRF systems for the measurement of Cd in bone phantoms showed that the cloverleaf hyperpure germanium system had the best LOD (2.1 μg g^{-1}) for 3 mm of overlying wax)	2.1 μg g^{-1}	118

Element	Sample	Technique	Description	Value	Ref
Cr	Rat liver	Enriched stable IT, SEC, INAA	Nine kinds of Cr-containing proteins were found		119
Fe	Rat brain	PIXE	Intracellular concentrations of Fe in the rat brain (approx. 30 µg g^{-1} dw) were determined by PIXE to assess differences between neurons with and without specialized structures (perineuronal net)	30 µg g^{-1} dw	120
Hg	Human hair and blood	XAS; ICP-MS	Provided quantitative speciation and structural information of Hg and S in hair samples and of Hg in erythrocyte and serum samples of people living in a mercury mine area. Established a method using nuclear XAS and ICP-MS techniques to provide speciation, structural and binding information for potential application in similar studies of other elements		88
Cu, Fe and Zn	Rat brain	NAA, SRXRF and XRF	The effect of iodine deficiency on the distribution of Cu, Fe and Zn in different regions and subcellular fractions of the brain of the developing rat was studied by means of nuclear analytical methods		121
As and Hg	Rabbit blood	LC-ICP-AES; XAS	Provided information on the molecular structure of compounds with As–Se and Hg–Se bonds, formed in the bloodstream simultaneously injected with As$_2$O$_3$ and SeO$_4^{2-}$ or HgCl$_2$ and SeO$_4^{2-}$		122
As and Se	Tuna fish	HPLC-ICP-MS	A method was developed for the simultaneous speciation of both arsenic and selenium. Thirteen arsenic and selenium species were studied using HPLC under elevated column temperatures followed by dual-element detection using ICP-MS	7–12 ng mL^{-1}	123

Table 1.6 (continued)

Element	Sample (matrix)	Method[a]	Main results	Detection limit	Ref.
Hg and Sn	Oyster tissue	CGC-ICP-ID-MS	A rapid, accurate, sensitive, and simple method for simultaneous speciation analysis of mercury and tin in biological samples has been developed. Demonstrated that isotope dilution analysis is a powerful method allowing the simultaneous speciation of TBT and MMHg with high precision and excellent accuracy. Analytical problems related to low recovery during sample preparation were minimized by SIDMS. Establish a performant routine method using CGC–ICPMS technique	Hg: 0.11–0.24 g kg^{-1}; Sn: 0.15–0.30 g kg^{-1}	124
Hg, Sn and Pb	Oyster tissue	SPME-MCGC-ICP-TOF-MS	A simple, rapid and accurate method on the basis of SPME in combination with MCGC linked to ICP-TOFMS was developed for simultaneous speciation analysis of 10 organometallic compounds of mercury (including inorganic mercury), tin and lead. Seven SPME fibers were compared in terms of extraction efficiency. Using MCGC separation, a total chromatographic run time below 200 s was obtained. With ICP-TOFMS detection peak widths at half height (FWHM) down to 0.3 s were measured without spectral skew because of the simultaneous character of the mass spectrometer	Sn, Pb: <pg g^{-1}; Hg: 1.3–2.0 pg g^{-1}	125

Analytes	Matrix	Technique	Description	Detection limit	Ref.
Hg, Pb and Sn	Human urine	SPME-GC/MS-MS	A GC/MS-MS method for the determination of Hg(II) and alkylated Hg, Pb, and Sn species in human urine is described. It offers a real multi-element/multi-species capability with low detection limits and a minimum of sample preparation	7–22 ng L^{-1}	126
Hg and Se	Human urine	HPLC-ICP-MS	Established a method for simultaneous speciation analysis of selenium and mercury. Batch-wise elution using two different mobile phases that are suitable for selenium and mercury speciation leads to successful determination of both selenium and mercury standards in 30 min with good efficiency and resolution	Se: 0.05–0.3 µg L^{-1}; Hg: 1.48–2.0 µg L^{-1}	127
Ge, Se, Sn, Sb, Hg	Human urine	HG/LT-GC/ICP-MS	Human urine samples after fish consumption have been investigated by HG/LT-GC/ICP-MS. This analytical technique enabled the identification of organometal(loid) compounds in human urine; species of the six elements germanium, arsenic, selenium, tin, antimony, and mercury were determined	2–12 pg L^{-1}	128
Se and As	Human urine	HPLC-ICP-MS	A procedure for the simultaneous determination of six selenium species and six arsenic species in human urine by ion-pair, reversed-phase liquid chromatography coupled to ICP-MS has been developed. The method was applied to the determination of arsenic and selenium species in urine, which only required filtering through a 0.45 µm membrane filter and dilution with mobile phase in order to measure arsenic and selenium urinary metabolites	As: 0.1–0.4 µg L^{-1}; Se: 0.7–2 µg L^{-1}	129

Table 1.6 (continued)

Element	Sample (matrix)	Method[a]	Main results	Detection limit	Ref.
EO-Cl, -Br, -I	Pine needles	NAA	Studied the concentrations and distribution of organochlorinated contaminants in pine needles (biomonitor to evaluate the levels of in the atmospheric environment) using instrument neutron activation analysis	Cl: 50 ng; Br: 8 ng; I: 3.5 ng	130

AES, atomic emission spectrometry; IC, ion chromatography; CGC, capillary gas chromatography; ID, isotope dilution; SPME, solid phase microextraction; MCGC, microcapillary gas chromatography; TOF, time of flight; HG, hydride generation; LT-GC, low temperature gas chromatography; NAA, Neutron activation analysis; HPLC, high-performance liquid chromatography; XAS, X-ray absorption spectroscopy; XRF, X-ray fluorescence spectrometry; SRXRF, synchrotron radiation X-ray fluorescence.

Introduction 35

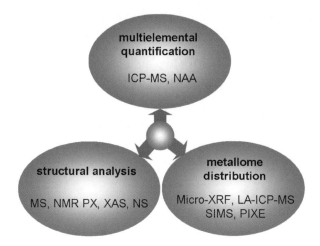

Figure 1.13 Selected analytical techniques used for metallomics studies. ICP-OES, inductively coupled plasma optical emission spectroscopy, ICP-MS, inductively coupled plasma mass spectrometry; LA-ICP-MS, laser ablation ICP-MS; XRF, X-ray fluorescence spectroscopy; PIXE, proton induced X-ray emission; NAA, neutron activation analysis; SIMS, secondary ion mass spectroscopy; GE, gel electrophoresis; LC, liquid chromatography; GC, gas chromatography; MS, mass spectrometry, which includes MALDI-TOF-MS, matrix-assisted laser desorption/ ionization time of flight mass spectrometry and ESI-MS, electron spray ionization mass spectrometry; NMR, nuclear magnetic resonance; PX, protein crystallography; XAS, X-ray absorption spectroscopy; NS, neutron scattering.

second aim is to provide a selected set of examples from recent literature of the latest results obtained with art-of-the-state techniques in metallomics and metalloproteomics. Thus, this book includes a number of illustrations, tables and documents relatd to the most up-to-date progress, as well as well-organized bibliographies of this novel emerging field.

The book is structured in two parts. In the first part, six chapters are included, with each devoted to a specific nuclear analytical technique and its application: Chapter 2, Neutron Activation Analysis (NAA, coordinated by Zhiyong Zhang); Chapter 3, X-ray Fluorescence (XRF, coordinated by Yuxi Gao); Chapter 4, Isotopic Techniques Combined with ICP-MS and ESI-MS (coordinated by Weiyue Feng); Chapter 5, Mössbauer Spectroscopy (coordinated by Chunying Chen); Chapter 6, X-ray Absorption Spectroscopy (XAS, coordinated by Chunying Chen) and Chapter 7, Protein Crystallography (PX, coordinated by Yuhui Dong). Each chapter has a dual structure: a first section devoted to an explanation of the basic concepts, providing the larger possible audience, and a second section devoted to a discussion of the most peculiar and innovative examples, allowing the book to have a longer shelf life.

In the second part, four chapters are included which show the integrated application of different analytical techniques, including nuclear analytical techniques in areas such as iron-omics, metallodrugs, and even metallic nanomaterials. Chapter 8 (coordinated by Guanjun Nie) introduces the application of nuclear analytical techniques in iron-omics while Chapter 9 (coordinated by Hongzhe Sun) shows the application of nuclear analytical techniques in metal-based drugs. Chapter 10 (coordinated by Chunying Chen) is devoted mostly to the imaging studies that use integrated nuclear analytical techniques. Last, but not least, Chapter 11 (coordinated by Chunying Chen) extends the concept of metallomics to the "hot" research field, nanoscience and nanotechnology, and, most importantly, proposes the new term "nanometallomics". Dedicated nuclear analytical techniques in nanometallomics are also introduced in this chapter.

1.6 Outlook

As an integrated biometal science, metallomics has received more and more attention. The complexity of metallomics in living organisms results in significant analytical challenges. The comprehensive quantification, distribution, speciation, identification, and structural characterization of metallomes require high-throughput and powerful analytical techniques. Thus, some of the analytical techniques which are or, at least, may be, promising for high-throughput in metallomics study have been reviewed. In general, high-throughput quantification of multielements can be achieved by ICP-MS and NAA while high-throughput multielement distribution mapping can be done by fluorescence-detecting techniques like SRXRF, XRF tomography, EDX, PIXE, and ion-detecting-based SIMS. LA-ICP-MS can also be used for multielement distribution mapping. When it comes to identification and structural characterization of metallomes, all the techniques are generally low-throughput. Therefore, this is likely to be the bottleneck in the study of metallomics. However, by automating arrays of small samples, rapid data collection of multiple low-volume low-concentration samples, and data reduction and analysis, high-throughput techniques may be achieved and, in fact, have partially been achieved.

Metallomics, understood as a holistic high-throughput metal speciation analysis in a biological system correlated with the genomics, proteomics, and metabolomics data, is taking shape as a multidisciplinary research field. While the performance of analytical techniques is steadily progressing,[34] advanced nuclear techniques are being used worldwide for metallomics and metallomproteomics studies in different biological and environmental systems. At present, these techniques are capable of providing speciation and structure data for analytical purposes. But, still, the speciation of some metals requires costly instrumentation, particularly for the use of elemental-selective spectroscopy, such as mass spectrometry (MS), which are not often feasible for conducting speciation analyses in developing and underdeveloped countries. Therefore, the role of analytical techniques is crucial in distributing speciation technologies worldwide. The techniques should involve inexpensive equipment that can be adapted to the needs of every nation, worldwide. And, as reproducibility in metal analysis is one of the serious and challenging issues for scientists, advancements are required in analytical techniques in terms of least interference and lowest limits of detection in unknown matrices. The techniques should be developed with good selectivity, sensitivity, reliability and reproducibility.

The future development of analytical techniques is a broad and opening topic. Applications of speciation methodologies may be increased by developing inexpensive and highly selective sample preparation devices, mobile and stationary phases and detectors. Briefly, growth in the advancement of advanced nuclear analytical techniques is increasing rapidly and continuously, and we hope that metallomics and metalloproteomics methodologies will be selective, sensitive, reproducible, and inexpensive in the near future. It is estimated that the prospects of metallomics and metalloproteomics are bright and will attract scientists and regulatory authorities for the betterment of human beings.

With the development of metallomics, especially high-throughput detection, distribution, speciation, and characterization of metallomes, more and more data will be obtained to gain a better understanding of the processes involved. *In silico* bioinformatics is therefore greatly needed to treat such a huge amount of data and perhaps the databases of metallome and metallome quantification, distribution, speciation, and structural information as well as methods to access, search, visualize, and retrieve the information. Thus, the study of metallomics and metalloproteomics will be not only for chemists involved in nuclear analytical techniques and speciation, but will also be important for environmental, nutritional and clinical researchers, and drug developers.

Acknowledgments

The authors acknowledge the financial support by the Ministry of Science and Technology of China as the National Basic Research Programs (2006CB705603, 2010CB934004 and 2009AA03J335), the Natural Science Foundation of China (10975040), the NSFC/RGC Joint Research Scheme (20931160430) and the Knowledge Innovation Program of the Chinese Academy of Sciences (KJCX2-YW-M02).

References

1. G. B. Smejkal, *Expert Rev. Proteomics*, 2006, **3**, 383–385.
2. R. Cornelis, J. Caruso, H. Crews and K. Heumann, *Handbook of Elemental Speciation: Techniques and Methodology*, Wiley, Chichester, 2003.
3. H. Bowen, *Trace elements in biochemistry*, Academic Press, London and New York, 1966.

4. J. Fraústo daSilva and R. Williams, *The Biological Chemistry of the Elements: The Inorganic Chemistry of Life*, Clarendon Press, Oxford, 1993.
5. G. W. Brummer, *The importance of chemical speciation in environmental processes*, Springer-Verlag, Berlin, 1986, **20**, 169–192.
6. W. Mertz, *Science*, 1981, **213**, 1332–1338.
7. N. Jakubowski, R. Lobinski and L. Moens, *J. Anal. At. Spectrom.*, 2004, **19**, 1–4.
8. R. Wagemann, E. Trebacz, G. Boila and W. Lockhart, *Sci. Total Environ.*, 1998, **218**, 19–31.
9. H. Harris, I. Pickering and G. George, *Am. Assoc. Advance. Sci.*, 2003, **301**, 1203.
10. E. Goldberg, *J. Geol.*, 1954, **62**, 249–265.
11. D. Templeton, F. Ariese, R. Cornelis, L. Danielsson, H. Muntau, H. Van Leeuwen and R. Lobinski, *Pure Appl. Chem.*, 2000, **72**, 1453–1470.
12. I. Ali and H. Aboul-Enein, *Instrumental Methods in Metal Ion Speciation*, CRC Press, Boca Raton, 2006.
13. F. Tack and M. Verloo, *Int. J. Environ. Anal. Chem.*, 1995, **59**, 225–238.
14. J. Caruso and M. Montes-Bayon, *Ecotoxicol. Environ. Safety*, 2003, **56**, 148–163.
15. R. Williams, *Coord. Chem. Rev.*, 2001, **216**, 583–595.
16. H. Haraguchi, *J. Anal. At. Spectrom.*, 2004, **19**, 5–14.
17. J. Szpunar, *Analyst*, 2005, **130**, 442–465.
18. D. W. Koppenaal and G. M. Hieftje, *J. Anal. At. Spectrom.*, 2007, **22**, 855–855.
19. S. Mounicou, J. Szpunar and R. Lobinski, *Chem. Soc. Rev.*, 2009, **38**, 1119–1138.
20. R. Lobinski, D. Schaumloffel and J. Szpunar, *Mass Spectrom. Rev.*, 2006, **25**, 255–289.
21. K. Hirschi, *Trends Biotechnol.*, 2003, **21**, 520–521.
22. K. Waldron and N. Robinson, *Nature Rev. Microbiol.*, 2009, **7**, 25–35.
23. C. Liu and H. Xu, *J. Inorg. Biochem.*, 2002, **88**, 77–86.
24. W. X. Shi, C. Y. Zhan, A. Ignatov, B. A. Manjasetty, N. Marinkovic, M. Sullivan, R. Huang and M. R. Chance, *Structure*, 2005, **13**, 1473–1486.
25. W. Shi and M. R. Chance, *Cell. Mol. Life Sci.*, 2008, **65**, 3040–3048.
26. M. de Bolster, R. Cammack, D. Coucouvanis, J. Reedijk and C. Veeger, *Pure Appl. Chem.*, 1997, **69**, 1251–1303.
27. I. Bertini, A. Sigel and H. Sigel, *Handbook on Metalloproteins*, CRC Press, Boca Raton, 2001.
28. S. Beranova-Giorgianni, *Trends Anal. Chem.*, 2003, **22**, 273–281.
29. E. Nagele, M. Vollmer, P. Horth and C. Vad, *Expert Rev. Proteomics*, 2004, **1**, 37–46.
30. K. A. Resing and N. G. Ahn, *FEBS Lett.*, 2005, **579**, 885–889.
31. Y. Gao, C. Chen and Z. Chai, *J. Anal. At. Spectrom.*, 2007, **22**, 856–866.
32. V. C. Wasinger, S. J. Cordwell, A. Cerpapoljak, J. X. Yan, A. A. Gooley, M. R. Wilkins, M. W. Duncan, R. Harris, K. L. Williams and I. Humpherysmith, *Electrophoresis*, 1995, **16**, 1090–1094.

33. P. Kahn, *Science*, 1995, **270**, 369–370.
34. S. Mounicou and R. Lobinski, *Pure Appl. Chem.*, 2008, **80**, 2565–2575.
35. C. W. Schmidt, *Environ. Health Perspect.*, 2004, **112**, A410–A415.
36. Y. F. Li, C. Y. Chen, Y. Qu, Y. X. Gao, B. Li, Y. L. Zhao and Z. F. Chai, *Pure Appl. Chem.*, 2008, **80**, 2577–2594.
37. N. Jakubowski and G. M. Hieftje, *J. Anal. At. Spectrom.*, 2008, **23**, 13–14.
38. D. E. Salt, I. Baxter and B. Lahner, *Annu. Rev. Plant Biol.*, 2008, **59**, 709–733.
39. A. Sanz-Medel, *Anal. Bioanal. Chem.*, 2005, **381**, 1–2.
40. F. Crick, *Nature*, 1970, **227**, 561–563.
41. A. Abbott, *Nature*, 1999, **402**, 715–720.
42. S. Srivastava, *J. Proteome Res.*, 2008, **7**, 1799.
43. A. D. Watson, *J. Lipid Res.*, 2006, **47**, 2101–2111.
44. H. Tweeddale, L. Notley-McRobb and T. Ferenci, *J. Bacteriol.*, 1998, **180**, 5109–5116.
45. I. Ascone, R. Fourme, S. Hasnain and K. Hodgson, *J. Synchrotron Radiat.*, 2004, **12**, 1–3.
46. R. Codd, *Chem. Commun.*, 2004, 2653–2655.
47. Z. Chai, *Nucl. Technol.*, 1999, **22**, 441–448.
48. A. K. Upadhyay, A. B. Hooper and M. P. Hendrich, *J. Am. Chem. Soc.*, 2006, **128**, 4330–4337.
49. M. Thompson and J. N. Walsh, *Handbook of Inductively Coupled Plasma Spectrometry*, Blackie, Glasgow, 1983.
50. I. Feldmann, N. Jakubowski, C. Thomas and D. Stuewer, *Fresenius' J. Anal. Chem.*, 1999, **365**, 422–428.
51. D. W. Koppenaal, G. C. Eiden and C. J. Barinaga, *J. Anal. At. Spectrom.*, 2004, **19**, 561–570.
52. Y. Li, C. Chen, B. Li, J. Sun, J. Wang, Y. Gao, Y. Zhao and Z. Chai, *J. Anal. At. Spectrom.*, 2006, **21**, 94–96.
53. Z. Chai, J. Sun and S. Ma, *Neutron Activation Analysis in Environmental Sciences, Biological and Geological Sciences*, Atomic Energy Press, Beijing, 1992.
54. Z. Chai and H. Zheng, *Introduction to Trace Element Chemistry*, Atomic Energy Press, Beijing, 1994.
55. J. Alexander, Y. Thomassen and J. Aaseth, *J. Appl. Toxicol.*, 1983, **3**, 143–145.
56. I. Falnoga, M. Tušek-Žnidaric, M. Horvat and P. Stegnar, *Environ. Res.*, 2000, **84**, 211–218.
57. F. Pinakidou, M. Katsikini, E. C. Paloura, P. Kavouras, T. Kehagias, P. Komninou, T. Karakostas and A. Erko, *J. Hazard. Mater.*, 2007, **142**, 297–304.
58. C. Chen, H. Yu, J. Zhao, B. Li, L. Qu, S. Liu, P. Zhang and Z. Chai, *Environ. Health Perspect.*, 2006, **114**, 297–301.
59. Y. Gao, N. Liu, C. Chen, Y. Luo, Y.-F. Li, Z. Zhang, Y. Zhao, Y. Zhao, A. Iida and Z. Chai, *J. Anal. At. Spectrom.*, 2008, **23**, 1121–1124.

60. R. Ortega, P. Cloetens, G. Devès, A. Carmona and S. Bohic, *PLoS ONE*, 2007, **2**, e925.
61. A. Snigirev, V. Kohn, I. Snigireva and B. Lengeler, *Nature*, 1996, **384**, 49–51.
62. A. Carmona, P. Cloetens, G. Devès, S. Bohic and R. Ortega, *J. Anal. At. Spectrom.*, 2008, **23**, 1083–1088.
63. L. Vincze, B. Vekemans, I. Szaloki, K. Janssens, R. Van Grieken, H. Feng, K. W. Jones and F. Adams, *Proc. SPIE 4503* (2002) 240–248.
64. S. D. Maind, S. A. Kumar, N. Chattopadhyay, C. Gandhi and M. Sudersanan, *Forensic Sci. Int.*, 2006, **159**, 32–42.
65. C. Hansel, M. La Force, S. Fendorf and S. Suttons, *Environ. Sci. Technol.*, 2002, **36**, 1988–1994.
66. M.-P. Isaure, B. Fayard, G. Sarret, S. Pairis and J. Bourguignon, *Spectrochim. Acta B.*, 2006, **61**, 1242–1252.
67. K. Kametani and T. Nagata, *Med. Mol. Morphol.*, 2006, **39**, 97–105.
68. M. Motelica-Heino, P. Le Coustumer, J. Thomassin, A. Gauthier and O. Donard, *Talanta*, 1998, **46**, 407–422.
69. J. J. D'Amore, S. R. Al-Abed, K. G. Scheckel and J. A. Ryan, *J. Environ. Qual.*, 2005, **34**, 1707–1745.
70. W. Przybyowicz, J. Mesjasz-Przybyowicz, C. Pineda, C. Churms, C. Ryan, V. Prozesky, R. Frei, J. Slabbert, J. Padayachee and W. Reimold, *X-Ray Spectrom.*, 2001, **30**, 156–163.
71. E. Johansson, U. Lindh, H. Johansson and C. Sundstrom, *Nucl. Instrum. Methods Phys. Res. Sect. B.*, 1987, **22**, 179–183.
72. K. Ishii, S. Matsuyama, Y. Watanabe, Y. Kawamura, T. Yamaguchi, R. Oyama, G. Momose, A. Ishizaki, H. Yamazaki and Y. Kikuchi, *Nucl. Instrum. Methods Phys. Res. Sect. A*, 2007, **571**, 64–68.
73. W. P. Meurer and D. T. Claeson, *J. Petrol.*, 2002, **43**, 607–629.
74. W. Devos, M. Senn-Luder, C. Moor and C. Salter, *Fresenius' J. Anal. Chem.*, 2000, **366**, 873–880.
75. T. Punshon, B. P. Jackson, P. M. Bertsch and J. Burger, *J. Environ. Monit.*, 2004, **6**, 153–159.
76. J. S. Becker, S. Mounicou, M. V. Zoriy, J. S. Becker and R. Lobinski, *Talanta*, 2008, **76**, 1183–1188.
77. K. Park and N. Kang, *Talanta*, 2007, **73**, 791–794.
78. A. Raith, J. Godfrey and R. C. Hutton, *Fresenius' J. Anal. Chem.*, 1996, **354**, 163–168.
79. A. Kindness, C. N. Sekaran and J. Feldmann, *Clin. Chem.*, 2003, **49**, 1916–1923.
80. A. Benninghoven, F. G. Rüdenauer and H. W. Werner, *Secondary Ion Mass Spectrometry: Basic Concepts, Instrumental Aspects, Applications and Trends*, Wiley, New York, 1987.
81. B. Y. D. Briggs, A. Brown and J. C. Vickerman, *Anal. Chem.*, 1988, **60**, 1791–1799.
82. R. G. Wilson, F. A. Stevie and C. W. Magee, *Secondary Ion Mass Spectrometry*, Wiley, New York, 1989.

83. D. B. Lazof, J. G. Goldsmith, T. W. Rufty and R. W. Linton, *Plant Physiol.*, 1994, **106**, 1107–1114.
84. V. A. P. Freitas, Y. Glories, G. Bourgeois and C. Vitry, *Phytochemistry*, 1998, **49**, 1435–1441.
85. S. Chandra and G. H. Morrison, *Biol. Cell.*, 1992, **74**, 31–42.
86. Z. Chai, X. Mao, Z. Hu, Z. Zhang, C. Chen, W. Feng, S. Hu and H. Ouyang, *Anal. Bioanal. Chem.*, 2002, **372**, 407–411.
87. Q. W. Wang and W. H. Liu, *X-ray absorption fine structure and its application*, China Science Press, Beijing, 1994.
88. Y.-F. Li, C. Chen, B. Li, W. Li, L. Qu, Z. Dong, M. Nomura, Y. Gao, J. Zhao, W. Hu, Y. Zhao and Z. Chai, *J. Inorg. Biochem.*, 2008, **102**, 500–506.
89. Y.-F. Li, C. Chen, L. Xing, T. Liu, Y. Xie, Y. Gao, B. Li, L. Qu and Z. Chai, *Nucl. Technol.*, 2004, **27**, 899–903.
90. S. S. Hasnain and R. W. Strange, *J. Synchrotron Radiat.*, 2003, **10**, 9–15.
91. I. Hazemann, M. T. Dauvergne, M. P. Blakeley, F. Meilleur, M. Haertlein, A. Van Dorsselaer, A. Mitschler, D. A. Myles and A. Podjarny, *Acta Crystallogr. Sect. D, Biol. Crystallogr.*, 2005, **61**, 1413–1417.
92. L. A. Feigin and D. I. Svergun, *Structure Analysis by Small-Angle X-Ray and Neutron Scattering*, Plenum Press, New York, 1987.
93. X. Liang, *Nuclear Magnetic Resonance*, Scientific Press, Beijing, 1976.
94. D. S. Wishart, B. D. Sykes and F. M. Richards, *Biochemistry*, 1992, **31**, 1647–1651.
95. A. Savchenko, A. Yee, A. Khachatryan, T. Skarina, E. Evdokimova, M. Pavlova, A. Semesi, J. Northey, S. Beasley and N. Lan, *Proteins Struct. Funct. Genet.*, 2003, **50**, 392–399.
96. R. A. Scott, J. E. Shokes, N. J. Cosper, F. E. Jenney and M. W. W. Adams, *J. Synchrotron Radiat.*, 2005, **12**, 19–22.
97. M. Necemer, P. Kump, J. Scancar, R. Jadmovic, J. Simcic, P. Pelicon, M. Budnar, Z. Jeran, P. Pongrac, M. Regvar and K. Vogel-Mikus, *Spectrochim. Acta Part B*, 2008, **63**, 1240–1247.
98. H. Verma, *Atomic and Nuclear Analytical Methods: XRF, Mössbauer, XPS, NAA and Ion-beam Spectroscopic Techniques*, Berlin, Springer Verlag, 2007.
99. A. Taylor, S. Branch, M. P. Day, M. Patriarca and M. White, *J. Anal. At. Spectrom*, 2009, **24**, 535–579.
100. N. Szoboszlai, Z. Polgari, V. G. Mihucz and G. Zaray, *Anal. Chim. Acta*, 2009, **633**, 1–18.
101. P. T. Palmer, R. Jacobs, P. E. Baker, K. Ferguson and S. Webber, *J. Agric. Food Chem.*, 2009, **57**, 2605–2613.
102. A. G. Karydas, *Ann. Chim. (Rome, Italy)*, 2007, **97**, 419–432.
103. R. Tertian and F. Claisse, *Principles of Quantitative XRF Analysis*, Heyden, London, 1982.
104. M. L. Carvalho, T. Magalhaes, M. Becker and A. von Bohlen, *Spectrochim. Acta Part B*, 2007, **62**, 1004–1011.
105. A. Montaser, *Inductively Coupled Plasma Mass Spectrometry*, Wiley-VCH, Frankfurt, 1998.

106. I. Ascone and R. Strange, *J. Synchrotron Radiat.*, 2009, **16**, 413–421.
107. F. Liu, A. Gentles and C. W. Theodorakis, *Chemosphere*, 2008, **71**, 1369–1376.
108. V. I. Slaveykova, C. Guignard, T. Eybe, H. N. Migeon and L. Hoffmann, *Anal. Bioanal. Chem.*, 2009, **393**, 583–589.
109. Z. Chai, Z. Zhang, W. Feng, C. Chen, D. Xu and X. Hou, *J. Anal. At. Spectrom.*, 2004, **19**, 26–33.
110. X. Sun, C. Tsang and H. Sun, *Metallomics*, 2009, **1**, 25–31.
111. K. Wuthrich, *The George Fisher Baker non-resident lectureship in chemistry at Cornell University*, Wiley, New York (USA), 1986.
112. W. B. Dunn, N. J. C. Bailey and H. E. Johnson, *Analyst*, 2005, **130**, 606–625.
113. B. Warren, *X-ray Diffraction*, Courier Dover Publications, New York, 1990.
114. B. Cullity and J. Weymouth, *Am. J. Phys.*, 1957, **25**, 394.
115. B. Kanngießer, M. Haschke, A. Simionovici, P. Chevallier, C. Streli, P. Wobrauschek, L. Fabry, S. Pahlke, F. Comin, R. Barrett, P. Pianetta, K. Lüning, B. Beckhoff, V. Rößiger, B. Nensel, S. Kurunczi, J. Osán, S. Török, M. Betti, D. Rammlmair, M. Wilke, K. Rickers, R. A. Schwarzer, A. Möller, A. Wittenberg, O. Hahn, I. Reiche, H. Stege, J. Engelhardt, J. Zieba-Palus, G. Weseloh, S. Staub, J. Feuerborn and P. Hoffmann, in *Handbook of Practical X-Ray Fluorescence Analysis*, Springer-Verlag, Berlin Heidelberg, 2006, pp. 433–833.
116. T. Hasegawa, J. Ishise, Y. Fukumoto, H. Matsuura, Y. Zhu, T. Umemura, H. Haraguchi, K. Yamamoto and T. Naoe, *Bull. Chem. Soc. Jpn.*, 2007, **80**, 498–502.
117. G. Pearson, G. Greenway, E. Brima and P. Haris, *J. Anal. At. Spectrom.*, 2007, **22**, 361–369.
118. M. Popovic, D. Chettle, F. McNeill and A. Pejovi-Mili, *J. Radioanal. Nucl. Chem.*, 2006, **269**, 421–424.
119. W. Feng, B. Li, J. Liu, Z. Chai, P. Zhang, Y. Gao and J. Zhao, *Anal. Bioanal. Chem.*, 2003, **375**, 363–368.
120. A. Fiedler, T. Reinert, M. Morawski, G. Brückner, T. Arendt and T. Butz, *Nucl. Instrum. Methods Phys. Res. Sect. B*, 2007, **260**, 153–158.
121. F. Zhang, N. Liu, W. Feng, X. Wang, Y. Huang, W. He and Z. Chai, *J. Radioanal. Nucl. Chem.*, 2006, **269**, 535–540.
122. J. Gailer, *Coord. Chem. Rev.*, 2007, **251**, 234–254.
123. X. Le, X. Li, V. Lai, M. Ma, S. Yalcin and J. Feldmann, *Spectrochim. Acta Part B*, 1998, **53**, 899–909.
124. M. Monperrus, R. Martin-Doimeadios, J. Scancar, D. Amouroux and O. Donard, *Anal. Chem*, 2003, **75**, 4095–4102.
125. P. Jitaru, H. Infante and F. Adams, *J. Anal. At. Spectrom.*, 2004, **19**, 867–875.
126. L. Dunemann, H. Hajimiragha and J. Begerow, *Fresenius' J. Anal. Chem.*, 1999, **363**, 466–468.

127. Y. Li, C. Chen, B. Li, Q. Wang, J. Wang, Y. Gao, Y. Zhao and Z. Chai, *J. Anal. At. Spectrom.*, 2007, **22**, 925–930.
128. J. Kresimon, U. Grüter and A. Hirner, *Fresenius' J. Anal. Chem.*, 2001, **371**, 586–590.
129. F. Pan, J. Tyson and P. Uden, *J. Anal. At. Spectrom.*, 2007, **22**, 931–937.
130. D. Xu, W. Zhong, L. Deng, Z. Chai and X. Mao, *Environ. Sci. Technol.*, 2003, **37**, 1–6.

CHAPTER 2
Neutron Activation Analysis

ZHIYONG ZHANG*

CAS Key Laboratory of Nuclear Analytical Techniques, Institute of High Energy Physics, Chinese Academy of Sciences, Beijing 100049, China

2.1 Introduction

Neutron activation analysis (NAA) is a sensitive analytical technique useful for performing both qualitative and quantitative multielement analysis of major, minor, and trace elements in samples from almost every conceivable field of scientific or technical interest. NAA was discovered in 1936 by Hevesy and Levi. They found that Dy in samples became radioactive when bombarded by neutrons. From this observation, they quickly recognized the potential of employing nuclear reactions for the determination of elements present in different samples. Neutron activation analysis was soon established as a method of qualitative and quantitative element analysis.

The most common type of nuclear reaction used for NAA is the neutron capture or (n,γ) reaction. When a neutron interacts with a target nucleus via a non-elastic collision, a compound nucleus forms in an excited state. The excitation energy of the compound nucleus is due to the binding energy of the neutron with the nucleus. The compound nucleus will almost instantaneously de-excite into a more stable configuration through emission of one or more characteristic prompt gamma rays. In many cases, this new configuration yields a radioactive nucleus which also de-excites (or decays) by emission of one or more characteristic delayed gamma rays which are unique in half-life and energy, but at a much slower rate according to the unique half-life of the radioactive nucleus. These distinct energy signatures provide positive identification of the targeted

---

*Corresponding author: Email: zhangzhy@ihep.ac.cn; Tel: +86-10-88233215; Fax: +86-10-88235294

Nuclear Analytical Techniques for Metallomics and Metalloproteomics
Edited by Chunying Chen, Zhifang Chai and Yuxi Gao
© Royal Society of Chemistry 2010
Published by the Royal Society of Chemistry, www.rsc.org

element(s) present in the sample, while quantification is achieved by measuring the intensity of the emitted gamma rays that are directly proportional to the amounts of the respective element(s) in the sample.

The probability of a neutron interacting with a nucleus is a function of the neutron energy. This probability is referred to as the capture cross-section, and each nuclide has its own neutron energy–capture cross-section relationship. For many nuclides, the capture cross-section is greatest for low-energy neutrons (referred to as thermal neutrons). Some nuclides have the greater capture cross-sections for higher energy neutrons (epithermal neutrons). For routine neutron activation analysis we are generally looking at nuclides that are activated by thermal neutrons.

With respect to the time of measurement, NAA falls into two categories: (1) prompt gamma-ray neutron activation analysis (PGNAA), where measurements take place during irradiation, or (2) delayed gamma-ray neutron activation analysis (DGNAA), where the measurements follow radioactive decay. The latter operational mode is more common.

2.1.1 Activation and Analysis

Although there are several types of neutron sources (reactors, accelerators, and radioisotope neutron sources) that can be used for NAA, nuclear reactors with their high fluxes of neutrons from uranium fission offer the highest available sensitivities for most elements.

The activity for a particular radionuclide, at any time, t, during an irradiation, can be calculated from the following equation:

$$A_t = \sigma \varphi N (1 - e^{-\lambda t}) \qquad (2.1)$$

where A_t is the activity in number of decays per unit time, σ is the activation cross-section, φ is the neutron flux (in number of neutrons $cm^{-2} s^{-1}$), N is the number of parent atoms, λ is the decay constant (number of decays per unit time), and t is the irradiation time. From this equation we can see that the total activity for a particular nuclide is a function of the activation cross-section, the neutron flux, the number of parent atoms, and the irradiation time. Note that for any particular radioactive nuclide radioactive decay is occurring during irradiation, hence the total activity is determined by the rate of production minus the rate of decay.

Figure 2.1 shows a schematic figure of a gamma spectroscopy system for NAA. After the sample has been activated, the resulting gamma ray energies and intensities are determined using a solid-state detector (usually Ge). Gamma rays passing through the detector generate free electrons. The number of electrons (current) is related to the energy of the gamma ray. Each radioactive nuclide is also decaying during the counting interval and corrections must be made for this decay. The standard form of the radioactivity decay correction is:

$$A = A_o e^{-\lambda t} \qquad (2.2)$$

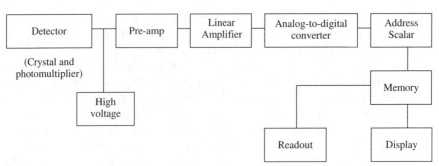

Figure 2.1 A gamma spectroscopy system for NAA.

where A is the activity at any time t, A_o is the initial activity, λ is the decay constant and t is time. Given the differences in half-lives for various nuclides, there are optimum times to count an activated sample. In general, nuclides with relatively short half-lives, on the order of hours to days, are determined within the first week of irradiation. Nuclides with half-lives on the order of weeks to months are determined 4–8 weeks after irradiation. Hence, activated samples are counted several times after irradiation.

Quantification of NAA can be accomplished by using the relative method or the absolute method. The relative method is based on the simultaneous irradiation of the sample with standards of known quantities of the elements in question in identical positions, followed by measuring the induced intensities of both the standard and the sample in a well-known geometrical position. In the absolute method, the efficiency of the detection system at several energies is determined and an efficiency-versus-energy curve is plotted. The efficiency of the detector is then used to determine the absolute activity of the sample.

With the use of automated sample handling, gamma ray measurement with solid-state detectors, and computerized data processing it is generally possible to simultaneously measure more than 30 elements in most sample types without chemical processing. The application of purely instrumental procedures is commonly called instrumental neutron activation analysis (INAA) and is one the most important advantages of NAA over other analytical techniques. If chemical separations of samples are carried out after irradiation to isolate one or more elements from the sample matrix or to concentrate the radioisotope of interest, the technique is called radiochemical neutron activation analysis (RNAA). Isolation of the element is performed to eliminate any spectral interferences and to lower the background. Using RNAA it is possible to achieve lower detection limits than are obtained using conventional INAA. RNAA has been used to analyze a wide variety of sample types. For example, RNAA has been used to determine the concentrations of tin and platinum in human liver tissue, and to determine cadmium and mercury in botanical reference materials. However, RNAA is performed infrequently due to its high labor cost.

2.1.2 Sensitivities

The sensitivities for NAA are dependent upon the irradiation parameters (i.e. neutron flux, irradiation and decay times), measurement conditions (i.e. measurement time, detector efficiency), nuclear parameters of the elements being measured (i.e. isotope abundance, neutron cross-section, half-life, and gamma-ray abundance). The accuracy of an individual NAA determination usually ranges between 1% and 10% of the reported value. Table 2.1 lists the approximate sensitivities for determination of elements assuming interference-free spectra.

2.1.3 Strengths and Limitations

In the decades after it became available in the mid-1940s, neutron activation analysis was considered to be the pre-eminent analytical method because few, if any, alternative techniques could match its high sensitivity (ppm or ppb) and accuracy. Even now that inductively coupled plasma mass spectrometry (ICP-MS) is available, NAA has the potential for superior accuracy because ICP-MS is more subject to matrix effects and interferences, at least for trace analysis.[2]

The particular advantages of INAA are that:

- It is a multielement technique capable of determining approximately 65 elements in many types of materials
- It is non-destructive and therefore does not suffer from the errors associated with sample digestion
- It has very high sensitivities for most of the elements that can be determined by INAA
- It is highly precise and accurate
- It permits the analysis of samples ranging in volume from 0.1 mL to 20 mL, and in mass from ~ 0.001 g to 10 g depending on sample density.
- Samples for INAA can be solids, liquids, gases, mixtures, and suspensions

Table 2.1 Estimated detection limits for INAA using decay gamma rays.[a1]

Sensitvity (pg)	Elements
1	Dy, Eu
1–10	In, Lu, Mn
10–100	Au, Ho, Ir, Re, Sm, W
100–1000	Ag, Ar, As, Br, Cl, Co, Cs, Cu, Er, Ga, Hf, I, La, Sb, Sc, Se, Ta, Tb, Th, Tm, U, V, Yb
1×10^3 to 1×10^4	Al, Ba, Cd, Ce, Cr, Hg, Kr, Gd, Ge, Mo, Na, Nd, Ni, Os, Pd, Rb, Rh, Ru, Sr, Te, Zn, Zr
1×10^4 to 1×10^5	Bi, Ca, K, Mg, P, Pt, Si, Sn, Ti, Tl, Xe, Y
1×10^5 to 1×10^6	F, Fe, Nb, Ne
1×10^7	Pb, S

[a]Assuming irradiation in a reactor neutron flux of 1×10^{13} n cm^{-2} s^{-1}. Reprinted with permission from the author M.D. Glascock.

In many applications INAA may be the best and only analytical technique required. In others it may comprise one of a suite of analytical methods or may serve as a primary calibration or reference method.

There are very few limitations. The major limitation is the detection limits of NAA are varied from one element to another (Table 2.1). The analytical sensitivities of some elements of great biological importance, such as lead, are low. Also, the number of elements that can be analyzed by this technique is another limitation.

2.2 Development of NAA for Metallomics and Metalloproteomics

The chemical speciation study of trace elements in life sciences has been paid more and more attention in recent years, mainly because it can provide more significant information on the pathway, distribution, accumulation, excretion, and functions of trace elements in biological systems of interest than the traditional bulk composition study. Almost all speciation techniques consist of two steps. The first step involves the separation of species from the sample followed by the second step of element specific detection. The so-called molecular neutron activation analysis (MoNAA)[3–6] or speciation neutron activation analysis (SNAA)[7] is, in fact, a combination of conventional NAA with physical, chemical, or biological separation procedures in order to meet the ever-increasing need for chemical species study.

For example, in order to study the chemical species of trace elements in biological samples, e.g. distribution patterns of trace elements in cells and subcellular fractions, and their combination with biological macromolecules (e.g. proteins, enzymes, or nuclear acids), the first step is to selectively separate various species fractions, followed by identification and determination. For this purpose the physical or chemical characteristics of biological macromolecules, e.g. size, charge, solubility, mobility or specificity of biological functions, are often utilized. Differential centrifugation, size exclusion chromatography, ion chromatography, polyacrylamide gel electrophoresis, and DNA electrophoresis, for example, are commonly employed methods that are combined with NAA.

2.2.1 Differential Centrifugation–NAA

It is well known that the cell is the unit of structure and function of living things. Within cells there is an intricate network of organelles that all have unique functions. These organelles allow the cell to function properly. Analysis of the intracellular distribution of trace elements of interests will shed some light on their physiological and pathological roles. Differential centrifugation is a common procedure in microbiology and cytology used to separate certain organelles from whole cells for further analysis of specific parts of cells. In the process, a tissue sample is first homogenized to break the cell membranes and mix up the cell contents. The homogenate is then subjected to repeated centrifugations, each time removing the pellet and increasing the centrifugal force. Finally, purification may be done through density gradient centrifugation, and the desired layer is extracted for further analysis.

Separation is based on size and density, with larger and denser particles pelleting at lower centrifugal forces. A scheme of differential centrifugation employed for isolation of subcellular fractions of animal cells is given in Figure 2.2. In general, the following cell components can be enriched, in the separating order in actual application: nuclei, mitochondria, lysosomes, microsomes, and cytosol.

Before differential centrifugation can be carried out to separate different portions of a cell from one another, the tissue sample must first be homogenized. The tissue sample is minced and a buffer solution is added to it, forming a liquid suspension of minced tissue sample. The buffer solution is a dense, inert, aqueous solution which is designed to suspend the sample in a liquid medium without damaging it through chemical reactions or osmosis. In most cases, the solution used is sucrose solution. The suspension is transferred into an all-glass tissue homogenizer. The cell membranes of the sample cells are broken by the grinding process, leaving individual organelles suspended in the solution. Then the homogenate is filtered to remove clumps of unbroken cells and connective tissue, and separated into different fractions by successive differential centrifugation. Finally, each fraction was lyophilized and subjected to NAA.

It should be noted that sucrose, which is used in isotonic solution, will remain in samples and affect the accurate weight of subcellular fractions. Chen et al.[8] suggested that ammonium acetate is a better isotonic solute than sucrose

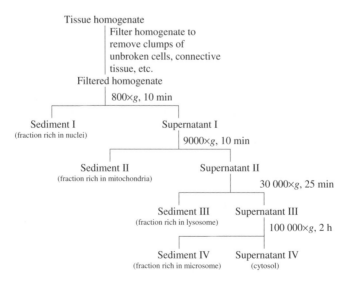

Figure 2.2 Scheme of differential centrifugation employed for isolation of subcellular fractions of animal cells.

because it can be volatilized by freeze drying. An alternative choice is to express the concentrations of trace elements in the organelles by weight of protein which can be determined by colorimetric methods.

Studies on subcellular distributions of trace elements could be greatly influenced by contaminants if proper cautions are not taken. All items of glassware, including the homogenizer, should be precleaned by soaking them in 4 mol L^{-1} nitric acid for more than 48 h followed by several rinsings with ultrapure water. The knife used for cutting samples should be titanium or Teflon® in order to avoid contamination. The reagent blanks also should be considered.

Subcellular distributions of trace elements in human liver,[8–14] human brain[15] and brain tumor,[16] beef heart,[17] bovine kidney,[18] rat tissues,[19–23] plant leaves,[24,25] have been studied by differential centrifugation–NAA. Tables 2.2 and 2.3 give the distribution of 24 elements in whole tissues and their subcellular fractions of human liver.

2.2.2 Liquid Chromatography–NAA

Liquid chromatography, or column chromatography, is an ideal technique for purification and analysis of biomacromolecules. If the sample solution is in contact with a second solid or liquid phase, the different solutes will interact with the other phase to differing degrees due to differences in adsorption, ion exchange, partitioning, or size. These differences allow the mixture components to be separated from each other by using these differences to determine the transit time of the solutes through a column. Simple liquid chromatography consists of a column with a fritted bottom that holds a stationary phase in equilibrium with a solvent. Typical stationary phases (and their interactions with the solutes) are: solids (adsorption chromatography), ionic groups on a resin (ion-exchange chromatography), liquids on an inert solid support (partitioning chromatography), and porous inert particles (size-exclusion chromatography). The mixture to be separated is loaded onto the top of the column followed by more solvent. The different components in the sample mixture pass through the column at different rates due to differences in their partitioning behavior between the mobile liquid phase and the stationary phase. The compounds are separated by collecting aliquots of the column effluent as a function of time.[26]

In order to study the chemical species of trace elements in biological samples, the elemental contents in the effluents will be determined after separation. The multielement character of NAA, as well as its high sensitivity for many elements, is ideal for the task. Since the column effluents are liquid, the fractions are usually freeze dried, then subjected to NAA.

Evans and Fritze[27] studied the distribution of protein-bound copper in human serum by a combination of gel chromatography and NAA. Norheim and Steinnes[28] described a method involving isolation of liver proteins followed by fractionation using gel filtration and subsequent determination of the elements

Table 2.2 Comparison of present results with some reported elemental contents in human liver of normal subjects in different areas of the world[9] ($\mu g \cdot g^{-1}$) © 2004 Springer-Verlag.

Area Year Number $\mu g\,g^{-1}$	USA 1983/1984 n=36 Wet weight	Canada 1985 n=143 Wet weight	UK 1972/1973 n=6–11 Wet weight	Japan 1981 n=9 Dry weight	Shanghai, China 1991 n=11–24 Dry weight	Tianjin, China 1996/2000 n=30–92 Dry weight
As	–	–	–	–	0.16±0.11 (24)	0.069±0.037 (43)
Br	2.62 (2)	–	4.0±0.03 (9)	–	4.33±1.10 (11)	2.32±0.93 (92)
Ca	–	33–181	54.3±6.7 (9)	610±380	217±103 (5)	320±128 (77)
Cd	≤0.2–5.20	0.1–9.1	4.3±1.0 (11)	7.1±4.1	8.6±5.3 (24)	5.92±3.45 (90)
Ce	–	–	–	–	–	0.267±0.165 (69)
Co	0.0468 (2)	–	–	0.31±0.10	0.155±0.056 (11)	0.120±0.045 (92)
Cr	0.0255 (2)	–	0.08±0.06 (11)	0.53±0.34	0.654±0.369 (7)	0.36±0.39 (53)
Cs	0.0149 (1)	–	–	–	–	0.038±0.014 (91)
Fe	313 (2)	51–997	176.0±28.9 (9)	1100±970	1160±1210 (11)	705±321 (92)
K	2860 (2)	815–3250	2400±100 (9)	7200±130	–	8150±1050 (30)
La	0.0205 (2)	–	0.08±0.03 (11)	–	–	0.135±0.088 (92)
Mo	354 (2)	–	0.4±0.2 (11)	1.50±0.59	–	3.35±1.10 (92)
Na	970 (2)	589–2660	–	5900±1900	2800±1000 (11)	3330±1300 (92)
Rb	9.52 (1)	–	7.0±1.0 (9)	–	44.3±15.2 (12)	27.2±6.4 (92)
Sb	11.4 (2)	–	0.010±0.002 (11)	–	–	0.050±0.049 (89)
Sc	0.00028 (2)	–	–	–	–	0.0056±0.0041 (92)
Se	0.34–0.69	0.20–1.02	0.3±0.1 (11)	–	1.45±0.38 (24)	0.76±0.21 (30)
Sm	–	–	–	–	–	0.0045±0.0021 (71)
Zn	97.2–27.0	20–183	75.8±9.6 (9)	210±89	210±55 (24)	186±41 (92)

Values in parentheses indicate the real number of samples measured.

Table 2.3 The elemental concentration in subcellular fractions of human liver (in $\mu g\, g^{-1}$ dry weight, unless indicated).[11]
© 2000 Akadémiai Kiadó, Budapest.

Element	Nuclei	Mitochondria	Lysosome	Microsome	Cytosol
Al	67.9 ± 38.5	34.0 ± 12.2	22.3 ± 11.0	53.0 ± 32.3	16.2 ± 1.7
As	0.69 ± 0.35	0.27 ± 0.18	0.18 ± 0.27	0.15 ± 0.07	0.08 ± 0.03
Au, ng g^{-1}	1.25 ± 0.91	0.93 ± 0.51	0.53 ± 0.35	0.86 ± 0.34	0.46 ± 0.23
Ba	9.13 ± 6.09	5.16 ± 2.21	3.22 ± 1.47	2.87 ± 1.47	1.57 ± 0.86
Br	1.78 ± 0.35	1.62 ± 0.49	0.97 ± 0.49	1.51 ± 0.56	3.65 ± 1.62
Ca	160 ± 40	128 ± 39	149 ± 116	249 ± 178	183 ± 84
Cd	1.02 ± 0.75	1.88 ± 1.43	0.92 ± 0.86	1.31 ± 1.01	4.37 ± 2.62
Cl	2030 ± 329	1320 ± 542	1350 ± 815	2340 ± 676	7850 ± 3390
Co	0.124 ± 0.040	0.127 ± 0.050	0.048 ± 0.030	0.068 ± 0.079	0.138 ± 0.076
Cs, ng g^{-1}	16.0 ± 4.4	14.2 ± 4.5	17.1 ± 2.9	19.2 ± 6.7	73.1 ± 17.8
Cu	24.8 ± 5.1	23.2 ± 6.7	17.4 ± 6.3	41.2 ± 36.8	43.2 ± 20.2
Fe	516 ± 304	410 ± 220	428 ± 240	3800 ± 3240	291 ± 122
Hg	0.043 ± 0.012	0.070 ± 0.025	0.037 ± 0.022	0.065 ± 0.050	0.065 ± 0.009
I	0.69 ± 0.25	0.58 ± 0.32	0.43 ± 0.28	0.33 ± 0.18	0.16 ± 0.06
In, ng g^{-1}	18 ± 10	13 ± 5	8.3 ± 5.2	20 ± 12	25 ± 7
K	5360 ± 1220	3760 ± 1670	3800 ± 1590	5880 ± 969	13500 ± 2770
Mg	499 ± 47	387 ± 151	247 ± 143	435 ± 179	523 ± 56
Mn	4.7 ± 1.9	3.8 ± 1.6	2.5 ± 1.3	5.5 ± 1.9	3.7 ± 1.7
Mo	4.5 ± 0.9	5.9 ± 2.6	2.0 ± 1.5	2.2 ± 0.7	1.9 ± 1.1
Na	2039 ± 827	1735 ± 1108	1426 ± 814	1836 ± 782	6347 ± 3600
Pb	4.81 ± 0.49	5.04 ± 0.98	6.35 ± 1.99	17.16 ± 9.61	7.98 ± 5.49
Rb	13.6 ± 4.6	11.3 ± 4.9	13.5 ± 2.2	14.1 ± 6.5	49.1 ± 18.4
Sb, ng g^{-1}	90.8 ± 46.8	101 ± 49.7	33.4 ± 17.3	27.3 ± 18.5	18.5 ± 8.4
Sc, ng g^{-1}	10.0 ± 6.1	10.8 ± 4.2	6.4 ± 4.9	3.1 ± 0.6	2.4 ± 1.7
Se	1.39 ± 0.19	1.48 ± 0.30	0.78 ± 0.24	0.74 ± 0.28	0.95 ± 0.54
Th, ng g^{-1}	8.9 ± 9.1	5.6 ± 1.1	3.9 ± 1.1	3.4 ± 0.6	4.2 ± 0.3
V, ng g^{-1}	<45	44 ± 23	41 ± 31	94 ± 46	137 ± 40
Zn	158 ± 19	167 ± 25	127 ± 35	133 ± 56	171 ± 66

Results are given as the average ± SD, $n = 7$.

Zn, Cd, Hg, Cu, As, and Se in the individual elution fractions by RNAA. Because of high levels of sodium and bromine, the background in the gamma spectrum in tissue samples are generally high after neutron activation and where many trace elements of interest are present in quite low concentration, displaying only very small peaks. RNAA is an ideal solution to this problem.

In research conducted by Sabbioni and Girardi,[29] NAA was developed for the simultaneous analysis of heavy metals in microsamples of potential metal binding components such as metalloenzymes (calf intestine alkaline phosphatase, cow milk xanthine oxidase, calf thymus deoxynucleotidyltransferase) and nucleic acid (calf thymus DNA), purified by size-exclusion chromatography and ion-exchange chromatography.

In 1980s, Rack and colleagues at University of Nebraska carried out a series study on speciation of iodine,[30] chlorinated pesticides,[3] and seleium[31,32] in urine using liquid chromatography and molecular neutron activation analysis. They developed procedures for trace level determination of iodoamino acids and hormonal iodine as well as DDT and its metabolites DDA [bis(*p*-chlorophenyl)acetic acid], DDD [l,l-dichloro-2,2-bis(*p*-chlorophenyl)ethane], and DDE [l,l-dichloro-2,2-bis-(*p*-chlorophenyl)ethylene] in a urine matrix employing high-performance liquid chromatography separation techniques with subsequent irradiation of the eluted fractions. By combining NAA with anion exchange chromatography, it was possible to detect and quantitify total selenium, trimethylselenonium (TMSe) ion, selenite (SeO_3^{2-}) ion, and total selenoamino acids in urine.

Chromium is an element which is considered to be related to glucose and ester metabolism. Feng *et al.*[33] investigated the fundamental distribution patterns of the chromium-containing proteins in the nucleic, mitochondrial, lysosomal, microsomal, and cytosolic subcellular fractions of rat liver by means of Sephadex G-100 gel chromatography combined with NAA via $^{50}Cr(n,\gamma)^{51}Cr$ reaction (Table 2.4). In total, nine kinds of Cr-containing proteins were found in the five subcellular fractions, whose relative molecular masses were 96.6 ± 6.2, 68.2 ± 1.4, 57.9 ± 4.7, 36.6 ± 1.2, 24.2 ± 1.8, 14.0 ± 1.5, 8.8 ± 0.6, 6.9 ± 0.4, and 4.2 ± 0.4 kDa. Approximately 64.5% of Cr proteins accumulated in the cytosolic fraction. The second enriched part was the nucleic fraction; about 12.2% Cr proteins were stored in this section. The 4.2 kDa molecular mass might contain the so-called low molecular weight chromium-containing substance; however, in this research, it was only observed in the mitochondria, lysosome, and microsome. In the mitochondrial fraction, most of the Cr proteins were present as relatively low molecular weight substances: about 56% of chromium-containing proteins had molecular masses ≤ 6.9 kDa. Nevertheless, more than 69% of Cr-containing proteins were observed with molecular masses ≥ 57.9 kDa in the liver cytosolic fraction.

Chen *et al.*[12] reported the chemical species of rare earth elements La, Sm, and Ce in human liver cytosol by size-exclusion chromatography combined with INAA. It was revealed that at least three La-binding proteins (MWs: 335 ± 50, 94.5 ± 15.4, 13.6 ± 3.8 kDa), three Ce-binding proteins (MWs: 335 ± 50, 85.1 ± 12.0, 22.8 ± 6.3 kDa) and about four Sm-binding proteins (MWs: 335 ± 70, 82.1 ± 5.4, 32.3 ± 5.8 and 13.6 ± 4.5 kDa) were present in the supernatant fraction of human liver. Most of La, Ce and Sm were found in the high-molecular-weight protein region.

2.2.3 Electrophoresis–NAA

Electrophoresis is a separation technique based on the mobility of ions in an electric field. Positively charged ions migrate towards a negative electrode and negatively charged ions migrate toward a positive electrode. For safety reasons one electrode is usually at ground and the other is biased positively or negatively. Ions have different migration rates depending on their total charge, size, and shape, and can therefore be separated.

An electrode apparatus consists of a high-voltage supply, electrodes, buffer, and a support for the buffer such as filter paper (filter paper electrophoresis),

Table 2.4 Distribution patterns of nine types of Cr-containing proteins in subcellular fractions.[33] © 2003 Springer-Verlag.

	Relative molecular weight (kDa)								
	96.6±6.2	68.2±1.4	57.9±4.7	36.6±1.2	24.2±1.8	14.0±1.5	8.8±0.6	6.9±0.4	4.2±0.4
Percentage in whole organelle Cr protein (%)[a]									
Nuclei	20.6	10.5	6.7	18.9	5.7	–	37.6	–	–
Mitochondrion	9.8	6.8	–	–	8.6	18.8	–	30.2	25.8
Lysosome	13.8	6.2	–	28.7	–	13.9	9.6	24.2	3.6
Microsome	20.5	26.2	–	6.1	8.3	–	12.1	18.5	8.4
Cytosol	15.8	15.9	38.2	–	3.9	–	26.2	–	–
Percentage in whole liver (%)[b]									
Nuclei	21.1	12.9	6.5	35.8	19.5	–	30.9	–	–
Mitochondrion	8.8	7.4	–	–	26.3	53.7	–	41.0	70.1
Lysosome	14.5	7.9	–	55.9	20.6	46.3	8.1	38.4	11.1
Microsome	15.1	23.0	–	8.3	–	–	7.1	20.6	18.8
Cytosol	40.5	48.8	93.5	–	33.6	–	53.9	–	–

[a]Data represent the ratio of different types of Cr-containing proteins distribution in each subcellular fraction.
[b]Data represent the ratio of different types of Cr-containing proteins distribution in the whole liver.

cellulose acetate strips (cellulose acetate electrophoresis), polyacrylamide gel (gel electrophoresis, PAGE), agarose gel (agarose gel electrophoresis), or a capillary tube (capillary electrophoresis). Open capillary tubes are used for many types of sample and the other supports are usually used for biological samples such as protein mixtures or DNA fragments. After a separation is completed the support is stained to visualize the separated components. Resolution of electrophoresis can be greatly improved using isoelectric focusing. In this technique the support gel maintains a pH gradient. As a protein migrates down the gel, it reaches a pH that is equal to its isoelectric point. At this pH the protein is netural and no longer migrates, i.e. it is focused into a sharp band on the gel.

Stone et al.[34] developed a procedure combing polyacrylamide gel electrophoresis (PAGE) with INAA to study biological macromolecules and their associated trace elements. The following protein samples were used: bovine albumin, carbonic anhydrase, casein, ceruloplasmin, phosvitin, trypsin inhibitor, and a human serum protein sample. After electrophoresis, the gels were removed from the glass cassettes and fixed with formalmehyde. Selected gels were stained by conventional Coomassie Blue techniques after fixing to make the separated proteins visible. Gel sections (5×10 mm) of both the stained and non-stained gels were identically sampled by removal from the gel. Individual gel sections were then freeze dried, packaged, and irradiated in a reactor. The samples were then counted after a suitable decay period.

The zinc results from this analysis are summarized in Figure 2.3. The individual gel sections referred to in the figure are locations in the gel where samples were punched for analysis. The shaded areas represent locations where protein bands occur. As indicated in the figure, protein bands for serum, phosvitin, and carbonic anhydrase in the non-stained gel show elevated zinc concentrations above the background. This indicates the expected association between these proteins and zinc, which is believed to be representative of the metal contained in each protein. This conclusion is supported by the background levels of zinc in the bovine albumin band, since bovine albumin has no known structural association with zinc. However, the stained gel sections that were analyzed showed no significant elevations of zinc concentrations above the background levels in the protein bands. The one exception was in the elevated zinc region above the carbonic anhydrase band. The results comparing zinc levels in stained and non-stained gel sections of the carbonic anhydrase band indicate possible removal of the trace element from the protein during the staining procedure.

Peng et al.[35] studied the incorporation of Se into protein as peroxidase isozymes in wheat seedlings by an optimized INAA after electrophoretic separation. After sodium dodecyl sulfate–polyacrylamide gel electrophoresis (SDS-PAGE), the gel sections containing corresponding protein bands were cut off and subjected to INAA. The lower detection limit of 0.02 ng was achieved through longer irradiation (8.3 days) with higher neutron flux and longer radioactivity counting time.

In some cases, several bioanalytical techniques were employed in conjunction with NAA to study the incorporation of trace elements with

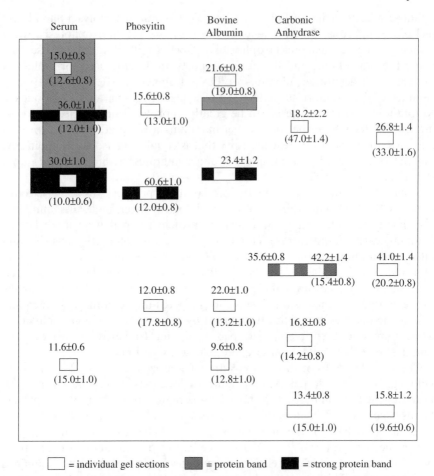

Figure 2.3 Zinc analysis of gel sections following SDS-PAGE of protein solutions. Results are given in ng cm^{-2}. Errors are 1 s counting statistics, ng Zn = results on unstained gel sections, (ng Zn) = results on Coomassie stained gel sections.[34] © 1987 Akadémiai Kiadó, Budapest.

biomacromolecules. Jayawickreme and Chatt[36] separated and purified proteins in bovine kidneys using ultracentrifugation, dialysis, gel filtration, ion exchange and hydroxylapatite chromatography, chromatofocusing, electrofocusing, isotachophoresis and ammonium sulfate precipitation. At the end of the experiments, all samples were freeze dried and subjected to INAA. The results indicates that in the bovine kidney microsomecytosol subcellular fraction, much of the copper is associated with a single protein of an isoelectric point around 5 and a molecular weight of about 30 kDa. Mn appears to be associated with several proteins. The results obtained from electrofocusing experiments indicated the possible existence of three manganese proteins with I_p values of 4.6, 7.4, and 8.1. The results from ammonium sulfate precipitation, hydroxylapatite chromatography and isotachophoresis also indicated the possible existence of at least two or three manganese proteins in the supernatant fraction.

Dicranopteris dichotoma is a hyperaccumulator of rare earth elements (REEs). A systemic study on the combination of REEs with biomacromolecules in *Dicranopteris dichotoma* has been carried out using several biochemical techniques and NAA.[37–39] The separation procedure of the REE-bound proteins (RBPs) from the fern leaves is shown in Figure 2.4 as an example.

Two REE-bound proteins (RBP-I and RBP-II) were obtained. The molecular weight of RBP-I on Sephadex G-200 gel colunm is about 8×10^5 Da and that of RB P-II is less than 12 400 Da, respectively. However, SDS-PAGE of the two proteins shows that they mainly have two protein subunits with MWs 14 000 and 38 700 Da. They are probably conjugated proteins, glycoproteins, with different glyco units.

Four polysaccharides (PSs) of different solubilities in the leaves of *Dicranopteris dichotoma*, which were named as cold-water-soluble, hot-water-soluble, acidic-soluble, and basic-soluble PSs, were obtained. After deproteinization and gel chromatography separation, the absorption curves of the PSs

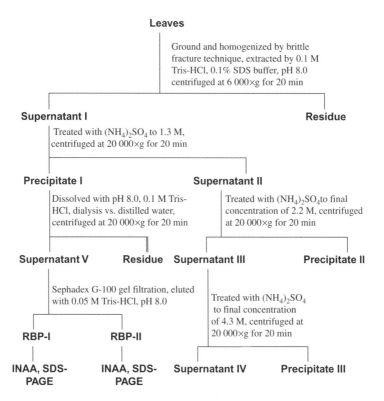

Figure 2.4 Separation procedure of the rare earth bound proteins (RBPs) from the fern leaves.[37] © 1996 Akadémiai Kiadó, Budapest.

Table 2.5 A summary of the application of NAA in chemical speciation study of trace elements.

Element	Matrix	Purpose	Method	Main results	References
Cr	Rat liver	Subcellular distribution of Cr and Cr-containing proteins	Enriched stable IT, SEC, INAA	Nine types of Cr-containing proteins were found	22,23,33
Se	Human liver	Subcellular distribution of Se and Se-containing proteins	SEC, INAA	Eight Se-containing proteins were identified	8,10
La, Ce, Sm, Yb	Human liver, Rat liver	Detection of REE bonding proteins	Enriched stable IT (in rat), SEC, INAA	About 12 REE-bonding proteins were detected	12
Hg	Rat brain	Study the affinity of Hg to metallothionein (MT) in the brain tissue of rats	GE, RNAA	About 80% of Hg was in the fraction of MT-like protein	41
Se	Wheat seedlings	Detection of selenoproteins	SDS-PAGE, INAA	Se incorporated into peroxidase isozymes	35
REE	Plant leaves	REE-bound biological macromolecules	SEC, INAA	Some new species of REE-bound proteins, DNA and polysaccharides have been found	24,25,37–39

SDS-PAGE, sodium dodecyl sulfate–polyacrylamide gel electrophoresis; REE, rare earth element; INAA, instrumental neutron activation analysis; SEC, size exclusion chromatography; IT, isotopic tracing; RNAA, radiochemical neutron activation analysis.

were obtained by a colorimetry method and the content of eight REEs (La, Ce, Nd, Sm, Eu, Tb, Yb, and Lu) were determined by NAA. The results indicated that REEs were bound firmly with the PSs. The molecular weight measurement on Sephadex G-200 demonstrated that those PSs bound with REEs were mostly low-molecular-weight polysaccharides (10–20 kDa).

The bonding of REEs with DNA in *Dicranopteris dichotoma* was also confirmed. The REE content of the DNA accounts for 0.087% of total REEs in the leaves. The molecular weight of the REE-DNA band is about 22 kb by agarose gel electrophoresis.

High selenium concentrations (7–12 g kg^{-1} dry mass) were found in the seeds of the selenium-accumulator plant coco de mono (*Lecythis ollaria*). In order to obtain information on the protein-bound part of selenium in extracts of these seeds, Hammel et al.[40] used dialysis and SDS-PAGE combined with INAA. Extractions were carried out at pH 4.5 and 7.5. In both cases about 90% of the element was dissolved. By analyzing the protein fractions separated by SDS-PAGE the element was found to be present in extremely selenium-rich proteins with molecular masses below 20 kDa.

2.3 Conclusion

The information on the chemical speciation of trace elements in biological systems is much needed to evaluate their biological significance. Although a number of analytical techniques based on atomic behavior are available for the analysis of chemical speciation of trace elements, neutron activation analysis, as a nuclear analytical technique, can be successfully used in chemical speciation studies, after appropriate fractionation steps. Table 2.5 lists some typical applications of NAA in chemical speciation analysis of metalloproteins. The main advantages of NAA are of its high sensitivity and the absence of matrix effects inherited from the conventional neutron activation analysis. It can, therefore, be used to analyze the chemical species of trace elements in very small samples or complicated matrices, which is often impossible for non-nuclear techniques.

Acknowledgments

The authors acknowledge National Natural Science Foundation of China (Grant No. 10505024, 20707027 and 10875136), the Knowledge Innovation Program of the Chinese Academy of Sciences (Grant No. KJCX3.SYW.N3), and the Ministry of Science and Technology of China (Grant No. 2006CB705605) for financial support.

References

1. M. D. Glascock, An overview of neutron activation analysis. www.missouri.edu/~glascock/archlab.htm.

2. J. W. Bennett, in *Proceedings of 15th Australian Conference on Nuclear and Complementary Techniques of Analysis, Melbourne, Australia, 21–23 November 2007*, The University of Melbourne, Australia, pp. 136–139.
3. L. R. Opelanio, E. P. Rack, A. J. Blotcky and F. W. Crow, *Anal. Chem.*, 1983, **55**, 677–681.
4. C. F. Chai, X. Y. Mao, Y. Q. Wang, J. X. Sun, Q. F. Qian, X. L. Hou, P. Q. Zhang, C. Y. Chen, W. Y. Feng, W. J. Ding, X. L. Li, C. S. Li and X. X. Dai, *Fresenius J. Anal. Chem.*, 1999, **363**, 477–480.
5. Z. F. Chai, *J. Nucl. Radiochem. Sci.*, 2000, **1**, 19–22.
6. Z. F. Chai, Z. Y. Zhang, W. Y. Feng, C. Y. Chen, D. D. Xu and X. L. Hou, *J. Anal. At. Spectrom.*, 2004, **19**, 26–33.
7. A. Chatt, in *Book of abstracts of the 7th International Conference on Nuclear Analytical Methods in the Life Sciences, Antalya, Turkey, 16–21 June 2002*, University of Bahçeşehir, Istanbul, Turkey, p. 10.
8. C. Y. Chen, P. Q. Zhang, X. L. Hou and Z. F. Chai, *Biol. Trace Elem. Res.*, 1999, **71/72**, 131–138.
9. P. Zhang, C. Y. Chen, M. Horvat, R. Jaćmović, I. Falnoga, B. Li and J. Zhao, Z. Chai. *Anal. Bioanal. Chem.*, 2004, **380**, 773–781.
10. C. Y. Chen, P. Q. Zhang, X. L. Hou and Z. F. Chai, *Biochim. Biophys. Acta*, 1999, **1427**, 205–215.
11. C. Y. Chen, X. L. Lu, P. Q. Zhang, X. L. Hou and Z. F. Chai, *J. Radioanal. Nucl. Chem.*, 2000, **244**, 199–203.
12. C. Y. Chen, P. Q. Zhang, X. L. Lu and Z. F. Chai, *Anal. Chim. Acta*, 2001, **439**, 19–27.
13. X. L. Hou, C. Y. Chen, W. J. Ding and C. F. Chai, *Biol. Trace Elem. Res.*, 1999, **69**, 69–76.
14. J. Zheng and G. S. Zhuang, *Nucl. Technol.*, [in Chinese], 1993, **16**, 111–114.
15. D. Wenstrup, W. D. Ehmann and W. R. Markesbery, *Brain Res.*, 1990, **533**, 125–131.
16. J. Zheng, G. S. Zhuang, Y. J. Wang and M. Dong, *Nucl. Technol.*, [in Chinese], 1993, **16**, 427–431.
17. P. O. Wester, *Biochim. Biophys. Acta*, 1965, **109**, 268–283.
18. C. K. Jayawickreme and A. Chatt, *J. Radioanal. Nucl. Chem.*, 1987, **110**, 583–593.
19. D. Behne, S. Scheid, H. Hilmert, H. Gessner, D. Gawlik and A. Kyriakopoulos, *Biol. Trace Elem. Res.*, 1990, **26/27**, 439–447.
20. D. Behne, C. Weiss-Nowak, M. Kalcklösch, C. Westphal, H. Gessner and A. Kyriakopoulos, *Biol. Trace Elem. Res.*, 1994, **43/45**, 287–297.
21. J. C. Lai, M. J. Minski, A. W. Chan, T. K. Leung and L. Lim, *Neurotoxicology*, 1999, **20**, 433–444.
22. W. Y. Feng, W. J. Ding, Q. F. Qian and Z. F. Chai, *Biol. Trace Elem. Res.*, 1999, **71/72**, 121–129.
23. W. Y. Feng, Q. F. Qian, W. J. Ding and Z. F. Chai, *Metabolism*, 2001, **50**, 1168–1174.

24. Z. Y. Zhang, Y. Q. Wang, J. X. Sun, F. L. Li, Z. F. Chai, L. Xu, X. Li and G. Y. Cao, *Chin. Sci. Bull.*, 2000, **45**, 1497–1499.
25. Z. Y. Zhang, Y. Q. Wang, F. L. Li and C. F. Chai, *J. Radioanal. Nucl. Chem.*, 2001, **247**, 557–560.
26. D. A. Skoog, *Principles of Instrumental Analysis*, 6th edn, Thompson Brooks/Cole, Belmont, CA, 2006.
27. D. J. Evans and K. Fritze, *Anal. Chim. Acta*, 1969, **44**, 1–7.
28. G. Norheim and E. Steinnes, *Anal. Chem.*, 1975, **47**, 1688–1690.
29. E. Sabbioni and F. Girardi, *Sci. Total Environ.*, 1977, **7**, 145–179.
30. M. L. Firouzbakht, S. K. Garmestanl, E. P. Rack and A. J. Blotcky, *Anal. Chem.*, 1981, **53**, 1746–1750.
31. A. J. Blotcky, G. T. Hansen, N. Borkar, A. Ebrahim and E. P. Rack, *Anal. Chem.*, 1987, **59**, 2063–2066.
32. A. J. Blotcky, A. Ebrahim and E. P. Rack, *Anal. Chem.*, 1988, **60**, 2734–2737.
33. W. Y. Feng, B. Li, J. Liu, Z. F. Chai, P. Q. Zhang, Y. X. Gao and J. J. Zhao, *Anal. Bioanal. Chem.*, 2003, **375**, 363–368.
34. S. F. Stone, D. Hancock and R. Zeisler, *J. Radioanal. Nucl. Chem.*, 1987, **112**, 95–108.
35. A. Peng, Y. Xu and Z. J. Wang, *Biol. Trace Elem. Res.*, 1999, **70**, 117–125.
36. C. K. Jayawickreme and A. Chatt, *J. Radioanal. Nucl. Chem.*, 1988, **114**, 257–279.
37. F. Q. Guo, Y. Q. Wang, J. X. Sun and H. M. Chen, *J. Radioanal. Nucl. Chem.*, 1996, **209**, 91–99.
38. Y. Q. Wang, J. X. Sun, F. Q. Guo, Z. Y. Zhang, H. M. Chen, L. Xu and G. Y. Cao, *Biol. Trace Elem. Res.*, 1999, **71/72**, 103–108.
39. Y. Q. Wang, P. Jiang, F. Q. Guo, Z. Y. Zhang and J. X. Sun, *Sci. China B*, 1999, **42**, 357–362.
40. C. Hammel, A. Kyriakopoulos, D. Behne, D. Gawlik and P. Brätter, *J. Trace Elem. Med. Biol.*, 1996, **10**, 96–102.
41. I. Falnoga, I. Kregar, M. Skreblin, M. Tusek-Znidaric and P. Stegnar, *Biol. Trace Elem. Res.*, 1993, **37**, 71–83.

CHAPTER 3
X-ray Fluorescence

YUXI GAO*

CAS Key Laboratory of Nuclear Analytical Techniques, Institute of High Energy Physics, Chinese Academy of Sciences, Beijing 100049, China

3.1 Background

In principle, X-ray fluorescence (XRF) *per se* is an atomic analytical technique, rather than a nuclear one, since it is based on electron transitions outside atomic nucleus. However, XRF is often defined as a nuclear-related technique, as frequently mentioned in many technical documents issued by the International Atomic Energy Agency, mainly because XRF is used where there are radioactive isotopic sources or accelerators, and even in large synchronous radiation facilities. Further, the detection of X-rays almost always needs nuclear radiation detectors, like Si(Li) or other planar detectors.

X-ray fluorescence is a non-destructive and multielemental analytical technique. Because of its excellent analytical sensitivity and spatial resolution under micro-beam conditions, the technique is capable of microscopic analysis, supplying information about two-dimensional (2D) distributions of trace elements. The technique can, thus, be used for imaging trace elements in biological specimens, and for the direct determination of trace elements in protein bands after slab-gel electrophoresis (GE), which is the benchmark for high-resolution protein separation, particularly in 2D format. Therefore, XRF is a useful technique for metallomics and metalloproteomics studies.

---

*Corresponding author: Email: gaoyx@ihep.ac.cn; Tel: +86-10-88233212; Fax: +86-10-88235294.

3.2 The Physics of X-ray Fluorescence

When materials are exposed to short-wavelength X-rays or to gamma rays provided by a primary radiation source, ionization of their component atoms may take place. One or more electrons are ejected from the atom via photoelectric absorption, creating a vacancy. X-rays and gamma rays can be energetic enough to expel tightly held electrons from the inner orbits of an atom. The ion is left in a highly excited state. As the atom returns to its stable state, electrons from the outer shells are transferred to the inner shells and in this process the excess energy of the ion can be dissipated through the emission of characteristic X-rays, whose energy is the difference between the two binding energies of the orbits involved. The process of emission of characteristic X-rays is called X-ray fluorescence (XRF). If the vacancy in a K shell is filled by an electron from the L shell, then the transition is accompanied by the emission of an X-ray line known as the $K\alpha$ line. The $K\alpha_1$ and $K\alpha_2$ X-rays are the products of transitions between the L subshells, L_{III} and L_{II}, and the K shell. If the vacancy in the K shell is filled by an electron from the M shell, then the line is known as the $K\beta$ line. The vacancies in the L or M shells can be filled by electrons from outer shells giving rise to the L and M series (Figure 3.1).[1] The number of permissible transitions between electronic shells is limited by the selection rules set by quantum mechanics. The wavelength (λ) of an X-ray characteristic line with the atomic number (Z) of the corresponding element is given by Moseley's law which plays an important role in systematizing X-ray spectra:

$$\frac{1}{\lambda} = k(Z - \sigma)^2 \tag{3.1}$$

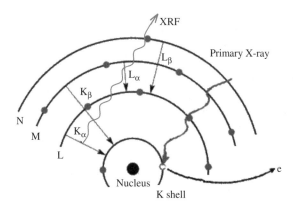

Figure 3.1 An electron in the inner shell (e.g. K shell) is ejected from the atom by an external primary excitation X-ray, creating a vacancy. An electron from outer shell "jumps in" to fill the vacancy. In the process, it emits a characteristic X-ray unique to this element and in turn, produces a vacancy in the outer shell.

where k is a constant for a particular spectral series and σ is a screening constant for the repulsion correction due to other electrons in the atom.

It is also possible for the excited ion to return to its ground state by transferring the excitation energy directly to one of the outer electrons, ejecting it from the atom. The ejected electron is called an Auger electron. This process is a competing process to XRF. Auger electrons are more probable in the low Z elements (elements of $Z<31$), than in the high Z elements. The probability that a vacancy is filled by a radioactive process is called the fluorescence yield (ω) and is a feature characteristic for each element. The fluorescence yield for the K shell is given by

$$\omega_K = \frac{I_K}{n_K} \qquad (3.2)$$

where I_K is the total number of characteristic X-ray photons emitted from a sample and n_K is the number of primary K-shell vacancies. For higher atomic shells, the definition of fluorescence yield becomes more complicated because (a) the shells above the K shell consist of more than one subshell, and (b) Coster–Kronig transitions occur.[1]

Because each element has a unique set of energy levels, each element produces X-rays at a unique set of energies. X-ray fluorescence from an irradiated sample should be composed of photons with their characteristic energies or wavelength. The fluorescent radiation can be analyzed either by sorting the energies of the photons (energy dispersive, ED) or by separating the wavelengths of the radiation (wavelength dispersive, WD). Once sorted, the intensity of each characteristic radiation is directly related to the amount of each element in the material. This is the basis of XRF analysis, which allows non-destructive measurement of the elemental composition of a sample. In most cases the innermost K and L shells are involved in XRF detection. A typical X-ray spectrum from an irradiated sample will display peaks of different intensities (Figure 3.2).

In principle, the lightest element that can be analyzed by XRF is beryllium ($Z=4$), but the fluorescence from lighter elements is of low energy (long wavelength) and has low penetrating power, and is severely attenuated if low-energy X-rays pass through air for any distance. Because of this attenuation, in high-performance analysis the path from an X-ray source to sample, then to the detector, is maintained under high vacuum (around 10 Pa residual pressure). This means, in practice, that most of the working parts of the instrument have to be located in a large vacuum chamber. Thus, instrumental limitations, low X-ray yields for the light elements, make it often difficult to quantify elements lighter than sodium ($Z=11$) unless the background corrections and very complicated inter-element corrections are made.

3.3 XRF Facilities

An XRF facility usually consists of a primary radiation source, a detector, an electronics system, and sometimes an optics system. Various XRF

X-ray Fluorescence

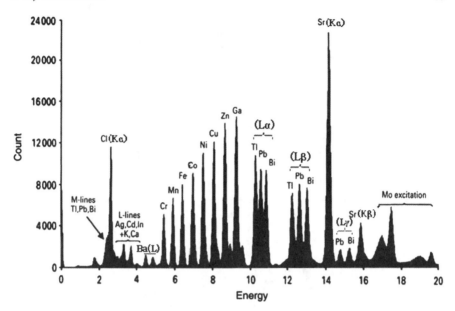

Figure 3.2 Typical energy dispersive XRF spectrum for a multielement standard. (Modified from http://webh01.ua.ac.be/mitac4/micro_xrf.pdf). Reprinted with permission from the author-Prof. Koen Janssens. (This figure was originally published in *The Handbook of Spectroscopy*, G. Gauglitz, T. Vo-Dinh, ed., Wiley-VCH, 2003, ISBN 3-527-29782-0.)

spectrometers based on different analytical principles have different structures and performances. An energy dispersive XRF (EDXRF) system will be introduced specifically for its extensive applications in metallomics and metalloproteomics fields. The geometrical setup for common EDXRF is shown in Figure 3.3.

3.3.1 Primary Radiation Source

In order to expel tightly held inner electrons, a source of radiation with sufficient energy is required. Conventional X-ray tubes are most commonly used for their output can readily be "tuned" for the application. The tube consists of a source of electrons as a cathode, which is usually a heated filament, and a thermally rugged anode (the "target"), which is enclosed in an evacuated glass envelope. The X-ray tube is operated at a high voltage (typically in the range 20–60 kV). The accelerating electrons from the cathode bombard the metallic surface of the target, hereby emitting X-rays, which consist of line emission and a continuous Bremsstrahlung (Figure 3.4). The Bremsstrahlung spectrum results from the rapid deceleration of the electrons as they hit the anode. The line emission is due to outer shell electrons falling into inner shell vacancies and hence is determined by the material used to construct the anode.

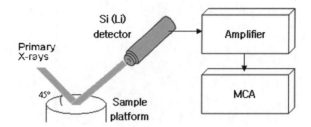

Figure 3.3 The geometrical setup for conventional EDXRF.

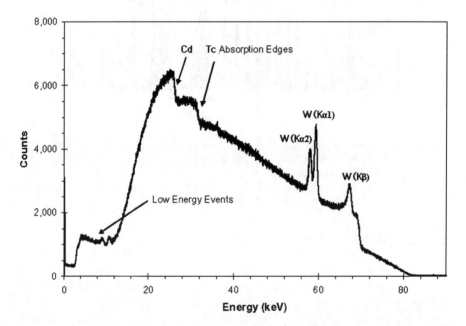

Figure 3.4 Plot showing the spectrum emitted from an $80\,kV_p$ X-ray tube with a tungsten (W) anode and a 15 keV filter, measured with a 1 mm thick CdTe detector. (Modified from http://www.amptek.com/pdf/ancdte1.pdf) Reprinted with permission from the publisher-John Pantazis.

Gamma-ray sources can also be used without the need for an elaborate power supply, allowing use in small portable instruments. The sources include a radioactive isotope which supply gamma rays with characteristic energy and can supply an X-ray or gamma-ray beam with the flux of 10^8 photons s^{-1} sr^{-1}. The main advantages of radioisotope excitation over X-ray tube excitation are the monoenergetic character of radioisotope-emitted X-rays, and that it is an inexpensive technique that is easily commercially available.

Synchrotron radiation (SR) is the electromagnetic radiation emitted by high-energy electrons (several GeV), circulating at highly relativistic speed in storage rings of a synchrotron accelerator. The SR spectrum is continuous and its energy depends on the energy and radius of curvature of the ring. The SR spectrum spans from microwaves to hard X-rays, and presents incomparable features like extremely high brightness and a very small divergence of the beam. During the last decade, the third-generation SR sources have been optimized for X-ray production by using undulators and wigglers[2] making it the brightest sources (12 orders of magnitude higher than the X-ray tubes), have a natural collimation in the vertical plane and are linearly polarized in the plane of the orbit. The advantages have greatly improved the sensitivity and space resolution of X-ray fluorescence (XRF) analysis. An absolute detection limit less than 10^{-15} g and a relative detection limit at ng g^{-1}, can be achieved with efficient excitation for low Z as well as high Z elements, and only micrograms of sample required. However, the number of synchrotron radiation facilities is limited due to the very high investment costs in equipment and manpower. Production of high-intensity X-ray beams using electron accelerators seems to be a promising alternative.[3] The technique is known as the electron microprobe, which might be used on a laboratory scale with the advantage of being less expensive than synchrotrons. Of course, the photon flux intensity obtained by this device is relatively low compared with synchrotron radiation sources.

It is also possible to create characteristic secondary X-ray emission using other incident radiation to excite the sample. When ions of sufficient energy (usually MeV protons) produced by an ion accelerator are employed as an excitation source, the technique is named particle-induced X-ray emission (PIXE), The sensitivity of the PIXE method is very high for Fe and neighboring elements with a similar Z number, i.e. approximately 20 essential metallic elements.[4] The lower detection limit for a PIXE beam is given by the ability of the X-rays to pass through the window between the chamber and the X-ray detector, which sharply decreases with the decrease of atomic number. The upper limit is dependent on the ionization cross-section, the probability of the K electron shell ionization, which decreases with the increase of atomic number.

3.3.2 Optics

In practice, a primary X-ray with specific photon energy and/or certain size of beam spot is often required to excite the samples being analyzed. Therefore, a high throughput adjustable optical system, such as collimators, monochromators, and focusing mirrors, becomes a necessary component for an XRF facility. X-rays are strongly absorbed by solid matter, so the optics used in the visible and near-infrared ranges of the electromagnetic spectrum cannot be used to focus or reflect the radiation. The radiation, however, can be "Bragg reflected" from a metal crystal when it irradiates the metal surface in a manner of grazing incidence, and each crystal plane acts as a weakly reflecting surface. Because the wavelengths of X-rays are comparable to the lattice spacing of the metal crystals, if the angle of incidence, θ, and crystal spacing, d, satisfy the Bragg condition,

$$n\lambda = 2d \sin \theta \tag{3.3}$$

where λ is the wavelength of X-ray and n is an integer, called the order of diffraction; the weak reflections can add constructively to produce nearly 100% reflection. Based on the principle, various X-ray optical elements have been designed, such as the multilayer monochromator, the double-crystal monochromator used for tuning the wavelength of the diffracted radiation, the Kirkpatrick–Baez mirror,[5,6] refractive lenses,[7] the Fresnel zone plate,[8] and (poly)capillary X-ray focusing optics[9] used to focus the X-rays to submicron spot sizes. The localized excitation makes XRF capable of analyzing a microscopically small area on the surface of a larger sample; the micro-analytical variant of bulk EDXRF is named microscopic X-ray fluorescence analysis (micro-XRF).

3.3.3 Detectors

In energy dispersive analysis, semiconductor crystals such as the PIN-diode, Si(Li), Ge(Li), and silicon drift detector (SDD) are generally used as X-ray detectors. They allow the count of an amount of photons and determination of the energy of the photon. An introduction of recent progress in X-ray detection can be seen in a published review.[10]

An incoming X-ray photon excites a number of electrons from the valence band to the conduction band. The electrons in the conduction band and the holes in the valence band are collected and measured. The amount of charge collected is proportional to the energy of the X-ray photon, and the process repeats itself for the next photon. The spectrum is then built up by dividing the energy spectrum into discrete bins and counting the number of pulses registered within each energy bin. A scheme for a Si(Li) detector is shown in Figure 3.5.

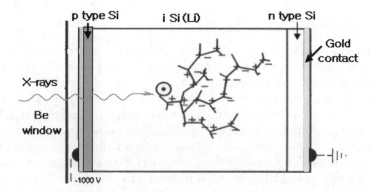

Figure 3.5 Schematic form of a Si(Li) detector. (Modified from http://webh01.ua.ac.be/mitac4/micro_xrf.pdf). Reprinted with permission from the author-Prof. Koen Janssens. (This figure was originally published in *The Handbook of Spectroscopy*, G. Gauglitz, T. Vo-Dinh, ed., Wiley-VCH, 2003, ISBN 3-527-29782-0.)

X-ray Fluorescence

The detector crystal itself consists essentially of a 3–5 mm thick silicon junction type p-i-n diode with a bias of $-1000\,\text{V}$ across it. Lithium ions drift (allowed to diffuse at elevated temperature) into the silicon crystal to neutralize the Si crystal defects and gold contacts are evaporated onto the crystal. In the crystal, the energy difference ΔE between the valence band and conduction band is 3.8 eV. To keep the leakage current as low as possible, the crystal is cooled with liquid nitrogen by placing it in a vacuum cryostat, in which almost all electrons remain in the valence band. The photons enter the Si crystal through a thin entrance window made of beryllium. By applying a reverse voltage to the charge-carrier-free intrinsic zone, an absorbed X-ray photon is converted into charge by ionization. Electrons are promoted from the valence to the conduction band, leaving "positive holes" in the valence band. The amount of electron–hole pairs, n, is $E/\Delta E$. The crystal thus temporarily becomes conducting. The electrons and holes are quickly swept to the contact layers by the electric field created by the applied reverse bias on the crystal. This causes a voltage pulse. The energy resolution of a Si(Li) detector optimally is of the order of 120 eV at 5.9 keV, and the energy range detected is 1–50 keV.

3.3.4 Electronics

The pulses generated by the detector are processed by pulse-shaping amplifiers. It takes time for the amplifier to shape the pulse for optimum resolution, and there is therefore a compromise between resolution and count rate: long processing time for good resolution results in "pulse pile-up" in which the pulses from successive photons overlap. Multi-photon events will prolong the processing time and pulse-length discrimination has to be used to filter out most of the overlap. Furthermore, pile-up correction should be built into the software being used for the application. To make the most efficient use of the detector, multi-photon events (before discrimination) should be kept at a reasonable level, e.g. 5–20% by adjusting the tube current or distance between detector and sample.

The signal from the amplifier is processed by a multi-channel analyzer (MCA), which converts these analog pulses to digital values and puts the numerical values in corresponding channels, producing an accumulating digital spectrum. The spectrum represents the distribution of the detected radiation intensity as a function of the energy in a whole energy range, and can be processed to obtain analytical data for all elements. Alternatively, a single-channel analyzer (SCA) can be used for selecting a range of output pulse amplitudes from the amplifier. It only counts the pulses of a given amplitude (equivalent to photons of a given energy) at a time, and can thus be used for single element analysis. The processing time can be sharply reduced in this manner.

Considerable computer power is dedicated to correcting for pulse pile-up and for extraction of data from poorly resolved spectra. These elaborated correction processes tend to be based on empirical relationships that may change with time, so continuous vigilance is required in order to obtain chemical data of adequate precision.

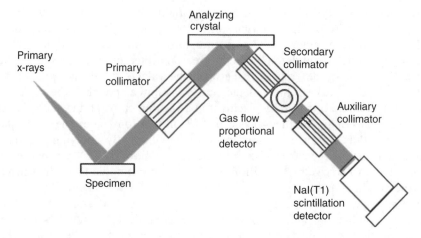

Figure 3.6 The experimental setup for WDXRF.

3.3.5 Wavelength Dispersive XRF

Wavelength dispersive X-ray fluorescence (WDXRF) is another type of X-ray fluorescence instrumentation. A WDXRF setup is displayed in Figure 3.6.

Collimators are employed to direct the beam in order to closely control the diffraction angle of all detected photons. The analyzing crystal or monochrometer angularly disperses incident radiation of wavelength according to the Bragg equation (3.3). The analyzing crystal may be rotated with the detector assembly simultaneously revolving around it to scan through the possible wavelengths for each element. Otherwise, a set of fixed detection systems is equipped, where each system measures the radiation of a specific element.

The system in the diagram utilizes two detectors in series. The first, a gas-flow proportional detector[11] is efficient for detecting long-wavelength radiation (>0.15 nm). Most high-energy X-rays pass through it, however, and are counted by the NaI(Tl) scintillation detector.[12] With the WDXRF system, an energy (spectral) resolution of 5 eV and 20 eV can be achieved. (EDXRF systems typically provide resolutions of more than 100 eV.) The higher resolution of WDXRF provides advantages in reduced spectral overlaps; therefore, the detector outputs can be used directly and heavy use of electronics and computer algorithms for evaluation of the spectra are generally not needed. In addition, high resolution backgrounds are reduced; thus, the detection limits and sensitivity are greatly improved. However, the additional optical components of a WDXRF system (e.g. diffracting crystal and collimators) means that it suffers from greatly reduced efficiency. Typically, this is compensated for by high-power X-ray sources. Another limitation of WDXRF lies in spectral acquisition. With an EDXRF system, an entire spectrum is acquired virtually simultaneously, so that many elements of the periodic table can be detected within a few seconds. But WDXRF spectrum acquisition is either made in a point-by-point mode (which is extremely time consuming), or by a very limited number of detectors. The method, therefore, can hardly satisfy the requirement of high throughput analysis in metallomics and metalloproteomics.

3.4 Analytical Procedures with EDXRF

XRF is widely used for elemental analysis in geochemistry, archaeology, forensic science, and life sciences. The method has the advantages of simple analytical procedure as well as being non-destructive with a relatively short testing time. As with any analytical procedure, the analytical procedure for EDXRF includes sample preparation, sample measurement, and data treatment.

3.4.1 Sample Preparation

Sample preparation is usually unavoidable in XRF analysis. It may contribute significantly to error in the total analysis, so all technical aspects related to sample preparation are of great importance. Whatever the kind of sample, no contaminating elements should be associated with it. The glassware should be cleaned with nitric acid and ultra-pure buffers/solutions should be prepared according to the experimental aim. Solid samples must be prepared to assure surface homogeneity, while powders are usually pressed into pellets. As matrix effects may be very important, the reproducibility of the sample thickness must be carefully checked. Furthermore, in the case of PAGE, gel drying must be well controlled to avoid dramatic thermal and radiative damage, which occurs under vacuum. Liquid samples should be deposited on an X-ray transparent supporting media (e.g. polyethylene, Kapton, Mylar, Nuclepore membrane) and dried before analysis. For micro-XRF, samples are cellular slices prepared for electron microscopy.

3.4.2 Element Measuring

The most simple EDXRF instrumental configuration is shown in Figure 3.3. The sample is usually mounted on a platform, which can be moved by computer-controlled stepping motors. A primary radiation and a Si(Li) detector are both placed at an angle of 45° with respect to the sample. Collimators are used to confine the excited and detected beam to a sample area. To make the most efficient use of the detector, the dead time should be controlled to a reasonable range by adjusting the X-ray tube current, beam size or distance between detector and sample. The counting time is determined on the base of the count error estimation: the probable error for a given count with a measured intensity value of I is $I^{1/2}$ because the emission of X-rays from sample is a random process that can be described by a Gaussian distribution. Thus, 10 000 counts have to be taken if a standard deviation is controlled below 1%.

When SR is used as a primary source, an additional advantage of this geometry setup is the high degree of polarization of SR, causing low spectral backgrounds due to scatter greatly reduced. The combination of a high primary beam intensity and low spectral background causes DL values of SRXRF to go down to the ppb level. It should be mentioned that the source intensity decreases with time (due to a gradual loss of orbiting particles in between ring refills), so that measurement of unknown samples must be bracketed between standards and samples by either introducing an internal marker, i.e. a known amount of an element, which is not likely to be found in the tested material (e.g. yttrium (Y) or rubidium (Rb)), or continuously monitoring the primary beam intensity, then normalize the intensity of XRF line to it.

3.4.3 Data Processing

The process to convert experimental XRF data into analytically useful information can be divided into two steps: first, the evaluation of the spectral data, and, second, the conversion of the net X-ray intensities into concentration data, i.e. quantification. In the latter step, especially, the appropriate correction of matrix effects is a critical issue.

In an EDXRF spectrum, the net number of counts under a characteristic X-ray peak (i.e. the integrated peak intensity) is proportional to the concentration of an analyte. At constant resolution, this proportionality also exists between concentration and net peak height. Because of the low and inconstant energy resolution of the solid detector used in EDXRF, the net peak area is preferred as analytical signal. It also results in a lower statistical uncertainty for the small peaks. Evaluation of the spectra is generally performed by computers. Many software packages have been developed for analysis of the collected spectra to obtain the net peak areas of all characteristic X-ray lines interested. Among them the computer package AXIL[13,14] is mostly suitable for the evaluation of energy-dispersive X-ray spectra and is commercially available from Canberra–Packard.

In practical X-ray fluorescence analysis, the observed XRF intensity is not linearly proportional to the concentration of an element (Figure 3.7), because the elemental composition of matrix can affect the intensity of the fluorescent X-ray lines by the following two methods:

1. *Matrix absorption*: All atoms of the specimen matrix will absorb photons from the primary source. The intensity/wavelength distributions of these photons available for the excitation of a given element may be modified by other matrix elements. On the other hand, other elements in the matrix can absorb the characteristic X-ray fluorescence of the element measured as the characteristic radiation passes out from the specimen. Matrix absorption will attenuate the XRF intensity of the given elements,
2. *Enhancement by multiple excitation*: In the case where the energy of a fluorescent photon (e.g. Ni Kα at 7.47 keV) is immediately above the absorption edge of a second element (e.g. the K edge of Fe at 7.11 keV), the fluorescence intensity of the second element (here Fe Kα and Fe Kβ radiation) will be enhanced as a result of the preferential excitation (here

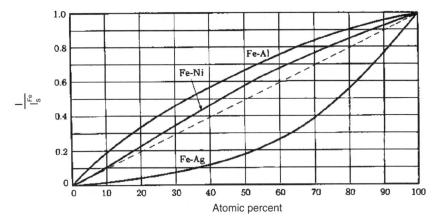

Figure 3.7 Effect of elemental composition of the matrix on the intensity of Fe (From http://www.postech.ac.kr/dept/mse/axal/note/xrsa/Chapter 5-1(X-ray).pdf).

by Ni Kα radiation) within the sample. For analysis of trace elements in biological sample, the magnitude of this effect is not significant because only a few heavier atoms exist in the matrix.

Matrix effects make quantitative analysis with XRF quite complicated. Many methodological processes are developed for calibration of the measuring arrangement, which may be performed by two major approaches: empirical and fundamental parameters (FP) calibration.

The empirical calibration is based on the analysis of standards with known elemental compositions. To produce a reliable calibration model, the standards must be representative of the matrix and target element concentration ranges of the sample. Maintaining the same sample morphology (particle size distribution, heterogeneity and surface condition) and source/sample geometry for both standard and sample measurements is essential in empirical calibrations.

Alternatively, "standardless" fundamental parameter (FP) techniques are based on built-in mathematical algorithms that describe the physics of the detector response to pure elements. In this case, the typical composition of a sample must be known, while the calibration model may be verified and optimized by one single standard sample. The techniques include the fundamental parameter method,[15] the influence coefficient method,[16] and the empirical coefficient method.[17]

3.4.4 Sensitivity, Limit of detection, and Precision

An important statistical consideration in XRF analysis is the capability of an instrument to detect whether an element is present or not in a specimen. The capability can be described by the detection limit. It is defined as being the minimum net peak intensity of an analyte, expressed in a concentration unit that can be detected by an instrument in a given analytical context with a 99.95% confidence level; that is

$$\text{LD} = \frac{6.58}{m_i}\sqrt{\frac{I_b}{T}} \quad m_i = \frac{I_p - I_b}{C_i} \tag{3.4}$$

where I_b and I_p are the background and peak intensity of analyte i, respectively; $I_p - I_b$ is net intensity, T is the counting time, C_i is the concentration of analyte, m_i is the sensitivity of an instrument for the analyte i, in a given analytical context.[18] From equation (3.4), it can be deduced that the limit of detection decreases as prolonging the counting time and reducing the background intensity. The LD also varies with the matrix composition of specimen and with the atomic number of the element to be determined.

The experimental analytical precision (EAP) is represented by the standard deviation, in terms of concentration unit, of 10 times measurements of the net intensity of the analyte in the same analytical context for a 95.4% confidence level.[18]

$$\text{EAP} = \frac{2}{m_i}\sqrt{\frac{\sum_{m=1}^{n}(I_m - \bar{I})^2}{n - 1}} \tag{3.5}$$

where I_m is the m-th measurement of the net intensity of the element i and $\bar{I}$ is the mean value of the n measured values of the net intensity. This experimental value (EAP) enables us to estimate the random errors due to the instrument and counting statistics. The counting statistical errors usually can be made very small by selecting appropriate counting times. It does not take into account the variability introduced by the sample preparation, which is the most important source of errors. The uncertainty introduced by this error source may limit considerably the accuracy and precision of the analytical results. A procedure was suggested by Rousseau[18] for evaluating the uncertainty introduced by random and systematic errors due to the sample preparation, which mainly includes:

1. Prepare 10 specimens from a single sample using the same preparation method.
2. Measure all elements of interest in each of the 10 specimens once using the same analytical program.
3. Measure all elements of interest in one of the specimens 10 times using the same analytical program.
4. Calculate the relative standard deviation of both series of measurements. If the relative standard deviation obtained in experiment 2 is sufficiently low, the method of sample preparation is suitable. If for a particular element the obtained relative standard deviation is too high, the results of step 2 can be compared with those of step 3 in order to determine whether the spread is due to the sample preparation or to the instrument and counting statistics.

3.5 Other XRF Techniques

Several variants of EDXRF with different performance and character are developed by changing the beam size of primary X-rays or the geometrical setup of the XRF instrument. Two of them, micro-XRF and total reflection XRF, will be introduced for their wide use or attractive prospects in metallomics and metalloproteomics fields.

3.5.1 Micro-XRF

When a beam of primary X-rays with (microscopically) small cross-section irradiates the sample, it induces the emission of fluorescent X-rays from a micro-spot, which carries information on the local composition of the sample. If the sample is moved either manually or under computer control in the X-ray beam path, spot or line analysis will be possible.

The key technique in the exploitation of this method lies in focusing the primary X-ray beam to a sufficiently small cross-section with enough intensity. Techniques to achieve this have only recently appeared. When using synchrotron sources as an exciting beam, combined with the excellent performance of X-ray focusing devices, the resulting microbeam technique could offer the possibility to obtain quantitative information on the elemental distributions in sample volume with trace level detection limits and on a (sub-)microscopic scale.[19] X-ray microprobes installed at the most advanced third-generation SR sources, such as the European Synchrotron Radiation Facility (ESRF) in Grenoble, offer absolute detection limits (DLs) below 10 ag and a potential lateral resolution level better than 50 nm.[20,21] The application of XRF to metallomics research for element distribution in biological samples is described in Chapter 11.

3.5.2 Total Reflection XRF

Total reflection XRF, like micro-XRF, is a variant of EDXRF based on the spatial confinement of the primary X-ray beam, so that only a limited part of the sample (+support) is irradiated. This is achieved in practice by use of dedicated X-ray sources, X-ray optics, and irradiation geometries. The principle of the setup of TXRF is shown in Figure 3.8.

When a monochromatic X-ray beam impinges upon an (optically) flat material under a very small angle (typically a few mrad), which is below the critical angle, σ_{crit}, for the substrate, total external reflection occurs. The critical angle is given by:

$$\sigma_{crit} = \sqrt{5.4 \times 10^{10} \times \frac{\rho Z \lambda^2}{A}} \qquad (3.6)$$

where Z is atomic number, A is the atomic weight, ρ, is the density of reflected substance, and λ is the wavelength of primary X-ray. Therefore, σ_{crit} is related to the characteristics of the reflected substance and X-ray beam.

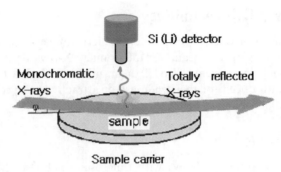

Figure 3.8 The principle of the setup of TXRF. A monochromatic X-ray beam irradiates the sample with a grazing manner. A Si(Li) detector is placed vertically to the surfaces of sample carrier.

In the TXRF condition, X-ray photons will only interact with the top few nm of material and then be reflected. Part of material that is present on top of the reflecting surface will be irradiated in the normal manner, and will interact with both the primary and the reflected X-rays. As a result, the double excitation of sample by both the primary and the reflected beam occurs. Hence, the fluorescent signal is practically twice as intense as in the standard EDXRF excitation mode.

Due to this grazing incidence, the primary beam is totally reflected. Absorption of the beam in the supporting substrate is largely avoided and the associated scattering is greatly reduced. This also reduces the background noise substantially. A further contribution to the reduction of the background noise is obtained by minimizing the thickness of sample. The scattering normally arising from sample and its matrix is greatly reduced. This makes the calibration and quantification independent from any sample matrix. Simplified quantitative analyses can be achieved with a single internal standard. All these advantages lead to lower limits of detection (LD) compared to the standard EDXRF mode. Depending on the X-ray source and the spectral modification devices, LD is in the picogram range for X-ray tubes and in the femtogram range for a synchrotron radiation source. The TXRF principle and techniques can be seen in published reviews (e.g. Streli et al.[22]).

TXRF, in particular, is applied for multielement determinations in water samples of various nature and for the routine analysis of Si-wafer surfaces employed in the micro-electronics industry. The samples with a few nanometers thickness are prepared by placing a small drop of the sample (5–100 μL of the substance dissolved in an appropriate solvent) on a silica carrier, then evaporating the solvent. In life sciences, the method is used for measuring amounts of trace elements in various tissues and body fluid.[23–25]

An obvious drawback of grazing incidence is poor lateral resolution because the incident X-rays irradiate the whole sample surface at small angles. An inverse arrangement, known as grazing-exit XRF, enables surface analysis of small regions to be carried out.

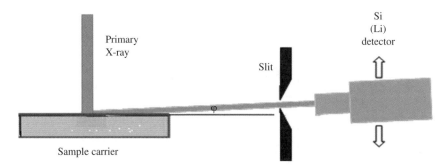

Figure 3.9 Experimental setup of the GE-XRF spectrometer.

3.5.3 Grazing-exit XRF

Grazing-exit XRF (GE-XRF) is a method related to TXRF, where the primary X-ray irradiates a sample perpendicularly to the reflector surface carrying the sample, while the X-ray fluorescence is collected under a grazing angle at small take-off angles. The experimental setup is shown in Figure 3.9.

It has been demonstrated that GE-XRF is theoretically equivalent to TXRF,[26] and the analyzing depth is also several nanometers. The method, therefore, has the same advantages for TXRF. An additional advantage of GE arrangement may be that it can be combined with a micro-XRF setup for elemental imaging. Furthermore, the depth information could be obtained by using this setup and changing the exit angle. Therefore, this instrument enables measurement of surface-sensitive line scanning and elemental mapping under grazing-exit conditions. In principle, measuring the elemental X-ray mappings at different exit angles enables the reconstruction of three-dimensional elemental distributions.[27]

Some important characteristics of XRF techniques for element analysis are list in Table 3.1.

3.6 Applications of EDXRF in Metallomics and Metalloproteomics

As mentioned above, EDXRF is a non-destructive multielemental analytical technique. Because the beam size of the primary X-ray can be monitored with a slit or focusing system, XRF is capable of microscopic analysis, supplying information about the 2D distribution of trace elements. The technique can, thus, be used for imaging of trace elements in a biological specimen, also for direct elemental determination in protein bands after slab-gel electrophoresis. Reviews concerning analysis of metalloproteins with XRF after electrophoresis separation have been published.[28–30]

Table 3.1 Some characteristics of XRF techniques for element analysis.

Method		Detection limit	Matrix effect	Detectable elements	Spectral interference	Preferred sample type	Lateral resolution	Multi-element
GEXRF	ED	ng mL^{-1}	Medium	$Z \geq 11$	High	Liquid (residue)	Good	Yes
	WD			$Z \geq 4$	Low			Limit
TXRF		ng mL^{-1}	Low	$Z \geq 11$	High	Liquid (residue)	Poor	Yes
EDXRF		1–10 µg g^{-1}	Medium to high	$Z \geq 11$	High	Solid	Good	Yes
WDXRF		1–10 µg g^{-1}	Medium to high	$Z \geq 4$	Low	Solid	Good	Limit
PIXE		0.2–3 µg g^{-1}	Medium to high	$Z \geq 11$	High	Solid	Good	Yes

3.6.1 XRF Methods Used for Elemental Analysis in Protein Fractions after Biochemical Separation

Proton-induced X-ray emission spectrometry (PIXE), using a high-energy photon beam as the excitation source, was first used in 1987 for metal analysis in proteins after electrophoresis separation by Szökefalvi-Nagy and coworkers.[31] They quantified the metals in high-potential iron–sulfur protein (HiPIP, containing four Fe–S clusters) band with a collimated 3 MeV proton beam. The method has also been used to show that Fe and Ni ions of the hydrogenase from *Thiocapsa roseopersicina* and *Desulfovibrio gigas* are located on the different polypeptides forming the enzyme.[32,33] Protein bands containing 1 mg HiPIP can be readily detected. Weber and his colleague quantified metals (Fe, Cu, Zn) in protein bands with PIXE. After a sodium dodecyl sulfate–polyacrylamide gel (SDS-PAGE) separation of a mixture of six proteins (ovotransferrin, carbonic anhydrase, cytochrome *c*, myoglobin, albumin, and ovalbumin) the gel is dried and each track is scanned with a 2.5 MeV proton beam which induces X-ray emission. The metal content in each band is obtained by comparing the characteristic X-ray peak area with those obtained with polyacrylamide gels doped with the same metal. Finally, the relative concentration of each protein is determined by densitometry in order to compute the protein/metal ratio. This procedure has proved to be a very useful multielementary method for the determination of the metal amounts inside proteins after separation by electrophoresis. Furthermore, it allows a check to be made to determine if metals remain bound to proteins.[34,35]

Solis *et al.* used PIXE to detect in a qualitative way the presence of Fe in some of the protein bands obtained by SDS-PAGE from a photosystems I complex, which was isolated from membranes of the thermophilic cyanobacteria *Synechochoccus* sp. After electrophoresis, the gels were stained for protein with Coomassie blue and dried on a layer of cellulose paper with a gel slab dryer. In-air PIXE analysis of Fe was performed at 2 MeV. Polyacrylamide gel electrophoresis of the complex shows eight subunits with molecular weights of 66, 60–65, 14, 13, 9, 8, and 7 kDa, respectively. Iron was consistently found in the bands of 60–65 kDa and 7 kDa. It was also occasionally found in the bands of 66, 8 and 6 kDa, but never in the bands of 13 and 14 kDa. Nevertheless, the results may not be trustworthy, because denaturing and staining processes may result in the loss of Fe from proteins.[36] In subsequent research, two Zn-containing enzymes (carbonic anhydrase, and cytoplasmic pyrophosphatase of *Rhodospirillum rubrum*) was purified by PAGE under non-denaturing conditions. After electrophoresis, one portion of the gel was removed and stained to locate the proteins with Coomassie brilliant blue, and the rest of the gel was dried without staining for PIXE analysis of Zn. PIXE analyses were carried out with an external proton beam of 3.7 MeV. In order to determine the amount of Zn detected by PIXE in each band, calibration was carried out by preparing gels with different amounts of $ZnSO_4 \cdot 7H_2O$ (between 0 and 300 µg g^{-1}). The results indicate that the metal/enzyme ratio is 0.7–1.3 for carbonic anhydrase, which is consistent with the acknowledged conclusion; and 1.9–2.1 for cytoplasmic pyrophosphatase. This is the first time that the association and stoichiometry of Zn has been reported for the cytoplasm pyrophosphatase of *R. rubrum*.[37] Additional studies concerning the use of PIXE to identify and quantify metals bound to proteins can be found in a published review.[28]

The feasibility of XRF for element detection in proteins separated by polyacrylamide gels was demonstrated by Stone *et al.* in 1994 using a bench-top microbeam X-ray system.[38] They evaluated bench-top microbeam X-ray spectrometry and energy-dispersive XRF analysis for Se detection in glutathione peroxidase after PAGE. The detection limits of Se in the gel matrix were 2.1 ng for the microbeam X-ray system and 30–60 ng when using energy-dispersive XRF. The poor sensitivity of the methods prevents their application in natural biological samples.

TXRF has an excellent detection limit due to its lower matrix effects. It can be used as an element-specific detector following an off-line separation step (e.g. gel electrophoresis thin layer chromatography, or high-performance liquid chromatography) in speciation studies. In early works, cytosols were investigated after gel permeation chromatography.[39] Cytosols of lamb's lettuce and cauliflower are separated by a Sephadex G50 column and the fractions obtained are analyzed directly by TXRF with an internal Co standard. Several metallic species could be distinguished, thereby elucidating the nature of metal complexing agents. Nickel complexation by Krebs cycle acids was investigated in xylem sap of cucumber plants grown in nutrient solutions contaminated with nickel by means of an off-line HPLC-TXRF method.[40] A primary HPLC separation was achieved with a size exclusion HPLC column employing ammonium acetate buffer as mobile phase; the fractions collected were freeze-dried and then the redissolved solids were injected onto a reversed-phase column. The fractions containing citric, malic or fumaric acids were analyzed by TXRF. Nickel could be identified in the fractions containing Krebs cycle acids.

TXRF has been used to quantify metals in protein bands in embryogenic callus (*Citrus sinensis L. Osbeck*) after separation by SDS-PAGE. The metal-containing bands were detected by micro-SRXRF. The gel was then decomposed in a microwave oven and the metal-binding protein was quantified by synchrotron radiation total reflection X-ray fluorescence (SR-TXRF). Signals that were almost free of background noise could be obtained.[41]

Recent efforts have focused on arsenic speciation in roots of cucumber plants grown in arsenic(III) containing hydroponics by thin layer chromatography and TXRF.[42,43] Separation of arsenic(III), arsenic(V), monomethyl arsonic acid (MMA) and dimethylarsinic acid (DMA) could be achieved on polyether imide cellulose plates. After scraping, the chromatographically developed plates were divided into 0.8 cm × 0.8 cm segments and microdigested in small PTFE vessels with about 100 μL nitric acid, arsenic in each segment was determined with TXRF. Better resolution for the separated peaks could be obtained by employing the automated version of TLC, called over-pressured liquid chromatography, Moreover, cucumber xylem sap and nutrient solutions containing arsenic(III) and arsenic(V) were investigated by X-ray absorption near-edge structure TXRF. Good agreement was observed with previous results achieved by HPLC-ICP-MS speciation.[44] In spite of the treatment with arsenate-containing nutrient solutions, cucumber plants reduce it to arsenite. However, these off-line methods are not competitive with the on-line coupled HPLC–inductively coupled plasma mass spectrometry.

In spite of the great efforts made to develop TXRF as a promising technique, its widespread implementation in speciation analysis, combined with HPLC, has been hampered by the lack of automation for (1) depositing the eluted fractions onto TXRF carriers, (2) the possible repeated collection for preconcentration purposes, (3) drying, and (4) element analysis. Such automation would speed up the investigations significantly.[25]

Special attention should be paid to XRF using synchrotron radiation (SR) as exciting sources. Combining advanced third-generation SR sources with excellent X-ray optics system, SRXRF offers absolute detection limits (DLs) of 10^{-18} g. Recently, the development of microprobe beamlines on third-generation synchrotrons has enabled spatially resolved X-ray fluorescence (XRF) at submicrometer levels, even nanometer levels.[20,21] These characteristics of SRXRF allow a multielemental analysis of major, minor, and trace elements in the microscopic region of biological specimens or in protein bands after electrophoretic separation. The technique has been applied to study the spatial distribution of elements in tissues with various physiological or pathological characters, even in a single cell to determine the pathogenesis of trace element-associated diseases and the biochemistry of the elements.[45–48] It is a suitable technique for direct element analysis in protein bands after electrophoresis separation. SRXRF was first reported as an elemental detector for electrophoresis separation in 1996.[49] The technique was used to monitor the interaction of metallothionein-II (MT-II) and Cu, Zn-superoxide dismutase (Cu, Zn-SOD) with mercury after incubation followed by isoelectric focusing-agarose or -polyacrylamide gel electrophoresis (IEF-AGE or IEF-PAGE). When MT-II reacted with mercuric chloride, an obvious change of isoelectric point (pI = 3.7–4.7) for the intact form to alkaline pI (9.4) was observed. This marked migration of MT-II by the metal was blocked by addition of glutathione, suggesting that sulfhydryl functions participate in the pI variation. In contrast, interaction of Cu, Zn-SOD with mercury did not cause any changes of its pI although the metal bound tightly to Cu, Zn-SOD after electrophoresis; however, the enzyme activity was drastically suppressed. These observations indicate that the combination of electrophoresis with SR-XRF analysis is a useful technique for detecting structural or functional alterations of proteins attributable to the binding of the mercury.

In order to investigate the suitability of a combination of SDS-PAGE and SRXRF in the detection of metal-containing proteins, Weseloh et al. have chosen metal-loaded apoazurin and selenoproteins in rat testis homogenate as examples for non-covalently bound metals and for covalently bound trace element, respectively, to obtain some basic information.[50,51] After SDS-PAGE separation, the proteins blotted onto a polyvinylidene difluoride (PVDF) membrane and then used for SRXRF measurements at the beamline L at HASYLAB. The pilot study showed that the combination of protein separation by SDS-PAGE and elemental analysis by SRXRF can be applied in the identification of the trace element-containing proteins at concentrations above 100 ng metalloprotein g^{-1} protein mixture, in which the metal or metalloid is either covalently bound or present in the form of a stable complex. In cases of weaker ionic metal–protein complexes where, due to the denaturing effect of the treatment with SDS, there is the danger of metal loss, native PAGE has to be employed. As the blotting membranes tend to be electrostatically charged, contamination problems have to be overcome by the application of clean-room conditions throughout the blotting membrane preparation and measurement to avoid contamination from air-borne particles.

The extraction of proteins from a biological sample is often the most problematic step in protein and metalloprotein analysis, because it is generally laborious and time consuming as well as frequently responsible for errors in the final results. Other problems to be considered are related to the instability of the analyte, when proteins and metalloproteins are being analyzed. A compromise between effective sample preparation and gentle procedures to maintain the integrity of the analytes is then imperative. Both the extraction procedure and the extraction medium are decisive to preserve species in the protein structure. To evaluate the efficiency of protein extraction from biological samples, 11 different procedures (Table 3.2) for protein and metalloprotein extraction from horse chestnuts (*Aescullus hippocastanum L.*) *in natura* were tested.[52] After each extraction, total protein was determined and, after protein separation through SDS-PAGE, those metals (Cr, Fe and Mn) belonging to the protein structure

Table 3.2 Protein extraction procedures.

Procedure no.	Remark
1	Samples manually shaken in water for ca. 5 min
2	Samples manually shaken in buffer[a] for ca. 5 min
3	Samples beside being shaken in water also dialysed[b] in water for 42 h
4	Samples manually ground in water, using a mortar and pestle at room temperature for 15 min
5	Samples were manually ground in buffer[a], using a mortar and pestle at room temperature for 15 min
6	Samples, beside being ground in water, were also dialysed[b] in water for 42 h
7	Samples were shaken in water for 5 min and then submitted to sonication[c] for 10 min
8	Samples were manually ground in water and then submitted to sonication[c] for 10 min
9	Samples were manually ground in buffer[a] and then submitted to sonication[c] for 10 min
10	Samples were prepared as in procedure 8, but at 40°C
11	Samples were prepared as in procedure 9, but at 40°C

The information is taken from de Magalhães and Arruda.[52] © 2006 Elsevier B.V.
[a] 1.0 mol L^{-1} Tris–HCl, pH 6.8 buffer.
[b] Membrane MWCO: 6–8000, Spectrum Laboratories Inc., USA.
[c] Unique 1400, Ultrasonik.

were mapped by SRXRF. After mapping the elements in the protein bands (approx. 33 and 23.7 kDa), their concentrations were determined using atomic absorption spectrometry (ET-AAS). The 11 different extraction procedures were compared in terms of their relative efficiencies for both total protein extraction and metal–protein binding preservation. The results indicate that procedure 9 in Table 3.2 (where Tris–HCl buffer, grinding and sonication were used) could be applied for protein analysis, but it was the worst in terms of metal–protein binding preservation and should not be recommended for metalloprotein analysis. Thus, the more gentle procedures (4 and 5 in Table 3.2) produced appropriate results, because they preserve the metal–protein structure more effectively, assuring more accurate results in terms of quantitative analysis of metalloproteins.

To investigate the application of a chemometric tool (Kohonen self-organizing maps) for an exploratory analysis of metalloproteins based on eight metallic descriptors (K, Ca, Cr, Mn, Fe, Co, Ni, Zn), the metal ions were detected by synchroton radiation X-ray fluorescence (SRXRF) in 43 bands of proteins from sunflower leaves after SDS-PAGE. The data were analyzed with a Kohonen neural network, reducing the data dimensionality from eight to only two without loss of information; therefore, the number of samples to be analyzed was optimized for sunflower metalloprotein identification as well as for realizing inferences concerning the metals and their relationship with the protein bands. Only six proteins represent the chemical behavior (related to metal ions) of those initial 43 protein samples evaluated. This benefits the choice of the significant metalloprotein for posterior characterization.[53]

The hybrid technique of electrophoresis and SRXRF was also used for metalloprotein speciation in our laboratory. The scheme for the workflow used is shown in Figure 3.10. Fe-, Cu-, and Zn-binding proteins in human liver cytosol were detected with SRXRF after separation by SDS-PAGE[54] or IEF.[55] More metal-containing bands can be detected after the non-denaturing IEF separation. To investigate the relationship between metalloprotein distributions and cancerization of human liver tissues, proteins in cytosol and microsome of tumor and surrounding non-tumor tissues from five individuals with hepatocellular carcinoma (HCC) were separated with IEF. Zn, Fe, Cu, and Mn contents in protein bands were measured with SRXRF. The results indicate that metal distributions among the metal-containing bands detected are distinguishable between tumor and non-tumor tissues. The differences in distribution may be involved in carcinogenesis of liver tissues.[56,57] In our resent work, the hybrid technique was used to explore the possible mechanism of antagonistic effects among Hg, Se, As in orgnisms.[58] Proteins in liver tissues of bighead carp and grass carp sampled from a mercury-polluted area of Wanshan, Guizhou Province, China, were separated by thin-layer IEF. The relative content of Hg, Se, and As in protein bands was measured with SRXRF. At least three Hg-containing bands in liver of bighead carp and one Hg-containing band in grass carp have were detected. Se and As were found in the Hg-containing bands with a pI of 3.7 in bighead carp and 6.2 in grass carp. The results imply that the bands containing Hg, As, Se simultaneously may correspond to

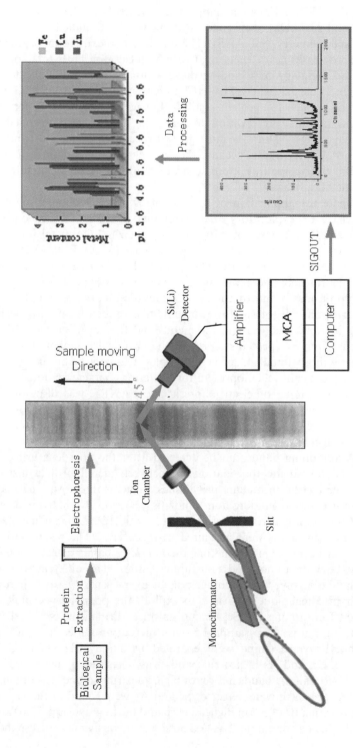

Figure 3.10 Scheme of the workflow for trace element speciation analysis in biological samples with electrophoresis and SRXRF.

the antagonistic effect of Se against the toxicity of Hg and As. In addition, Hg and As often coexist in the same band, suggesting that the two elements may be involved in the same metabolic processes.

Reports concerning to the application of XRF to direct analysis of trace elements associated with proteins in electrophoresis gel are listed in Table 3.3.

3.6.2 Electrophoresis and Sample Preparation for SRXRF Measurement

Flatbed gel electrophoresis (GE) is used extensively in the analysis and characterization of proteins with a variety of protocols such as isoelectric focusing (IEF) and polyacrylamide gel electrophoresis (PAGE), but has not been specifically developed for the separation of metal-containing proteins. Therefore, before using the method for speciation of biological samples, some important parameters have to be checked. It has to be investigated in detail if speciation data are hampered by contamination or alteration of native species due to the influence of the electric field and/or denaturing buffer systems.[60]

A commonly used electrophoresis technique is SDS-PAGE, which is a denaturing electrophoresis technique. The proteins are first treated with the detergent, SDS, in a ratio of 1.4 g detergent g^{-1} protein. The excess negatively charged detergent around proteins transforms all of them into macropolyanions with a uniformly random coil shape. The proteins, having a similar charge-to-mass ratio, are separated in the electric field according to their individual molecular weights (MWs). After separation, SRXRF can be used for the detection of metalloprotein bands. To reduce the matrix effects and improve the signal-to-noise ratio, the gel should be dried before XRF measurement without loss of protein-bound metal or cracking of the gel.[54] This was successfully achieved by incubation in glycerol followed by heating.[61] The protein contained in individual electrophoresis bands can be quantified by densitometry after coloration with Coomassie blue.[35] The metals in protein bands can be quantified according to a calibration curve, which is established by doping polyacrylamide gels with metals at different concentrations,[34] or by electrophoresis of metalloproteins containing known amounts of metals with stacking gel.[62] Thus the molecular weights, metal contents, and amounts of metalloproteins and the metal-to-protein ratios could be obtained.

Following the separation, proteins are generally visualized by staining the gel with one of a range of stains (typically silver or Coomassie blue). It has been demonstrated that staining the gel using silver or Coomassie blue can result in loss of most metals from their bound proteins.[61] To avoid the loss of metals from protein bands and the contamination of sample resulting from the use of colorant and decolorant, two electrophoresis gel must be loaded with the same sample and run simultaneously, the first is used for XRF analysis; and the second is colored and used for protein quantification. The inability to stain the gel before SRXRF measurement requires the use of scan mode for XRF

Table 3.3 Some applications of hybrid technique of XRF electrophoresis on chemical speciation analysis of trace elements.

Sample	Purpose	Excitation source	Electrophoresis	Ref.
Hydrogenase enzyme HiPIP	Detection of Fe, Ni	3 MeV proton beam	SDS-PAGE	31–33
Mixture of 6 proteins	Quantification of Fe, Cu, Zn	2.5 MeV proton beam	SDS-PAGE	34,35
Photosystems PSI RC in thermophilic cyano-bacteria *Synechoccocus*	Quantification of Fe	2 MeV proton beam	SDS-PAGE	36
Carbonic anhydrase, cytoplasmic Pyrophosphatase	Quantification of Zn	3.7 MeV proton beam	PAGE	37
Glutathione peroxidase	Detection of Se	X-ray tube	PAGE	38
Metallothionein II, Cu/Zn-superoxide dismutase	Interaction of Hg and the proteins	SR	IEF	49
Rat testis, and metal-loaded apoazurin	Estimation of the suitability of SDS-PAGE and SRXRF in the detection of metal-containing proteins	SR	SDS-PAGE	50,51
Metalloprotein extracted from horse chestnuts	Mapping of Cr, Mn, and Fe in gel to evaluate the efficiency of metalloprotein extraction	SR	SDS-PAGE	52
Sunflower leaves	Evaluation of the potentiality of qualitative/quantitative metallomics analysis with electrophoresis, SRXRF, and Kohonen neural network	SR	SDS-PAGE	53
Human liver cytosol	Distribution of Fe, Cu, Zn-containing proteins	SR	SDS-PAGE, IEF	54,55
Cytosol of hepatocellular carcinoma tissues	Distribution pattern of metalloproteins	SR	IEF	56,57
Fish live cytosol	Detection of As, Se, Hg-containing proteins	SR	IEF	58
Cu/Zn-superoxide dismutase	Speciation of oxidation states of Cu and Zn	SR-XAS	IEF	59

analysis, since the location of the protein spot on an unstained gel for spot analysis is ambiguous.

As SDS triggers dramatic changes in protein conformation, the ternary and quaternary structures are lost, causing the metals that are loosely bound to proteins to dissociate. As an example, after SDS-PAGE, no Cu or Zn can be detected in the band of superoxide dismutase, a Cu,Zn-containing protein, whereas after non-denaturing electrophoresis, both metals are detected in the corresponding protein band.[35] Therefore, the comparative analysis of metals from proteins separated either under denaturing or non-denaturing conditions can be used to study the affinity of the metal binding to proteins.

Nevertheless, the absence of detergent modifies the conditions of migration; the charge, size, and shape of and protein influences the migration velocity and proteins are not easily separated. For analysis of an unknown biological sample, a protocol was suggested by Bertrand and colleagues,[28] starting with classical denaturing electrophoresis, first to obtain sharp protein bands. Then, less and less detergent is used, the bands become broader, but the protein-bound metals are preserved. Finally, XRF analysis confirms it.

Isoelectrofocusing (IEF) is a native electrophoresis technique, which provides sharp protein bands in non-denaturing conditions. Gels used in IEF are usually made of appropriate amounts of acrylamide ($\sim 5\%$), ampholyte ($\sim 2\%$), and double-distilled water. The ampholytes existing in polyacrylamide gels can form a pH gradient between anode and cathode in an electric field. The charged protein molecules migrate through the pH gradient until their respective isoelectric point (pI). Proteins with characteristic pI values can thus be separated. They do not experience a denaturing process; metal species can be conserved to a larger degree. More metalloproteins could be detected by SRXRF analysis of proteins separated with gel filtration chromatography and thin layer IEF[55] compared to SRXRF analysis of the same protein fraction separated by SDS-PAGE.[54]

A number of amphoteric buffer solutions and pre-made gels are available covering broad and narrow pH ranges. High resolution is obtained when narrow pH ranges are employed. A resolution of 0.01 pH unit can be achieved under optimal conditions. The technique of IEF represents a remarkable advantage compared with PAGE and SDS-PAGE for metalloprotein research. However, it is necessary to be aware that ampholytes may capture metals from proteins competitively. The affinity between the ampholytes and metal ions depends on the pH environment.[63] It is necessary to find the optimum spot of sample application at which the metalloproteins are stable and to pre-focus the gel, which allows the pH gradient partly to be established prior to application.

Whatever format of electrophoresis is used for separating metalloproteins, the influence of the electric field on the stability of metal-containing proteins must be taken into account. Lustig demonstrated that In-, Ga-, and V-transferrin complexes can be stripped off during a PAGE or IEF process, while the binding of Cr- and Pt-serum protein complexes seem to be strong enough to withstand the electric field.[60] Therefore, it is necessary to check the capability of electrophoresis as a speciation method before starting the application.

As for the elemental analysis techniques, apart from XRF, metal in an electrophoresis gel can also be analyzed directly by following three techniques: autoradiography,[64] X-ray absorption spectroscopy (XAS),[59] and laser ablation inductively coupled plasma mass spectrometry (LA-ICP-MS).[65–69] Some important characteristics of these techniques are listed in Table 3.4.

3.6.3 XRF as an On-line Detector of Capillary Electrophoresis and Other Separation Techniques

The combination of XRF with electrophoresis can constitute a quasi-on-line coupling technique. Great efforts have been made to develop on-line analysis techniques with XRF. Mann et al.[70] reported a technique of on-line SRXRF detection of metal ions, such as Fe, Co, Cu, and Zn, in their high binding-constant complexes for capillary electrophoresis (CE) separation. An X-ray transparent polymer coupling is used to create a window for the on-line X-ray detection. In contrast to ICP-MS, this detection technique is not limited by sample or the buffer volatility or atomization efficiency. Simultaneous XRF and UV absorbance detection can be used to provide on-line determination of metal/chelate ratios. A bench-top energy-dispersive micro X-ray fluorescence system was also combined with the CE apparatus constructed by a thin-walled fused-silica capillary for elemental analysis of the species containing Fe, Co, and Cu, for example.[71,72] This coupled technique used for metalloprotein speciation analysis can avoid the compromise between optimal separation and sensitive detection, which must be taken into account in the HPLC-ICP-MS procedure. In addition, this detection scheme is non-destructive, so the separated material can be recovered for additional characterization.

Preliminary average detection limits obtained by this system were at the order of 10^{-4} M. The poorer detection limits are mainly caused by a high Bremsstrahlung background resulting from scatter by the silica capillary. The peak definition and the ratio of S/N are mainly limited by the low instrument sampling rate (10 s per spectrum is necessary for CE-micro-XRF system). Improvements in the detecting window materials and detection technology should be able to decrease spectral dwell times, therefore increasing the sampling rate, which would enhance sample peak definition and increase S/N, as well as the sensitivity of the procedure.

An exploratory study has been carried out for on-line use of GE-XRF as an element-specific detector for chemical microchips. The microchip is specially designed. It includes an analyzing region in which the sample solution can be dried and concentrated in a suitable area corresponding to the size of the primary X-ray beam. In GE-XRF, the background intensity in the XRF spectrum is reduced at grazing-exit angles. In addition, a good relationship between the X-ray fluorescence intensities and the concentrations of standard solutions that are introduced into the microchip can be obtained. This indicates that the GE-XRF method is feasible for trace elemental analysis in chemical microchip systems.[73] However, the X-ray spectra observed had considerably high

Table 3.4 The main analytical techniques for direct metalloprotein analysis on electrophoresis gels.

Analytical technique	Detection limit ($\mu g\,g^{-1}$)	Selectivity	Quantification	Analytical depth (μm)
Autoradiography	<0.01	Monoelemental (radio-labeled isotope)	Quantitative	—
LA-ICP-MS	0.01	Multielemental and isotopic	Quantitative (matrix effects)	200
XAS	100	Chemical species	Quantitative	>100
PIXE	1	Multielemental	Quantitative	10–100
SR-XRF	0.1	Multielemental	Quantitative	>100

The information is taken from Ortega.[30] ©The Royal Society of Chemistry 2009.

background intensity, suggesting that the grazing-exit condition was not actually achieved. Further improvement of microchips will be necessary.

3.7 Outlook and Challenges

XRF, combined with electrophoresis, is a powerful and non-destructive method for speciation analysis of metals in biological samples. It can give element-specific information and identify proteins simultaneously, which is difficult for other hybrid techniques. This XRF technique is also capable of studying the synergistic and antagonistic effects among bio-elements for its multielement character. For a high throughput analysis in metallomics and metalloproteomics, quick, precise and multielement analysis should be essential, combined with high-resolution separation procedure. Two dimensional gel electrophoresis is a powerful and widely used technique for proteome analysis. This technique begins with IEF electrophoresis and then separates molecules by a second property, usually molecular weight, in a direction 90° to the first. The result is that molecules are spread out across a 2D gel. Because it is unlikely that two molecules will be similar in two distinct properties, molecules are more effectively separated in 2D electrophoresis than in 1D electrophoresis. However, such a protein separation protocol is not used very often to study metalloproteins because of the denaturing process in the second dimension of separation. Development of native 2D electrophoresis with native PAGE, such as blue native PAGE or clear native PAGE, is significant for metallomics and metalloprotemics studies. Lustig et al. have described a native and powerful 2D method for the separation of platinum proteins.[74] In the first dimension IEF was performed using immobilised pH gradients (IPGs). Native PAGE was done in the second dimension. Mild conditions were selected, particularly for the second dimension, e.g. avoiding buffer systems with platinophile N- or S-donor groups. The separation reagents were checked to determine the concentrations that could be used for this purpose. However, non-denaturing conditions are known to alter the resolving power of 2D PAGE. One solution to this problem could be the separation of protein fractions under native conditions before 2D PAGE by using, for example, steric exclusion chromatography, or other related chromatographic methods that would reduce the number of proteins in the electrophoresis gel.[30]

For elemental analysis, the application of SR-XRF and XAS to the direct analysis of proteins on 2D gels will simultaneously allow the identification, quantification, and speciation of proteins and inorganic elements. Chevreux and colleaques have developed a new method for the speciation analysis of inorganic elements bound to proteins with X-ray absorption near edge structure (XANES) after separation using polyacrylamide gel electrophoresis.[59] They separated the copper–zinc superoxide dismutase (CuZnSOD) isoforms with IEF, and then investigated the copper and zinc oxidation states in the protein bands with XANES directly. Results of the study indicate that zinc is present in its Zn(II) oxidation state in all analysed isoforms. Copper is present in the Cu(II) oxidation state in the main acidic isoform, while it is found in both Cu(II) and Cu(I) states in the main basic isoform. This method enables the direct speciation analysis of metals in metalloproteins without any protein purification step that could modify their oxidation state. Nevertheless, the detection limit must be improved for the analysis of metalloproteins expressed in low amounts. This may be achieved by enhancing the brightness of SR, the efficiency of X-ray detectors, and the performance of the X-ray optics. Meanwhile, the enhancement will reduce the analysis time, which is a crucial issue for high-throughput analysis.

Acknowledgments

The authors acknowledge the Natural Science Foundation of China for financial supports (Grant No.: 20777075 and 10475092).

References

1. R. E. Van Grieken and A. A. Markowicz, *X-ray physics, Handbook of X-ray Spectrometry,* Marcel Dekker, New York, 2002, ch. 1.
2. M. Kocsis and A. Snigirev, *Nucl. Instrum. Methods Phys. Res. Sect. A,* 2004, **525**, 79–84.
3. W. Mondelaers, P. Cauwels, B. Masschaele, M. Dierick, P. Lahorte, J. Jolie, S. Baechler and T. Materna, International Linac Conference, Monterey, Calif. THC, 2000, www.slac.stanford.edu/econf/C000821/THC16.pdf pp. 890–892.
4. S. A. E. Johansson and J. L. Campbell, ed. *PIXE. A Novel Technique for Elemental Analysis,* Wiley, New York, 1988.
5. S. Matsuyama, H. Mimura, H. Yumoto, Y. Sano, K. Yamamura, M. Yabashi, Y. Nishino, K. Tamasaku, T. Ishikawa and K. Yamauchi, *Rev. Sci. Instrum.,* 2006, **77**, 103102.
6. R. Ortega, S. Bohic, R. Tucoulou, A. Somogyi and G. Devès, *Anal. Chem.,* 2004, **76**, 309–314.
7. A. Snigirev, V. Kohn, I. Snigireva and B. Lengeler, *Nature,* 1996, **384**, 49–51.
8. E. Di Fabrizio, F. Romanato, M. Gentili, S. Cabrini, B. Kaulich, J. Susini and R. Barrett, *Nature,* 1999, **401**, 895–898.
9. G. Hirsch, *X-ray Spectrom.,* 2003, **32**, 229–238.
10. I. Szaloki, S. B. Török, C. U. Ro, J. Injuk and R. E. Van Grieken, *Anal. Chem.,* 2000, **72**, 211–234.
11. R. Jenkins, *X-Ray Fluorescence Spectrometry,* Wiley, New York, 1988, p. 61.
12. G. F. Knoll, *J. Radioanal. Nucl. Chem.,* 2000, **243**, 125–131.
13. P. Van Espen, K. Janssens and J. Nobels, *Chemom. Intell. Lab. Syst.,* 1987, **1**, 109–114.
14. B. Vekemans, K. Janssens, L. Vincze, F. Adams and P. Van Espen, *X-Ray Spectrom.,* 1994, **23**, 278–285.
15. H. A. van Sprang, *Adv. X-Ray Anal.,* 2000, **42**, 1–10.

16. G. R. Lachance, *Spectrochim. Acta Part B*, 1993, **48**, 343–357.
17. J. W. Criss, L. S. Birks and J. V. Gilfrich, *Anal. Chem.*, 1978, **50**, 33–37.
18. R. M. Rousseau, *Rigaku J.*, 2001, **18**, 33–47.
19. L. Vincze, B. Vekemans, F. Brenker, G. Falkenberg, K. Rickers, A. Somogyi, M. Kersten and F. Adams, *Anal. Chem.*, 2004, **76**, 6786–6791.
20. F. Pfeiffer, C. David, M. Burghammer, C. Riekel and T. Salditt, *Science*, 2002, **297**, 230–234.
21. C. G. Schroer, O. Kurapova, J. Patommel, P. Boye, J. Feldkamp, B. Lengeler, M. Burghammer, C. Riekel, L. Vincze, A. van der Hart and M. Küchler, *Appl. Phys. Lett.*, 2005, **87**, 6951–6953.
22. C. Streli, P. Wobrauschek, F. Meirer and G. Pepponi, *J. Anal. At. Spectrom.*, 2008, **23**, 792–798.
23. B. Ostachowicz, M. Lankosz, B. Tomik, D. Adamek, P. Wobrauschek, C. Streli and P. Kregsamer, *Spectrochim. Acta Part B*, 2006, **61**, 1210–1213.
24. T. Magalhães, A. von Bohlen, M. L. Carvalho and M. Becker, *Spectrochim. Acta Part B*, 2006, **61**, 1185–1193.
25. N. Szoboszlai, Z. Polgári, V. G. Mihucz and G. Záray, *Anal. Chim. Acta*, 2009, **633**, 1–18.
26. R. S. Becker, J. A. Golovchenko and J. R. Patel, *Phys. Rev. Lett.*, 1983, **50**, 153–156.
27. K. Tsuji and F. Delalieux, *Spectrochim. Acta Part B*, 2003, **58**, 2233–2238.
28. M. Bertrand, G. Weber and B. Schoefs, *Trends Anal. Chem.*, 2003, **22**, 254–262.
29. R. Ma, C. W. McLeod, K. Tomlinson and R. K. Poole, *Electrophoresis*, 2004, **25**, 2469–2477.
30. R. Ortega, *Metallomics*, 2009, **1**, 137–141.
31. Z. Szökefalvi-Nagy, I. Demeter, C. Bagyinka and K. L. Kovacs, *Nucl. Instrum. Methods Phys. Res. Sect. B*, 1987, **22**, 156–158.
32. Z. Szökefalvi-Nagy, C. Bagyinka, I. Demeter, K. L. Kovacs and L. H. Quynh, *Biol. Trace Elem. Res.*, 1990, **26–27**, 93–101.
33. Z. Szökefalvi-Nagy, C. Bagyinka, I. Demeter, K. Hollos-Nagy and I. Kovaks, *Fresenius' J. Anal. Chem.*, 1999, **363**, 469–473.
34. D. Strivay, B. Schoefs and G. Weber, *Nucl. Instrum. Methods Phys. Res. Sect. B*, 1998, **136–138**, 932–935.
35. G. Weber, D. Strivay, C. Menendez, B. Schoefs and M. Bertrand, *Int. J. PIXE*, 1996, **6**, 215–225.
36. C. Solis, A. Oliver and E. Andrade, *Nucl. Instrum. Methods Phys. Res. Sect. B*, 1998, **136–138**, 928–931.
37. C. Solis, A. Oliver, E. Andrade, J. L. Ruvalcaba-Sil, I. Romero and H. Celis, *Nucl. Instrum. Methods Phys. Res. Sect. B*, 1999, **150**, 222–225.
38. S. F. Stone, G. Bernasconi, N. Haselberger, M. Makarewicz, R. Ogris, P. Wobrauschek and R. Zeisler, *Biol. Trace Elem. Res.*, 1994, **43–45**, 299–307.
39. K. Günther and A. von Bohlen, *Spectrochim. Acta Part B*, 1991, **46**, 1413–1419.

40. V. G. Mihucz, E. Tatár, A. Varga, G. Záray and E. Cseh, *Spectrochim. Acta Part B*, 2001, **56**, 2235–2246.
41. F. M. Verbi, S. C. C. Arruda, A. P. M. Rodriguez, C. A. Pérez and M. A. Z. Arruda, *J. Biochem. Biophys. Methods*, 2005, **62**, 97–109.
42. C. Streli, G. Pepponi, P. Wobrauschek, C. Jokubonis, G. Falkenberg, G. Záray, J. Broekart, U. Fittschen and B. Peschel, *Spectrochim. Acta Part B*, 2006, **61**, 1129–1134.
43. F. Meirer, G. Pepponi, C. Streli, P. Wobrauschek, V. G. Mihucz, G. Záray, V. Czech, J. A. C. Broekaert, U. E. A. Fittschen and G. Falkenberg, *X-Ray Spectrom.*, 2007, **36**, 408–412.
44. V. G. Mihucz, E. Tatár, I. Virág, E. Cseh, F. Fodor and Gy. Záray, *Anal. Bioanal. Chem.*, 2005, **383**, 461–466.
45. K. Geraki, M. J. Farquharson and D. A. Bradley, *Phys. Med. Biol.*, 2002, **47**, 2327–2339.
46. A. Ide-Ektessabi, S. Fujisawa, K. Sugimura, Y. Kitamura and A. Gotoh, *X-Ray Spectrom.*, 2002, **31**, 7–11.
47. B. S. Twining, S. B. Baines, N. S. Fisher, J. Maser, S. Vogt, C. Jacobsen, A. Tovar-Sanchez and S. A. Sanudo-Wilhelmy, *Anal. Chem.*, 2003, **75**, 3806–3816.
48. R. Ortega, S. Bohic, R. Tucoulou, A. Somogyi and G. Deves, *Anal. Chem.*, 2004, **76**, 309–314.
49. S. Homma-Takeda, M. Shinyashiki, I. Nakai, C. Tohyama, Y. Kumagai and N. Shimojo, *Anal. Lett.*, 1996, **29**, 601–611.
50. G. Weseloh, M. Kühbacher, H. Bertelsmann, M. Özaslan, A. Kyriakopoulos, A. Knöchel and D. Behne, *J. Radioanal. Nucl. Chem.*, 2004, **259**, 473–477.
51. M. Kuhbacher, G. Weseloh, A. Thomzig, H. Bertelsmann, G. Falkenberg, M. Radtke, H. Riesemeier, A. Kyriakopoulos, M. Beekes and D. Behne, *X-Ray Spectrom.*, 2005, **34**, 112–117.
52. C. S. de Magalhães and M. A. Z. Arruda, *Talanta*, 2007, **71**, 1958–1963.
53. J. S. Garcia, G. A. da Silva, M. A. Z. Arruda and R. J. Poppi, *X-Ray Spectrom.*, 2007, **36**, 122–129.
54. Y. X. Gao, C. Y. Chen, P. Q. Zhang, Z. F. Chai, W. He and Y. Y. Huang, *Anal. Chim. Acta*, 2003, **485**, 131–137.
55. Y. X. Gao, C. Y. Chen, Z. F. Chai, J. J. Zhao, J. Liu, P. Q. Zhang, W. He and Y. Y. Huang, *Analyst*, 2002, **127**, 1700–1704.
56. Y. X. Gao, Y. B. Liu, C. Y. Chen, B. Li, W. He, Y. Y. Huang and Z. F. Chai, *J. Anal. At. Spectrom.*, 2005, **20**, 473–475.
57. Y. B. Liu, L. N. Li, Y. X. Gao, C. Y. Chen, B. Li, W. He, Y. Y. Huang and Z. F. Chai, *Hepatogastroenterology*, 2007, **54**, 2291–2296.
58. L. N. Li, G. Wu, J. Sun, B. Li, Y. F. Li, C. Y. Chen, Z. F. Chai, A. Iida and Y. X. Gao, *J. Toxicol. Environ. Health*, 2008, **18**, 1266–1269.
59. S. Chevreux, S. Roudeau, A. Fraysse, A. Carmona, G. Devès, P. L. Solari, T. C. Weng and R. Ortega, *J. Anal. At. Spectrom.*, 2008, **23**, 1117–1120.

60. S. Lustig, D. Lampaert, K. De Cremer, J. De Kimpe, R. Cornelis and P. Schramelb, *J. Anal. At. Spectrom.*, 1999, **14**, 1357–1362.
61. A. Raab, B. Pioselli, C. Munro, J. Thomas-Oates and J. Feldmann, *Electrophoresis*, 2009, **30**, 303–314.
62. Y. X. Dong, Y. X. Gao, C. Y. Chen, B. L. Xing, H. W. Yu, W. He, Y. Y. Huang and Z. F. Chai, *Chin. J. Anal. Chem.*, 2006, **34**, 443–446.
63. A. Oratore, A. M. D. Alessandro and G. D. Andrea, *Biochem. J.*, 1989, **259**, 909–912.
64. J. S. Garcia, C. Schmidt de Magalhaes and M. A. Zezzi Arruda, *Talanta*, 2006, **69**, 1–15.
65. J. L. Neilsen, A. Abildtrup, J. Christensen, P. Watson, A. Cox and C. W. McLeod, *Spectrochim. Acta Part B*, 1998, **53**, 339–345.
66. T. W. M. Fan, E. Pruszkowski and S. Shuttleworth, *J. Anal. At. Spectrom.*, 2002, **17**, 1621–1623.
67. M. R. Binet, R. Ma, C. W. McLeod and R. K. Poole, *Anal. Biochem.*, 2003, **318**, 30–38.
68. G. Ballihaut, L. Tastet, C. Pecheyran, B. Bouyssiere, O. F. X. Donard, R. Grimaud and R. Lobinski, *J. Anal. At. Spectrom.*, 2005, **20**, 493–499.
69. J. Sa. Becker, M. Zoriy, J. Su. Becker, C. Pickhardt, E. Damoc, G. Juhacz, M. Palkovits and M. Przybylski, *Anal. Chem.*, 2005, **77**, 5851–5860.
70. S. E. Mann, M. C. Ringo, G. Shea-McCarthy, J. Penner-Hahn and C. E. Evans, *Anal. Chem.*, 2000, **72**, 1754–1758.
71. T. C. Miller, M. R. Joseph, G. J. Havrilla, C. Lewis and V. Majidi, *Anal Chem.*, 2003, **75**, 2048–2053.
72. J. Vogt and C. Vogt, *Nucl. Instrum. Methods Phys. Res. Sect. B*, 1996, **108**, 133–135.
73. K. Tsuji, T. Emoto, Y. Nishida, E. Tamaki, Y. Kikutani, A. Hibara and T. Kitamori, *Anal. Sci.*, 2005, **21**, 799–803.
74. S. Lustig, J. De Kimpe, R. Cornelis and P. Schramel, *Fresenius' J. Anal. Chem.*, 1999, **363**, 484–487.

CHAPTER 4
Isotopic Techniques Combined with ICP-MS and ESI-MS

MENG WANG, WEIYUE FENG* AND ZHIFANG CHAI

CAS Key Laboratory of Nuclear Analytical Techniques, Institute of High Energy Physics, Chinese Academy of Sciences, Beijing 100049, China

4.1 Inductively Coupled Plasma–Mass Spectrometry

4.1.1 Introduction

Inductively coupled plasma–mass spectrometry (ICP-MS) appeared as a commercial instrument in the early 1980s,[1] and nowadays it has become a powerful tool for analysis of trace, minor, and major elements in a variety of samples, including environmental, geological, semiconductor, biomedical, and nuclear application fields.[2,3] As the name suggests, ICP-MS is an ionization source of inductively coupled plasma attached to a mass spectrometer. ICP-MS uses a high-temperature plasma (partly ionized argon) and can analyze almost all elements in the periodic table. With recent advances in instrumental techniques, different mass analyzers have been introduced in ICP-MS, such as a quadrupole, a double-focusing magnetic sector,[4] the time-of-flight technology,[5] and a multipole collision/reaction cell.[6] On the other hand, a large variety of sample introduction systems can be combined with ICP-MS, such as high-performance liquid chromatography (HPLC), gas chromatography (GC), capillary electrophoresis (CE), and laser ablation (LA).[7–9]

In many publications of biomedical studies, mass spectrometry is usually associated with a soft, low-temperature ionization source, such as matrix-assisted laser desorption ionization (MALDI) or electrospray ionization (ESI).

---

*Corresponding author: Email: fengwy@ihep.ac.cn; Tel: +86-10-88233209; Fax: +86-10-88235294.

Nuclear Analytical Techniques for Metallomics and Metalloproteomics
Edited by Chunying Chen, Zhifang Chai and Yuxi Gao
© Royal Society of Chemistry 2010
Published by the Royal Society of Chemistry, www.rsc.org

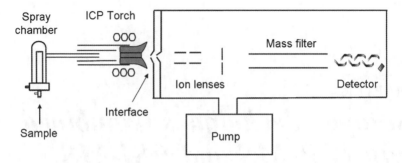

Figure 4.1 The schematic of ICP-MS.

Table 4.1 Characteristic of ICP-MS.

1 High analytical throughput
2 Excellent detection limits ($pg\,g^{-1}$ to $ng\,g^{-1}$) for most elements
3 Minimal matrix effects
4 Specific response to most elements (metals, semimetals or metalloids)
5 The capability of up to eight magnitudes of linear dynamic range
6 Information of isotope ratios
7 Simple coupling to another separated method (e.g. high performance liquid chromatography, HPLC)

However, ICP-MS and its capabilities for biomedical applications are still not well-known. In fact, ICP-MS has played and undoubtedly will continue to play an important role in many aspects of analytical chemistry, especially in the analysis of elements and isotope ratios. Nowadays, ICP-MS has been considered as a promising analytical method for metallomics and metalloproteomics study, due to its outstanding characteristics described above.

Although there are different ICP-MS instruments today, an ICP-MS instrument is usually composed of a sample introduction system, a plasma source, an interface, a mass analyzer, a detector and a vacuum system (see Figure 4.1). Most of elements in the periodic table can be fully ionized in a high-temperature ionization source like ICP, and thus can be analyzed sequentially with a mass spectrometer. The outstanding characteristics of ICP-MS are summarized in Table 4.1.[10,11] In this section, an overview of ICP-MS will be given briefly with an emphasis on the unique characterization and potential of ICP-MS, in comparison with the molecular mass spectrometry that is widely used in proteomics studies.

4.1.2 ICP as a High-temperature Ionization Source

Argon plasma is the most common type in plasma source mass spectrometry. Generally, the sample is introduced into ICP-MS as a liquid solution via a nebulizer, where a fine aerosol is generated. Then the aerosol is transported to the ICP ionization source. In contrast to soft-ionization sources, such as ESI and MALDI, the ICP operates at a high temperature between 6000 and 10 000 K;[11] the energy is enough for sample volatilization, desolvation, and atomization, and ionization of the atom to occur. All chemical bonds in the sample are broken readily, and the isotopes are ionized into ions with a positive charge. Therefore, the response of an element in ICP-MS is independent of the molecular environment of the element,[12] implying that the quantitative analysis of an element in a protein can use any species of the element as a calibrated standard. For example, Svantesson *et al.* demonstrated that as long as the molecule concentration is sufficiently low, the matrix interference from the surrounding molecule is almost negligible and inorganic elemental standards can be used for quantification with an accuracy of 10% or better.[13] Moreover, the atomic mass spectra are more straightforward than molecular mass spectra, because the number of isotopes in nature is very limited. This unique characteristic of ICP-MS is suitable for accurate measurement of elements in proteins regardless of the diverse matrixes of the samples.

The high-temperature plasma mainly consists of electrons and argon ions and thus has a very high ion density. It also results in a high collision rate in the plasma. The ionization efficiency of an element is dependent on its ionization potential. Because the first ionization potential of argon is about 15.8 eV, it is high enough to ionize most of elements in the periodic table, the majority of which have the first ionization potentials in the range of 4–12 eV.[11] Unfortunately, the first ionization potentials of many abundant elements in proteins, such as carbon, oxygen, and hydrogen, are too high to be ionized in argon ICP. Thus these elements have poor detection limits in ICP-MS.

In summary, the ionization efficiency and analytical sensitivity of ICP is higher than other ion sources, and the matrix effect in ICP-MS is much smaller. It can be found that the ICP and other soft ion sources (e.g. ESI or MALDI) are really complementary techniques. Structural information is preferably obtained by means of ESI- or MALDI-MS, whereas ICP-MS is ideal to quantify elements in samples, even in the very low concentration ranges.

4.1.3 Mass Analyzers for ICP-MS

The ions generated in the plasma are introduced into the mass analyzer, where they are separated by their mass-to-charge ratio (m/z). There are mainly three mass analyzers in commercial ICP-MS: quadrupole, double-focusing magnetic sector (high-resolution),[4] and time-of-flight (TOF).[2,11] The quadrupole-based analyzer was first developed in the early 1980s and now represents most of all ICP-MS. The resolution of a quadrupole mass analyzer is normally kept at 0.3–1 AMU (atomic mass unit) and can offer a resolving power of about 300,[11] which is enough for most routine applications. However, the quadrupole analyzer has proved to be inadequate for many elements that are liable to spectral interference derived from argon, solvents, and sample matrixes, such as the interferences of $^{40}Ar^{16}O$ for ^{56}Fe and $^{40}Ar^{35}Cl$ for ^{75}As determination and so forth. The resolving power of a commercial TOF analyzer in ICP-MS is typically in the range of 500–2000,[5] which is also inadequate to solve most polyatomic interferences in ICP-MS. A TOF analyzer, however, is ideally suited for analysis of rapid transient signals, high precision of isotope ratio analysis, and rapid data acquisition because ions to be determined are sampled at the exact same moment. Compared to the quadrupole and TOF analyzers, a magnetic-sector analyzer has a resolving power of up to 10 000.[4] The great improvement in resolving power allows the elements mentioned above to be analyzed, even in complex sample matrixes, but the application of high-resolution ICP-MS is limited due to its high cost and complex operation.

Based on a quadrupole ICP-MS, another technique, termed collision/reaction cell, has frequently been used to reduce polyatomic ion interferences.[14] These techniques provide simple, efficient, and low-cost methods in the face of many difficult interference problems. In these methods, ions to be analyzed first enter a radio-frequency-only multipole (e.g. a quadrupole, hexapole, or octapole), in which the analytes react with the collision/reaction gas, which is usually oxygen, ammonia, xenon, or methane,[14] to remove polyatomic interference or generate a new analyte ion of m/z showing less interference. The RF-only multipole does not separate ions like a traditional quadrupole, but it has profound influence on collisional focusing of ions, both of the energy and spatial distributions. An example of removing the polyatomic interference is shown below, which uses ammonia gas to reduce any $^{40}Ar^+$ interference in the measurement of $^{40}Ca^+$:

$$Ar^+ + NH_3 \rightarrow NH_3^+ + Ar \quad \text{(fast reaction)}$$

$$Ca^+ + NH_3 \rightarrow NH_3^+ + Ca \quad \text{(slow reaction)}$$

Because of the disparity of the reaction rates of the two neutralization reactions, the analyte can be efficiently determined after the introduction of ammonia as a reactive gas into the multipole. There are many excellent reviews about the development and applications of collision/reaction cell in ICP-MS.[6,14] In order to eliminate the new isobaric interferences produced by secondary reactions, two methods are commonly used in the commercial instrument: the discrimination of kinetic energy or mass filtering.[14] The former mainly utilizes the post-cell kinetic energy discrimination (KED) to suppress transport of the produces of the side reactions to the analyte in the hexapole and octapole cell instruments. Whereas in the latter, the quadrupole cell has a capability to reduce the formation of the unwanted side product ions by selecting an appropriate mass bandpass. The details of the KED and bandpass approaches can refer to many excellent books and reviews.[6,11,14]

4.1.4 ICP-MS Coupled Techniques

Using different types of sample introduction system, ICP-MS can be easily combined with various separation methods, including gas chromatography (GC), high-performance liquid chromatography (HPLC), and capillary electrophoresis.[8] Among the above methods, the coupling of HPLC to ICP-MS appears to be one of the most common methods for metallomics and metalloproteomics study because of its ease of sample preparation and simplicity of the interface to the detector.[15,16] The separation strategies commonly used in HPLC include reversed-phase partition (RP), ion exchange (IE), and size exclusion (SE).[15] However, this combined technique possesses some problems when a liquid chromatography column is coupled to ICP-MS. The traditional nebulizer (e.g. Meinhard and cross-flow) has a very low nebulized efficiency (less than 3%) and a large dead volume,[11] thus causing a deterioration of the detection limits. On the other hand, the gradients of organic modifiers and inorganic salts used in the HPLC mobile phase may significantly decrease the signal of ICP-MS.

To overcome these problems, many kinds of micronebulizers, such as a total consumption micronebulizers with a low-volume spray chamber[17] and direct-injection high-efficiency nebulizer (DIHEN),[18,19] have been developed for interfaces of HPLC to ICP-MS. In comparison with the traditional interface, these devices can minimize the dead volume, improve the nebulization efficiency, and allow efficient coupling of HPLC to ICP-MS. Another method in HPLC/ICP-MS is to use post-column sheath flow, which ensures the constant sensitivity of ICP-MS during HPLC gradient elution.[20,21]

Capillary electrophoresis (CE), as a complement to other separation methods, is widely applied in analytical laboratories.[22] This method is versatile for diverse analytes because of its extraordinary resolution power of up to 500 000 theoretical plates, and its ability to separate anions, cations, and neutral molecules in a single analysis. Because the coupling of CE to ICP-MS is introduced into speciation analysis, especially in recent years, several interfaces of CE-ICP MS have been developed, thus the approach has been proven to be potential for protein quantification.[22]

Equipped with laser ablation (LA), ICP-MS can be directly accessed to analytes in solids matrices. Protein identification and semi-quantification directly from a one-dimensional (1D) or 2D gel combined with LA-ICP-MS is used to determine metal distributions in a proteome.[23] Moreover, with the measurement of LA-ICP-MS, an *in situ* element-related image can be obtained in a tissue section.[24] Figure 4.2 shows Cu distributions in the cross-section of a rat brain with a tumor region.[24]

Nowadays, a large number of proteins that contain heteroelements such as S, P, and Se and natively binding metals (e.g. Zn, Fe, Mn, Cu) have been detected and quantified with ICP-MS coupled techniques,[12] with the emphasis on its potential in life sciences, including the emerging metallomics and metalloproteomics. The available coupled techniques for metallomics and metalloproteomics are shown in Figure 4.3.[7]

4.2 ESI-MS

4.2.1 Introduction

Over the past years, great advances of ICP-MS-based coupled techniques have been made for the purpose of trace and ultra-trace element speciation

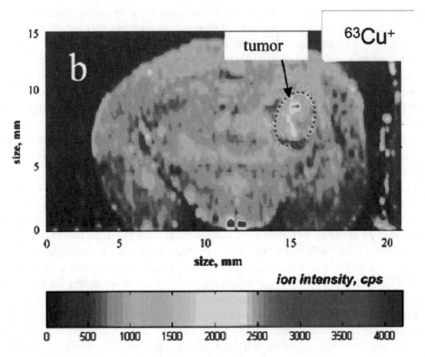

Figure 4.2 Cu distributions in the section of a rat brain with tumor region.[24] © 2007 The Royal Society of Chemistry.

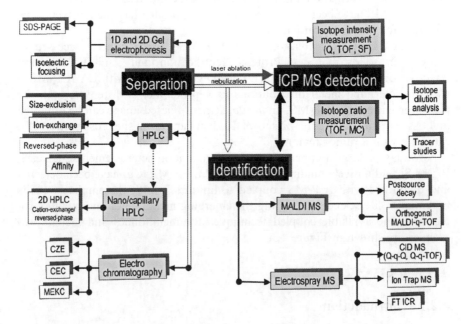

Figure 4.3 The hyphenated techniques for metallomics and metalloproteomics[7] © 2005 Wiley Periodicals, Inc.

analysis.[12] These techniques are mainly based on high-resolution separation and on-line element-specific detection. In order to further elucidate the biological functions of elements in organisms, the studies of metallomics and metalloproteomics have been proposed recently.[25] More advanced analytical techniques are required to explore the new research fields. Because the structural information is lost during the ionization process in ICP-MS, the analytes can be identified only when they have been fully isolated and their chromatographic retention times are matched with known standards. Unfortunately, only a limited number of standards are commercially available. Therefore, the methods that can provide both direct structural information and element specific information are highly desirable. These capabilities are offered by a synergistic method of ICP-MS and soft ionization mass spectrometry, such as electrospray ionization–mass spectrometry (ESI-MS) and matrix-assisted laser desorption ionization–time of flight (MALDI-TOF), which have become the most powerful analytical methods for accurate detection and structural characterization of various types of biomolecules. In fact, it is now considered that ICP-MS, electrospray and a MALDI ionization source followed by mass spectrometry are truly complementary techniques for the acquisition of information on metallomics and metalloproteomics.[7] Multi-element specificity and quantification are the key features of ICP-MS, whereas molecular mass determination and structural information can be supplied by ESI- or MALDI-MS. This section will briefly describe the theory and instrumentation of ESI-MS.

4.2.2 Electrospray and Related Ionization Techniques

A mass spectrometer usually consists of two major parts: the ionization source and mass analyzer. In the ionization source, the samples are ionized and then come into the gas phase. Gas-phase ions are chosen by the mass analyzer and introduced to the detector. Traditionally, small organic molecules are ionized into gas phase by thermal vaporization and electron ionization (EI) or chemical ionization (CI).[26] These methods have difficult problems for generating ions from large, non-volatile analytes such as proteins and peptides without significant fragmentation. During the 1990s, two technical breakthroughs for ionization methods, MALDI and ESI, changed protein analysis.

Fenn and co-workers described electrospray ionization for mass spectrometry analyzing large biomolecules.[27] ESI greatly enhanced MS ability for analysis of proteins or peptides. The mechanism of the ESI source is relatively simple. In an electrospray ionization source, the solution of analytes is nebulized into fine droplets via a capillary tube under a high electric field. The positive charges are accumulated on the surface of the droplets in this field. Later, because of evaporation of droplets, the surface coulombic forces exceed the surface tension and the droplets are dissociated into smaller droplets. This process continues until nanometer-sized droplets are formed. In this way, the ions pass from the source into the mass analyzer, whereas the bulk solvent is pumped away by a vacuum system.[28] The stability limit (Rayleigh limit) of droplets is determined by the Coulomb forces of the accumulated positive

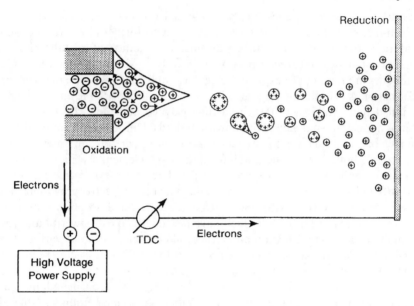

Figure 4.4 Sketch of the electrospray ionization process.[29] © 2000 John Wiley & Sons, Ltd.

charges and the surface tension of the solvent. The liquid cone formed at the capillary tip is called the "Taylor cone". The whole process is illustrated in Figure 4.4.[29]

In comparison with other ionization sources, ESI represents an even softer ionization technique and causes no fragmentation of analyte ions. A major benefit of the generation of multiply charged ions from polypeptides is that they can be readily analyzed with a less sophisticated instrument with limited mass range, such as a quadrupole. A serious problem is its poor tolerance to matrix, such as mobile-phase buffers, when ESI-MS is used for metallomics and metalloproteomics studies. Since the MS response significantly depends on solvent and sample composition, ion signal intensities of a given analyte do not necessarily correlate with its concentration in samples. Therefore, the internal standards are essential for the quantitative analysis. Another solution is the use of ICP-MS as complement.[30] The detailed methods will be described later.

ESI is commonly coupled to a quadrupole or ion-trap mass analyzer, where the molecular mass of samples can be determined. The basic principle is based on measuring the mass-to-charge ratios of analytes. To obtain information on the primary structure of polypeptides, however, more sophisticated devices are used in ESI tandem MS, e.g. triple quadrupole, the ion trap, and quadrupole-time of flight.[31] These different tandem mass analyzers with ESI sources offer excellent flexibility. From a mixture of peptide ions generated by the ESI source, the tandem MS analyzers select a single m/z species. This ion is then subjected to collision-induced dissociation (CID), which induces fragmentation of the peptide into fragment ions and neutral fragments. The fragment ions are then analyzed on the basis of their m/z to form a product ion spectrum. The information contained in this tandem or MS–MS spectrum permits the sequence of the peptide to be deduced. Moreover, the nature and sequence location of peptide modification can also be established from an MS–MS spectrum.

Matrix-assisted laser desorption ionization (MALDI) is another soft ionization source that generates ions by irradiating a solid mixture with a pulsed laser beam.[32] The solid mixture is comprised of the samples dissolved in an organic matrix compound. The positive ions are formed by accepting a proton as they are ejected from the matrix. Then the samples are indirectly ionized and desorbed from the mixture by the laser pulse. One of the advantages is that MALDI produces mostly single charge ions with little fragmentation and thus mass spectra are relatively easy to interpret.[31] Each peptide molecule tends to pick up a single proton. MALDI is less vulnerable to matrix effects than ESI. However, the information on the metal–protein interactions is often lost during the sample ionization from the solid mixture into the gas phase because preservation of metal complexes in MALDI is difficult. Therefore, molecular mass spectrometry alone seems to be unable to solve many questions that are relevant to metallomics and metalloproteomics.[7] An increasing number of reports have emphasized the synergy between ICP-MS, ESI-MS, and MALDI-MS.[30,33,34] The comparison of ICP-MS and ESI-MS is listed in Table 4.2.

4.3 Isotopic Tracer Techniques Combined with ICP-MS in the Study of Metallomics

4.3.1 Introduction

The use of isotopic tracer techniques cannot be regarded as very recent, as it dates from at least the 1930s when Hevesy and Paneth used radio-lead as a tracer to facilitate the determination of lead in rocks.[35] Up to now, isotopic tracer techniques have been applied for *in vivo* or *in vitro* studies in nearly every field of life sciences, such as medicine, biology, physiology, nutrition, toxicology, and biotechnology; in fact, it is difficult to find anywhere it does not reach.[36–38] In comparison with other techniques, the isotopic tracer method has many unique advantages, such as high sensitivity, good accuracy and precision, and time saving. Moreover, the use of isotopic tracer techniques can easily

Table 4.2 Comparison of ICP-MS and ESI-MS.

	ICP-MS	ESI-MS
Application	Atomic information	Molecular information
Structure determination	Possible with coupling to separation and standards	Bimolecular fragmentation
Quantitative capability	Direct quantification via element signal	Indirect quantification via isotope labeling
Matrix effect	Weak	Strong

distinguish endogenous and exogenous sources of elements to be analyzed in samples, even at an extremely low concentration.[38] It can also provide dynamic information of the elements or compounds of interest in organisms.

Usually, the detectable isotopes, including radioactive and enriched stable isotopes, are added to a system then a pathway or a mode of distribution in biological systems can be determined. The systems cover a wide range, such as the blood flow in an animal or a manufacturing process. All these methods depend on isotopic properties, *i.e.* the chemical identity of all isotopes of any given element. At the same time, radioisotopes or enriched stable isotopes can readily be determined by virtue of their emitted radiations or different mass-to-charge ratios, respectively.

The emission of alpha, beta and gamma beams can be measured in experiments of radioactive tracers. The majority of radioisotopes used in the biomedical analysis involve beta-ray emitters, such as ^{3}H, ^{14}C, ^{35}S, ^{33}P, and ^{32}P. In addition, a gamma-ray emitter, such as ^{131}I, is also used but there is a difficulty in shielding.[39] The activity of an isotope emitting beta-rays can be analyzed with a liquid-scintillation counter, whereas a crystal-scintillation counter can be used to measure a gamma-ray emitter. However, the radioactive experiments require specialized safe handling and laboratory equipment for radiation protection. Before performing a possible application of radioisotopes, it is important to assess the justification. In addition, the isotopic effect and radiation effect must be considered. This means that a radioisotope may not behave in the same manner analogous to its stable counterpart, e.g. the different behaviors of tritium and hydrogen.

Owing to the great development of modern analytical techniques, especially mass spectrometry, the use of enriched stable tracers has rapidly increased recently. The use of enriched stable-isotope tracers is very similar to radioisotope ones and more general acceptance by the scientists who are not familiar with nuclear techniques. Compared with the common radioactive isotope tracers, it possesses many merits and also drawbacks. Enriched stable isotopes are non-radioactive and thus can be used in experiments involving living organisms and even humans without any concerns of radiation hazard. Thus, it is especially suitable for the radiation-sensitive population, e.g. children, pregnant women, the elderly, and immuno-deficient subgroups. The main drawback of stable tracers is that they are often not true tracers like radioisotopes, because the enriched stable isotopes usually exist as part of the natural abundance of an element constituting the biological systems. Therefore, in comparison with radioactive tracers, a greater amount of the stable tracers with a highly enriched abundance should be used in the experiments, which is expensive in many cases.

Generally, determination of stable isotopes is more complicated than radioactivity counting. Activation analysis is useful for isotopic tracer technique. In this method, radioactive isotope is not added to the sample but produced in it by irradiation. The common used nuclear reaction is the (n,γ) reaction that is brought about by thermal neutrons. After the radioactive isotopes are produced, they can be determined simply by virtue of their irradiations. In addition, traditional methods for analysis of isotope ratios include thermal ionization mass spectrometry and fast atom bombardment mass spectrometry. Since ICP-MS instruments were commercialized in the 1980s, more applications for the analysis of isotope ratios are carried out by ICP-MS, which has become one of the most widely used methods for isotopic tracer experiments. The reliable results in stable tracer experiments depend on the accurate and precise determination of isotope ratios by using ICP-MS. The unique characterization of ICP-MS opens a new way to investigate different aspects of metallomics and metalloproteomics, such as the identity, metabolism, chemical species, and functions of metal-binding proteins in biological systems. In the following section, the applications of radioisotopes or enriched stable isotopes to metallomics and metalloproteomics study are selectively reviewed.

4.3.2 Examples of Applications

There are many successful applications in literature for the use of stable isotopes as tracers combined with ICP-MS determination for the study of metallomics. In the most common applications, a highly isotopically enriched element used as a tracer is administered to a living organism. Then metallomics studies can be focused on this element, which is detected by ICP-MS. Rodriguez-Cea *et al.* used the stable isotope ^{111}Cd as a tracer for studying the *de novo* incorporation of cadmium in fish (European eel) liver and kidney.[40] The exposure to ^{111}Cd (100 ng L^{-1}) gave rise to the *in vivo* dilution of the natural existing Cd in the selected tissues and therefore a change of Cd isotope ratios can be measured by ICP-MS. By use of mathematical calculations based on the isotope dilution methodology, the previously accumulated natural Cd and the isotopically enriched Cd (*de novo* incorporated) can be determined. This method allows the quantitative discrimination of the Cd binding metallothionein isoforms induced by *de novo* incorporation of Cd in fish liver and kidney.[40]

It is well-known that mercury is a global pollutant in the environment. One of main mechanisms for mercury toxicity derives from the formation of highly stable complexes with sulfhydryl groups of proteins. In order to investigate trace mercury-containing proteins in maternal rats and their offspring, a method of enriched stable isotopic tracer (^{196}Hg and ^{198}Hg) combined with size-exclusion chromatography (SEC) coupled to inductively coupled plasma–isotope dilution mass spectrometry (ICP-IDMS) was developed.[41] The qualitative and quantitative information on mercury-containing proteins in the organisms indicate that the detection sensitivity could be increased by the tracer method. Additionally, the tracer technique could be used to identify the "artifact" species via isotopic ratio measurement. Figure 4.5 shows two results of the tracing-rat tissue samples after using HPLC/ICP-MS. In the figure, for the measurement of ^{198}Hg, five significant mercury fractions were observed at retention times of 12.5, 14.2, 16.9, 19.5 and 27.6 min, while ^{202}Hg showed one low broad peak from 14 to 22 min and another obvious peak at 27.6 min, and ^{196}Hg showed three significant peaks at 12.5, 16.9 and 19.5 min. Since the ratios

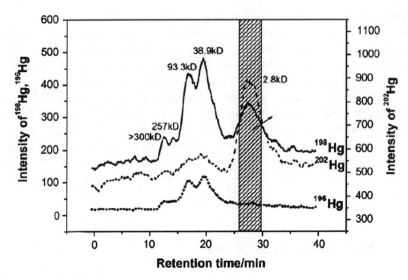

Figure 4.5 Mercury-containing protein separation and ^{196}Hg, ^{198}Hg, ^{202}Hg measurement in a rat samples. The shadow peak is identified as artifact mercury species.[41] © 2006 Elsevier B. V.

of ^{196}Hg to ^{202}Hg and ^{198}Hg/^{202}Hg in the tracer are already known, thus by measurement of the isotopic ratios of each fraction, it can be identified that the last peak in the figure is the artifact or contamination fraction. The limitations of the traditional techniques are mainly related to the artifact elimination, contamination control and peak identification, especially in mercury species with strong "memory" effect in the instrumental system. The isotopic tracer technique seems to be a nice solution for eliminating the above limitations.[41]

Selenium is an essential trace element and plays important roles in mammalian organisms. For example, many publications report that selenium is involved in the prevention of diseases, such as cancer, Alzheimer's disease, and cardiovascular disease.[42,43] The distribution and speciation of selenium in organisms are complicated, therefore isotopic tracer techniques are often imperative in this aspect. Both enriched stable isotopes (e.g. ^{82}Se) and radioactive isotopes (^{75}Se) were used to study the metabolic pathway and the selenium speciation in organisms. In the early 1970s, the radioactive isotope ^{75}Se was used as a tracer in rat liver and plasma and an important selenium-containing protein, selenoprotein P, was found.[44] Burk et al., using ^{75}Se as a tracer,[45] reported that selenium administered to selenium-deficient rats was rapidly incorporated into selenoprotein P in preference to glutathione peroxidase. More than 20 selenoproteins or subunits in rats after replenishment with ^{75}Se-labeled selenite were reported later.[46] Suzuki et al. examined the metabolism of different selenium species in rats by using enriched stable-isotope labeling. The rats received the ^{82}Se label via the drinking water. Then the rat samples were analyzed with HPLC coupled to ICP-MS. Information on the metabolism of selenium compounds can be obtained. The advantage of HPLC coupled to ICP-MS in combination with enriched stable isotopes is the simultaneous detection of exogenous and endogenous selenium. In addition, because ^{82}Se is in low abundance and little background exists in rats, the stable tracer can be viewed as a true radioactive tracer.[47–49]

Thus, the isotopic tracer technique has been proven as an effective and powerful method for analysis of element speciation in various kinds of samples. This technique will continually play an important role in the emerging field of metallomics or metalloproteomics. In future, the application of isotopic tracers will probably increase and contribute significantly to the quest for new knowledge.

4.4 Isotope Dilution Analysis in the Quantitative Study of Proteins

4.4.1 Introduction

In recent years, a remarkable trend can be observed; *i.e.* the utilization of inductively coupled plasma-mass spectrometry (ICP-MS) as an attractive complement for protein quantification. If the heteroatom-containing proteins are known and the standards are available, the absolute quantification of proteins can be easily obtained via element analysis by ICP-MS. However, when ICP-MS is interfaced to a separation system, such as HPLC, the organic solution and inorganic salts introduced into ICP-MS usually decrease the instrumental stability and detection limits. To overcome the problems, isotope dilution analysis is introduced to an ICP-MS-based linked system.[50]

Unlike other calibration methods, isotope dilution analysis (IDA) is an analytical technique based on the measurement of the change of isotope ratios in samples after the mixture of a spike, which is the same element with different isotopic compositions from the sample that is introduced in a controlled manner. For an IDA, equilibration of the isotopically labeled spike with the sample is a prerequisite. Generally, the element to be analyzed must have at least two isotopes without spectral interferences. In isotope dilution analysis, there is the only parameter, *i.e.* isotope ratio, to be determined. Therefore, isotope dilution has many unique advantages.[51] First, the final results obtained by isotope dilution analysis are independent of the shift of instrumental signals and the matrix effects of samples. Second, the results of isotope dilution analysis can be traceable and achieve very high accuracy and precision using a mass spectrometer. Last, but not least, once complete equilibration is achieved between the sample and the spike, possible loss of isotope-diluted analyte has no influence on the analytical results, because any fraction of isotope-diluted sample contains the same isotope ratio. Therefore, isotope dilution analysis is internationally regarded as a highly qualified primary method. It is obvious that when isotope ratios are determined with a mass spectrometer, corrections for isotopic bias are essential. The accuracy of the results is critically dependent on instrumental calibration. A large number of publications, including many excellent reviews about applications of ICP-MS in IDA can be found.[51–54]

The equation for isotope dilution can be expressed as:

$$N_s = N_{sp} \frac{A_{sp}^b}{A_s^b} \frac{R_{sp} - R_m}{R_m - R_s} \qquad (4.1)$$

where:

- N_s and N_{sp} are the number of atoms in the sample and spike, respectively
- A_s^b and A_{sp}^b are the abundance of isotope b in the sample and spike, respectively
- R_{sp}, R_s and R_m are the isotope ratio of isotope a to isotope b in the spike, sample and mixture, respectively

Isotope dilution analysis for protein quantification with ICP-MS is usually performed by linked techniques based on the coupling to an effective separation method, such as HPLC or CE. Generally, proteins to be analyzed should contain ICP-MS-detectable elements, including both naturally occurring and artificially labeled. After introduction of the spike that has different isotopic composition from the analyte, the elemental isotope ratio can be measured in ICP-MS and thus the detectable elements can be absolutely quantified by use of the isotope dilution equation. If proteins have been identified with biological mass spectrometry, or the elemental stoichiometric ratios of proteins have been known, the amount of proteins can be deduced from the elemental concentration of the proteins.

The application of isotope dilution analysis to ICP-MS-based hyphenated techniques, for example HPLC-ICP-MS, can be divided into two different modes: the species-unspecific mode and species-specific mode.[52] The two modes are illustrated in Figure 4.6.[51] Because IDA-ICP-MS relies on the measurement of isotope ratios by mass spectrometry, the traditional interference to the accuracy and precision of IDA should all be considered and some biases need further correction in this case, such as spectral interferences, dead time of detector, mass discrimination, and statistic and stability of ion counting.

4.4.2 Species-specific Method

In the species-specific method, the spike is usually an isotopically labeled analyte and supposed to simultaneously elute with the analyte in the entire separation procedure. The method is only possible when the composition and structure of the analyte are known in order to obtain the spike that is labeled with an enriched isotope. The premise of the method is that the labeled spikes have enough thermodynamic and kinetic stability and no isotopic exchange occurs between the spike and analyte. In this mode, the loss of the analyte after isotope dilution step has no influence on the analytical results. In fact, this

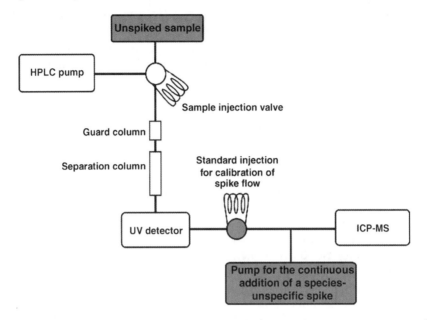

Figure 4.6 Schematic diagram for isotope dilution analysis by HPLC-ICP-MS.[51] © 2005 Elsevier B. V.

mode is very similar to the isotope dilution analysis in the quantitative proteomics, such as PROTEIN-AQUA that applies synthetic, stable-isotope-labeled peptide analogues as internal standards.[55] However, because the detector is ICP-MS, the labels should be ICP-MS detectable nuclides, rather than ^{2}H, ^{13}C, for example. On the other hand, the ratios of the isotopes are measured by ICP-MS, instead of the ratios of molecules or fragments in biological mass spectrometry.

The main challenge when applying this method for protein analysis is the lack of commercial isotopically labeled proteins. Thus, most applications are focused on small molecules, such as organic mercury,[56,57] organic tin[58–60] and so on. However, there is increasing interest in the use of the ICP-MS linked system and species-specific isotope dilution for quantification of peptides or proteins due to the outstanding performance of ICP-MS.

Encinar *et al.* developed a method for the accurate determination of selenoamino acids in human serum by species-specific isotope dilution analysis. A human serum was enzymatically digested, and then the selenoamino acid and carboxymethylated selenocysteine were separated and quantified by HPLC-ICP-MS. Quantification of selenomethionine was carried out by isotope dilution using a synthetic ^{77}Se-labeled counterpart. The selenomethionine in samples was also measured by using selenomethionine as an internal standard. The instrumental detection limit was down to 75 fg Se and the precision was better than 5% RSD.[61]

Harrington *et al.* labeled a copper-containing protein rusticyanin (Rc) with ^{65}Cu for use as a spike material in species-specific isotope dilution analysis, and indicated the potential of the development of a new method for the absolute quantification of metalloproteins by HPLC-ICP-MS.[62] They found that a pH 7.0 buffer could afford the most appropriate conditions. Moreover, the stability of the copper center could be verified by analysis of mixtures of different isotopic solutions.[62]

Later, Busto *et al.* used synthesized ^{57}Fe-labeled transferrin to determine individual transferrin isoforms in human serum samples for the species-specific mode. The stability of the prepared proteins was tested for 1 week and no iron exchange had occurred.[63] They concluded that the results are in good agreement with other calibration methods, e.g. the species-unspecific method; however, the species-specific mode can offer better precision.[63] Hoppler *et al.* also synthesized and characterized ^{57}Fe-labeled *Phaseolus vulgaris* ferritin for isotope dilution analysis.[64]

More recently, Deitrich *et al.* described the chemical preparation and characterization of an isotopically enriched metalloenzymes (superoxide dismutase, SOD) containing two different metal isotopes. The metal co-factors in enzyme were firstly removed and then replaced with isotopically enriched ^{65}Cu and ^{68}Zn. This application shows the potential of using isotopically enriched SOD for the accurate quantification of that enzyme in real samples.[65]

The use of species-specific spiking greatly enhances the traceability of the analytical procedure and improves the reliability of the results. On the other hand, the use of isotopically enriched metalloproteins provides a means to evaluate the analytical conditions, under which the metal center is most stable. Despite all the advantages of species-specific spiking for ICP-MS, it has to be mentioned that huge effort is necessary to produce and characterize an isotopically enriched protein spike. Therefore, this approach might be sensible in practice for the assay of only a few proteins or biomarkers.[65]

4.4.3 Species-unspecific Method

In the species-unspecific method, the spike, the chemical species of which is different from the analyte, is added to the analyte after the complete separation by chromatography or electrophoresis. This mode is especially useful when the structure and composition of the analytes is not exactly known or isotopically labeled spike is not available. In contrast with the species-specific mode, this spiking mode cannot correct for any loss in the separation procedure until the complete mixing between the spike and analyte. However, the errors derived from instrumental instabilities and matrix effects can be corrected because of its determination of the relative isotope ratio rather than the absolute signal. For example, Rottmann and Heumann compared the matrix effect of humic substances on the accuracy of molybdenum determination by isotope dilution and by external calibration (see Figure 4.7).[52] They concluded that external calibration strategies are not suitable for a coupling system (such as HPLC/ICP-MS) and accurate results can be obtained by isotope dilution analysis.[52] The unique feature is helpful for the samples with complicated matrices or for the gradient separation by use of HPLC. One prerequisite for successful

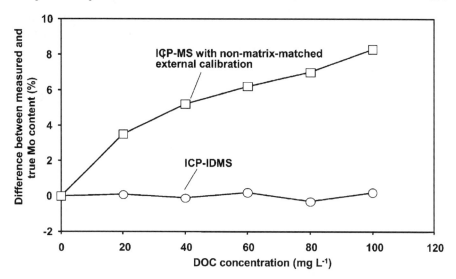

Figure 4.7 Matrix effect of humic substances the accuracy of molybdenum determination by IDMS and by external calibration with a standard solution.[52]
© 1994 Springer-Verlag.

application of the method is that mass balance should be accurately calculated during the whole procedure.

The method of the species-unspecific modes is well described in literature.[51–54] In brief, isotope ratios are monitored during the whole chromatographic procedure and corrected by the isobaric interference, dead time, and mass bias. Then the chromatogram of isotope ratios is transformed into the chromatogram of mass flow (mass versus time) using the following isotope dilution equation:

$$\mathrm{MF_s} = c_{sp} d_{sp} f_{sp} \frac{M_s}{M_{sp}} \frac{A^b_{sp}}{A^b_s} \frac{R_m - R_{sp}}{R_s - R_m} \qquad (4.2)$$

where:

- $\mathrm{MF_s}$ is the mass flow of the sample eluting from the column
- c_{sp} is the concentration of the element in the spike (ng g^{-1})
- d_{sp} is the density of the spike (g mL^{-1})
- f_{sp} is the flow rate of the spike (mL min^{-1})
- M_s and M_{sp} are the atomic weights of the element in the sample and spike, respectively
- A^b_{sp} and A^b_s are the abundance of isotope b in the spike and sample, respectively
- R_m, R_s, and R_{sp} are the isotope ratio (isotope a/isotope b) in the mixture, sample and spike, respectively

The absolute amount of the element in different species can be obtained by the integration of the corresponding peaks in the chromatogram of mass flow. From recent literature, a trend can be found that the species-unspecific mode has been widely applied in measurement of different metalloproteins, such as metallothioneins (MTs).[40,66,67] It should be mentioned that molecular mass spectrometry is usually required to identify the analyte with the complicated matrices.

Prange et al. analyzed the isoforms of metallothioneins in human brain cytosol with capillary electrophoresis (CE) coupled to ICP-SFMS (inductively coupled plasma-sector field mass spectrometry). In the study, three metallothionein isoforms were separated by CE, and the enriched isotopes of ^{65}Cu, ^{68}Zn, ^{116}Cd, and ^{34}S were mixed as a specific–unspecific spiking solution; finally, the Cu, Zn, Cd, and S in metallothioneins were quantified with IDA. They found that the levels of MT-3 and MT-1 were lower in the brain samples from the patients with Alzheimer's disease in comparison with the controls (see Figure 4.8). The results indicated that quantification of metalloproteins gave additional information on its relationship with Alzheimer's disease.[68]

Muniz et al. quantitatively analyzed the metal containing proteins in human serum by anion-exchange chromatography coupled to post-column isotope dilution analysis with double focusing ICP-MS. Parallel analysis of human serum from healthy volunteers and patients on hemodialysis was also undertaken. They found different amounts of iron-containing proteins between healthy and renal disease individuals and indicated that Fe bound to albumin decreased in patients with renal disease.[69]

Sulfur exists in cysteine and methionine and is one of the most abundant elements in natural proteins. The cumulative abundance of sulfur in proteins is about 5%.[70] The relative amount of proteins can be obtained by the isotope-labeling method with the sulfhydryl group and molecular mass spectrometry measurement, such as the isotope-coded affinity tags (ICAT) method.[71] On the other hand, when the amino acid sequence of a protein has been identified, the absolute amount of a protein can be obtained via the determination of sulfur with ICP-MS. In comparison with other naturally existing elements in proteins, sulfur is preferable as an internal standard for protein quantification due to its high abundance.[70,72]

However, sulfur measurement by ICP-MS is hampered by the low ionization efficiency. Because sulfur has a relatively high first ionization potential, the ionization of sulfur in the argon plasma is only 10%.[10] In addition, serious polyatomic ions, such as $^{16}O^{16}O^+$, $^{14}N^{18}O^+$ and $^{16}O^{18}O^+$, $^{16}O^{17}OH^+$, interfere with the main isotopes, ^{32}S and ^{34}S, respectively. These spectral interferences at m/z 32 and 34 can be overcome by a sector field ICP-MS (ICP-SF MS) with a mass resolution of about $m/\Delta m = 4000$.[73]

Sulfur has four stable isotopes: ^{32}S, ^{33}S, ^{34}S, and ^{36}S, of which ^{32}S and ^{34}S have relative high natural abundance and are suitable for ICP-MS determination. Therefore, IDA can be introduced for the accurate and precise determination of sulfur in proteins. This strategy was first implemented with ICP-SFMS. With the method of on-line IDA in combination with CE-ICP-SFMS,

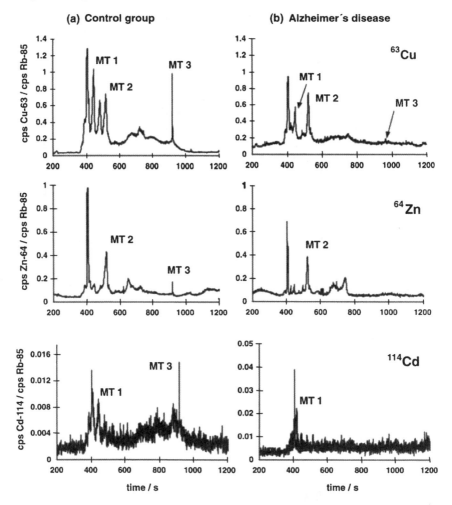

Figure 4.8 Comparison of electropherograms from a brain affected by Alzheimer's disease (b) and those from a control brain (a) detected on ^{63}Cu, ^{64}Zn and ^{114}Cd (temporal region).[68] © 2001 Springer-Verlag.

Schaumlöfel et al. have characterized and quantified metallothionein (MT) isoforms.[74]

Although the ICP-SF MS has been proved to be suitable for sulfur determination, the expensive and complicated method of ICP-SF MS limits the application in a wide range. For a quadrupole ICP-MS, a simple method for sulfur analysis is to adjust the ICP-MS at a condition of high yield of oxides and thus the sulfur content can be measured by SO$^+$ signals.[75,76] Comparatively, with the development of the dynamic reaction cell (DRC) or collision cell (CC) the quadrupole ICP-CC-MS or ICP-DRC-MS can perform more efficient ion-molecule reaction, e.g. the oxidation of S$^+$ to SO$^+$, and thus can achieve a lower detection limit.[14] Wang and colleagues developed a method for absolute quantification of proteins via sulfur measurement with SEC coupled to ICP-CC-MS and post-column isotope dilution analysis.[77] In the study, the mixture of three standard proteins, bovine serum albumin (BSA), superoxide dismutase (SOD), and metallothionein-II (MT-II), were separated and quantified. The isotope-enriched ^{34}S, ^{65}Cu, and ^{67}Zn were used as a spiking solution. The oxygen was added as a reactive gas into the collision cell, where sulfur reacts with oxygen to form a sulfur-oxygen ion; thus, the ratio of $^{32}S^{16}O^+/^{34}S^{16}O^+$ that represented $^{32}S^+/^{34}S^+$ was measured with ICP-MS. The detection limits for BSA, SOD, and MT-II were 8, 31, and 15 pmol, respectively. The RSDs for the proteins were less than 3%.[77] In order to obtain efficient separation, multidimensional HPLC have to be used when analyzing complex protein samples.

Recently, pre-column sulfur isotope dilution analysis in nano HPLC-ICP-MS was introduced where the sulfur spike (^{34}S) was added directly into the chromatographic eluents for measurement of sulfur-containing peptides.[78] In contrast to the post-column isotope dilution in which the spike is added after the chromatographic column by a three-way connection, pre-column isotope dilution is propitious for reducing broadening of the chromatographic peaks and for maintaining stability of the spike flow. In the study, xenon was utilized as the collision gas to eliminate the interference of ^{32}S and ^{34}S. Figure 4.9 shows 24 peaks in the mass flow chromatography of a human serum albumin (HSA) digest. By integration of these peak areas, the amount of sulfur in each peak could be obtained. After identification by ESI-MS (see Table 4.3), the peptide can be quantified based on the amount of sulfur. The combination of nano

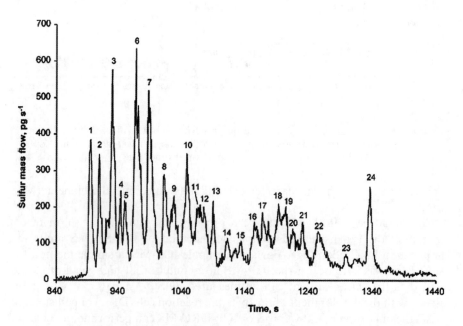

Figure 4.9 Sulfur mass flow chromatogram obtained from HSA tryptic digest by the pre-column isotope dilution technique.[78] © 2007 American Chemical Society.

Table 4.3 Assignment of peaks in Figure 4.9[a,78] © 2007 American Chemical Society.

Peak no.	Sulfur mass, pg	Peptide amino acid sequence[b]	Position in the HAS amino acid sequence[c]	Peptide mass, ng
1	2427.2	TCVADESAENCDK	52–64	52.4
2	1868.6	ETCFAEEGKK	565–574 mc	66.5
3	4382	CCAAADPHECYAK	360–372	62.9
4	1035.2	CASLQKFGER	200–209 mc	36.7
5	1397.1	ADDKETCFAEEGK	561–573 mc	62.8
6	5615.1	AAFTECCQAADK	163–174	
		YICENQDSISSK	263–274	
7	4929.2	CCTESLVNR	476–484	
		ETYGEMADCCAK	82–93	
8	2259.8	QEPERNECFLQHK	94–106 mc	116.8
9	2571.8	ECCEKPLLEK	277–286	47.7
10	3435.1	NECFLQHK	99–106	
		LKECCEKPLLEK	275–286 mc	
11	2150.3	VHTECCHGDLLECADDR	241–257	42.8
12	1254.7	NECFLQHKDDNPNLPR	99–114 mc	75.9
13	1305.1	AACLLPK	175–181	29.1
14	999.6	LCTVATLR	74–81	27.3
15	772	VHTECCHGDLLECADDRADLAK	241–262 mc	19.4
16	1770.7	RPCFSALEVDETYVPK	485–500	102.3
17	1126	QNCELFEQLGEYK	390–402	56.2
18	1765.5	LVRPEVDVMCTAFHDNEETFLKK	115–137 mc	74.9
19	1737.6	SLHTLFGDKLCTVATLR	65–81 mc	101.6
20	1093.4	LVRPEVDVMCTAFHDNEETFLK	115–136	44.2
21	1436.1	EFNAETFTFHADICTLSEK	501–519	98.6
22	2364.1	AVMDDFAAFVEK	546–557	98.9
23	665.8	SHCIAEVENDEMPADLPSLA ADFVESK	287–313	30.3
24	2134.8	TYETTLEKCCAAADPHECYAK	352–372 mc	
		MPCAEDYLSVVLNQLCVLHEK	446–466	
		ALVLIAFAQYLQQCPFEDHV K	21–41	

[a] Absolute quantification of sulfur-containing peptides in 1 mL of HSA tryptic digest via sulfur quantification.
[b] Sulfur-containing amino acids: C cysteine, M methionine.
[c] mc: miscleavage.

HPLC-ICP-MS with nano HPLC-ESI MS/MS allowed the precise quantification and identification of sulfur-containing peptides in tryptic digests of human serum albumin and salt-induced yeast protein (SIP18) at the picomole level.[78]

4.5 Isotope Tagging and Labeling Techniques for Protein Quantification

4.5.1 Introduction

Molecular mass spectrometers, such as ESI-MS and MALDI-MS, can be utilized to identify proteins because only one unique peptide from a protein would be sufficient to unambiguously identify the parent protein.[79] However, the proteome is extremely dynamic and constantly changing and thus an understanding of biological functions will not only rely on protein identification but also on protein quantification. Quantification of proteins is expected to provide new insights into biological processes, disease diagnoses, and the discovery of therapeutic targets. Unfortunately, molecular mass spectrometry is not a directly quantitative method, because the relationship between the quantity of proteins or peptides and the signal intensity obtained in ESI-MS and MALDI-MS is complex and ambiguous due to the diverse ionization efficiency of different proteins or peptides.[31] Although the label-free method has emerged as a promising option in quantitative proteomics analysis,[80] nowadays most of quantitative proteomics methods are based on the molecular mass spectrometry in combination with stable isotope labeling, as well reviewed in recent literatures.[81,82] This kind of method relies on the fact that a protein or peptide is chemically identical to its counterpart labeled with enriched stable isotopes and these analytes can be considered to behave identically in a chromatography column and produce the same response signals in a mass spectrometer. In the method, one sample is labeled with a heavy isotope and the other is labeled with a light isotope. The two samples are then mixed and determined with mass spectrometry. The ratios between the peptides or proteins derived from different biological conditions can be accurately determined by ESI-MS.

The stable-isotope tags can be introduced into proteins or peptides by using different methods. These methods can be divided into the following categories: (1) enzymatic labeling, such as labeling with ^{18}O from heavy-oxygen water; (2) chemical labeling with enriched isotopes; and (3) metabolic labeling through cell growth and protein turnover.[83] Figure 4.10 shows schematic representations of methods for stable-isotope protein labeling for quantitative proteomics.[31] In this section, we will selectively discuss the widely used and promising methods based on isotope labeling technology in quantitative proteomics.

4.5.2 Chemical Labeling Methods for Protein Quantification

Isotopic labeling can be introduced by chemical reactions with a wide variety of isotopically labeled reagents, which are targeted toward different amino groups in proteins or peptides. Several strategies have been developed using the derivate of the peptides containing unique (N- or C-terminal) or rare amino acids (e.g. Cys, Met, His).[83] Each amino acid chain of a protein or peptide contains a number of reactive sites which can covalently bind to the isotopically labeled reagents. The peptides labeled with heavy or light isotopes are chemically identical and their ionization efficiencies are the same as in a mass spectrometer. Thus the relative abundance ratios in the mass spectrometer are directly proportional to relative ratios of proteins or peptide. The representative chemical labeling strategies for quantitative proteomics are summarized in Table 4.4.

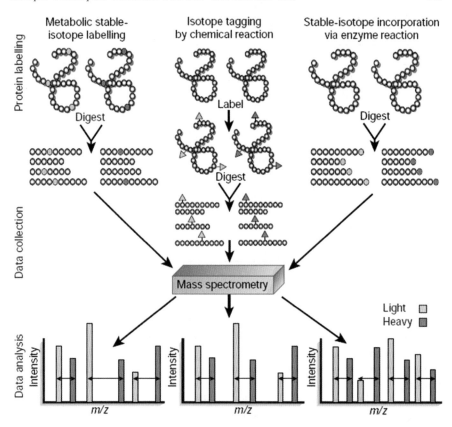

Figure 4.10 Schematic representations of methods for stable-isotope protein labeling for quantitative proteomics.[31] © 2003 Nature Publishing Group.

Gygi *et al.* first introduced a method termed isotope-coded affinity tagging (ICAT) to determine the relative expression levels of proteins from two different samples.[71] The original ICAT reagent consists of a thiol-specific reactive group, a linker incorporating either hydrogen (light) or deuterium (heavy) at eight positions, and a biotin group for selective capture and enrichment of the labeled peptides, shown in Figure 4.11a. The proteins of yeast grown in two different conditions are separately alkylated by the light or heavy reagents. Then the isotopically labeled analytes are mixed, digested, and then separated by an avidin monomer column. Peptides recovered from the column are further separated by reversed-phase chromatography and the relative amount of the two samples can be simultaneously measured by ESI-MS. Nowadays, the method has been widely applied in quantitative proteomics, such as applications in low abundance proteins[84] and membrane proteins.[85] An advantage of this method is that the affinity tag (biotin) is utilized to isolate cysteine-containing peptides, and thus the peptide mixture was less complex and more advantageous for quantification. However, the proteins that do not contain

Table 4.4 Some applications of chemical labeling strategies for quantitative proteomics.

Targeting residue	Name of labeling method	Isotopes	Reference
Sulfhydryl	Isotope-coded affinity tagging (ICAT)	D	71
	Cleavable ICAT	^{13}C	88, 102
	Catch and release (CAR)	D	103
	MeCAT	Metal	100
	Iodoaceanilide	D	104
Amines	Isotope-coded protein labeling (ICPL)	D	90
	Tandem mass tag (TMT)	D	105
	Isotope-coded n-terminal sulfonation	^{13}C	106
	Isobaric tag for relative and absolute quantification (iTRAQ)	^{13}C, ^{15}N and ^{18}O	107
Lysines	Guanidination	^{13}C and ^{15}N	108
Carboxyl	Methyl esterification	D	109
	C-terminal isotope-coded tagging using sulfanilic acid	^{13}C	110

cysteine cannot be analyzed with this method. In addition, the large ICAT tag severely influences the interpretation of mass spectra. Another problem is that deuterium labeling influences the retention time; deuterium-labeled and natural hydrogen peptides fail to co-elute during reversed-phase chromatography.[86] At last, because there are only two versions of the tag, only relative amount of two samples is restricted.

To overcome the problems of the original ICAT reagent, many relative methods have been developed. More isotopes, including ^{13}C, ^{15}N, and ^{18}O, are incorporated into the labeled reagents in order to minimize isotope effects.[87] A method called cleavable ICAT (cICAT) uses ^{13}C and ^{12}C as heavy or light tags, instead of ^{1}H or ^{2}H, and eliminates the problem of co-elution of peptide isoforms on reversed-phase column.[88] This reagent also contains an acid-cleavable linker that allows the removal of the affinity tag before mass spectrometric analysis. Another reagent named HysTag contains a six-histidine tag for enrichment, a 2-thiopyridyl disulfide group to react with thiols, a deuterium-labeled alanine and a tryptic cleavage site.[89] The HysTag offers advantages compared with the original ICAT reagent because the HysTag reagent is easy to synthesize and economical due to use of deuterium instead of a ^{13}C isotope label, and allows robust purification and flexibility through the affinity tag, which can be extended to different peptide functionalities.[89]

Apart from sulfhydryl group, the primary amines or the carboxylic groups have been successfully applied as the labeling sites. These amino and carboxyl group based methods can be suitable for quantification of every observed peptide in principle. For example, the reagents in isotope-coded protein labeling (ICPL) target all amino groups using nicotinoyloxy succinimide (Nic-NHS).[90] The ^{13}C labeling step is performed before protein digestion and results in a mass shift of about 6 Da. Because the labeling process is at protein level, information on post-translational modification can be obtained. Another popular method is termed

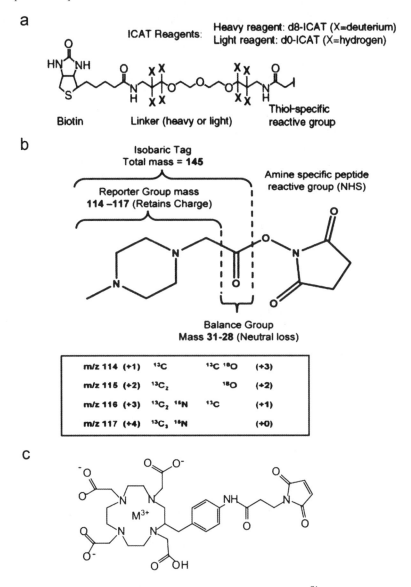

Figure 4.11 Schematic structures of the reagents of ICAT[71] (a) © 1999 Nature America Inc.; iTRAQ[91] (b) © 2004 The American Society for Biochemistry and Molecular Biology, Inc.; and MeCAT[100] (c) © 2007 The American Society for Biochemistry and Molecular Biology, Inc.

iTRAQ (isobaric tagging for relative and absolute quantitation). The iTRAQ reagent uses the same chemical reaction mentioned above and consists of three parts: an amine-targeting reactive group, a reporter group containing ^{13}C or ^{15}N atoms, and a mass balance group (see Figure 4.11b).[91,92] The iTRAQ is identical in the single MS mode, but the reporter groups can generate a specific reporter ion in MS/MS fragmentation spectra. This method allows analysis of four separately labeled samples simultaneously and thus increases analytical throughout. The multiplexing capability, the better protein sequence coverage, and more sensitive quantification make a success of iTRAQ in quantitative proteomics.

Isotopic labeling via chemical reactions allows great flexibility and selectivity for different reactive groups in proteins or peptides, including side chains or terminals of amino acids. In the labeling reactions, specific functional groups or affinity tags can be introduced into samples to facilitate the selective isolation and analysis of targeted proteins. Although a large number of chemical-labeling methods have been developed, relatively few of these methods have been applied in real experimental samples. This is mainly attributed to the non-specific or side reactions during labeling. Future chemical reaction for protein or peptide labeling should be extremely specific, complete and involve minimal sample handling.

4.5.3 Metabolic Labeling for Protein Quantification

Metabolic labeling is an *in situ* method which employs ^{15}N or isotope tagging (^{13}C or ^{2}H) amino acids.[93] In this strategy, proteins are labeled by use of cell culture in media that are isotopically enriched or isotopically depleted. In fact, this method has been used for decades for quantification and trace monitoring of metabolites with radioactive isotopes. Unlike the experiment with a small amount of a radioactive tracer, metabolic labeling for quantitative proteomics replaces all unlabeled proteins with stable isotopes before the start of the experiment. In contrast to chemical labeling, metabolic labeling does not require any chemical reaction. Therefore, metabolic labeling is easy to apply and mainly used for the relative quantification of protein expression in cell culture.

The first reports of metabolic labeling appeared in 1999. In this procedure, yeast cells were grown in two media, one of which used ^{15}N-enriched media.[94] The two yeast cultures were combined and the proteins of interest were digested with trypsin before mass spectrometry analysis. The labeling by use of an ^{15}N-ammonium salt results in the complete labeling of amino acid and could be quantified with mass spectrometry. The corresponding mass shift between the unlabeled and the labeled form of the peptide requires high resolution mass spectrometry for analysis.

Later, Ong *et al.* introduced a method termed SILAC (stable isotope labeling by amino acids in cell).[95] In SILAC, two cell-culture populations are grown under identical conditions, except that one is supplied with the labeled amino acids (e.g. arginine with six ^{13}C atoms) and the other is with the non-labeled. After five or six doublings, two kinds of amino acids have fully incorporated into proteins. Every peptide pair is separated by the mass shift by the labeled amino acid. This approach cannot be applied to tissues or body fluids, and is limited to cultured cells. More recently, a method termed QCAT[96] or Qcon-CAT[97] has been developed for multiplexed absolute quantification of proteins. The artificial proteins (QconCATs), which are concatamers of tryptic peptide for a group of proteins under study, are expressed in *Escherichia coli*, and can be metabolically labeled with stable isotopes. The QconCAT proteins are then purified, quantified, and added to samples as internal standards for absolute quantification.

4.5.4 Hetero-elements used as Elemental Tags

There is an interesting trend of using elements as the chemical labeling for quantitative proteomics in literature. Whetstone *et al.* described a cysteine-targeted strategy termed ECAT (element-coded affinity tag).[98] The ECAT reagent consists of bi-functional chelating agent DOTA (1,4,7,10-tetraazacyclododecane-N,N',N'',N'''-tetraacetic acid), which contains a reactive group targeted to cysteine residue and stable chelation with lanthanides. Liu *et al.* also developed a similar method called MECT (metal element chelate tag).[99] As an inexpensive, well-known reagent used in the field of radio-pharmacology, the bi-cyclic anhydride diethylenetriamine-N,N,N',N'',N''-pentaacetic acid (DTPA) is covalently coupled to primary amines of peptides, and the ligand is then chelated to the rare earth metals Y and Tb. Peptides can be quantified by the relative signal intensities for the Y and Tb tag pairs in ESI-MS. The method can offer convenient and rapid sample preparation and good sensitivity at the femtomole level. Moreover, all primary amines are labeled, and clearer fragmentations increased the overall coverage.

The above methods have opened a new door to tackle quantitative proteomics by element mass spectrometry, although it is still quite difficult to quantify proteins by element mass spectrometry detection after element labeling. In fact, element labeling is not a new concept. For example, a silver saturation assay was applied for measurement of metallothionein in tissue. But this kind of metal-binding assay is based on certain properties of a protein and, being less selective, is applied mainly for a known and isolated protein. Comparatively, bi-functional chelating reagents are established in selective, reproducible reactions and thus are more suitable for protein quantification by element mass spectrometry detection.

Ahrends *et al.* described a metal-coded affinity tag (MeCAT) for the relative and absolute quantification of peptides and proteins (Figure 4.11c).[100] A macrocyclic metal chelate complex (1,4,7,10-tetraazacyclododecane-1,4,7,10-tetraacetic acid, DOTA) loaded with different lanthanide ions as the essential part of the tag. If required, the structures of peptides can be determined using ESI or MALDI MS/MS data, but the quantitative information comes from ICP-MS measurements. Peptides tagged with the reagent loaded with different metals co-elute in liquid chromatography. The calculated detection limit for the example of bovine serum albumin is 110 amol. They have also used MeCAT to analyze proteins of the *Sus scrofa* eye lens as a model system. These data showed that MeCAT allowed quantification not only of peptides but also of proteins in an absolute fashion at low concentrations and in complex mixtures.

Patel et al. derivatised bradykinin and substance P with cyclic diethylenetriaminepentaacetic anhydride (cDTPA) and subsequently labeled with natural and isotopically enriched Eu^{3+}.[101] Relative quantification was achieved by differentially labeling two peptide sources, after derivatisation with cDTPA, using natural and enriched ^{151}Eu respectively. The $^{151}Eu/^{153}Eu$ isotope ratio was measured and used to calculate the original peptide ratio. The measured ratios came within 5.2% of the known ratio.

Obviously, the combination of element labeling and then detection by element mass spectrometry is still in its infancy for quantitative proteomics. However, this is such a promising method for metallomics and metalloproteomics, because of the unique advantages: (1) results can be validated due to the traceability of ICP-MS measurement; (2) multiplexed analysis is easy to implement because of the availability of so many isotopes for labeling; and (3) absolute quantification at a very low concentration allows comparison between the results from different laboratories.

4.6 Conclusions

At present, isotope techniques combined with ICP-MS and ESI-MS have been successfully applied for metallomics and metalloproteomics studies, because of their unique characterization. It should be stressed that element mass spectrometry (e.g. ICP-MS) and molecular mass spectrometry (e.g. ESI-MS) are really two complementary techniques. The integrated roadmap of this combination is shown in Figure 4.3. Element mass spectrometry has a unique quantitative ability and an unmatched sensitivity for elemental detection, whereas molecular mass spectrometry can provide information on protein sequence. Isotopic techniques combined with ICP-MS and ESI-MS are very promising for metallomics and metalloproteomics. More successful advances are expected in the future.

Acknowledgments

The authors acknowledge the Natural Science Foundation of China for financial supports (Grant No.: 20805048, 10975148 and 20931160430).

References

1. R. S. Houk, V. A. Tassel, G. D. Flesch, H. J. Svec, A. L. Gray and C. E. Taylor, *Anal. Chem.*, 1980, **53**, 2283–2289.
2. A. A. Ammann, *J. Mass Spectrom.*, 2007, **42**, 419–427.
3. D. Beauchemin, *Anal. Chem.*, 2006, **78**, 4111–4135.
4. N. Jakubowski, L. Moens and F. Vanhaecke, *Spectrochim. Acta Part B*, 1998, **53**, 1739–1763.

5. M. Guilhaus, *Spectrochim. Acta Part B*, 2000, **55**, 1511–1525.
6. D. W. Koppenaal, G. C. Eiden and C. J. Barinaga, *J. Anal. At. Spectrom.*, 2004, **19**, 561–570.
7. R. Lobinski, D. Schaumlöffel and J. Szpunar, *Mass Spectrom. Rev.*, 2006, **25**, 255–289.
8. J. Szpunar and R. Lobinski, *Hyphenated Techniques in Speciation Analysis*, The Royal Society of Chemistry, Cambridge, 2003.
9. R. E. Russo, X. L. Mao, H. C. Liu, J. Gonzalez and S. S. Mao, *Talanta*, 2002, **57**, 425–451.
10. K. E. Jarvis, A. L. Gray and R. S. Houk, ed. *Handbook of Inductively Coupled Plasma Mass Spectrometry*, Blackie, Glasgow, 1992.
11. S. M. Nelms, ed. *Inductively Coupled Plasma Mass Spectrometry Handbook*, Blackwell, Oxford, 2005.
12. J. Szpunar, *Analyst*, 2005, **130**, 442–465.
13. E. Svantesson, J. Pettersson and K. E. Markides, *J. Anal. At. Spectrom.*, 2002, **17**, 491–496.
14. S. D. Tanner, V. I. Baranov and D. R. Bandura, *Spectrochim. Acta Part B*, 2002, **57**, 1361–1452.
15. B. Michalke, *TrAC, Trends Anal. Chem.*, 2002, **21**, 142–153.
16. M. Montes-Bayon, K. DeNicola and J. A. Caruso, *J. Chromatogr. A*, 2003, **1000**, 457–476.
17. D. Schaumlöffel, J. R. Encinar and R. Lobinski, *Anal. Chem.*, 2003, **75**, 6837–6842.
18. B. W. Acon, J. A. McLean and A. Montaser, *J. Anal. At. Spectrom.*, 2001, **16**, 852–857.
19. M. Wind, A. Eisenmenger and W. D. Lehmann, *J. Anal. At. Spectrom.*, 2002, **17**, 21–26.
20. P. Giusti, D. Schaumlöffel, J. R. Encinar and J. Szpunar, *J. Anal. At. Spectrom.*, 2005, **20**, 1101–1107.
21. A. P. Navaza, J. R. Encinar and A. Sanz-Medel, *Angew. Chem. Int. Ed.*, 2007, **46**, 569–571.
22. S. S. Kannamkumarath, K. Wrobel, K. Wrobel, C. B'Hymer and J. A. Caruso, *J. Chromatogr. A*, 2002, **975**, 245–266.
23. R. L. Ma, C. W. McLeod, K. Tomlinson and R. K. Poole, *Electrophoresis*, 2004, **25**, 2469–2477.
24. J. S. Becker, M. Zoriy, J. S. Becker, J. Dobrowolska and A. Matusch, *J. Anal. At. Spectrom.*, 2007, **22**, 736–744.
25. H. Haraguchi, *J. Anal. At. Spectrom.*, 2004, **19**, 5–14.
26. F. Turecek, *Org. Mass Spectrom.*, 1992, **27**, 1087–1097.
27. J. B. Fenn, M. Mann, C. K. Meng, S. F. Wong and C. M. Whitehouse, *Science*, 1989, **246**, 61–71.
28. R. Westermeier and T. Naven, *Proteomics in Practice*, Wiley-VCH, Weinheim, 2002.
29. P. Kebarle, *J. Mass Spectrom.*, 2000, **35**, 804–817.
30. M. Wind and W. D. Lehmann, *J. Anal. At. Spectrom.*, 2004, **19**, 20–25.
31. R. Aebersold and M. Mann, *Nature*, 2003, **422**, 198–207.

32. R. Aebersold and D. R. Goodlett, *Chem. Rev.*, 2001, **101**, 269–295.
33. N. Jakubowski, R. Lobinski and L. Moens, *J. Anal. At. Spectrom.*, 2004, **19**, 1–4.
34. J. Szpunar, *Anal. Bioanal. Chem.*, 2004, **378**, 54–56.
35. G. Hevesy and F. Paneth, *Z. Anorg. Chem.*, 1913, **82**, 322.
36. Z. Y. Zhang and Z. F. Chai, *Radiochim. Acta*, 2004, **92**, 355–358.
37. Z. F. Chai, Z. Y. Zhang, W. Y. Feng, C. Y. Chen, D. D. Xu and X. L. Hou, *J. Anal. At. Spectrom.*, 2004, **19**, 26–33.
38. Y. X. Gao, C. Y. Chen and Z. F. Chai, *J. Anal. At. Spectrom.*, 2007, **22**, 856–866.
39. R. J. Slater, *Radioisotopes in Biology*, Oxford University Press, New York, 2002.
40. A. Rodriguez-Cea, M. D. F. de la Campa, J. I. G. Alonso and A. Sanz-Medel, *J. Anal. At. Spectrom.*, 2006, **21**, 270–278.
41. J. W. Shi, W. Y. Feng, M. Wang, F. Zhang, B. Li, B. Wang, M. T. Zhu and Z. F. Chai, *Anal. Chim. Acta*, 2007, **583**, 84–91.
42. M. P. Rayman, *Lancet*, 2000, **356**, 233–241.
43. L. C. Clark, G. F. Combs, B. W. Turnbull, E. H. Slate, D. K. Chalker, J. Chow, L. S. Davis, R. A. Glover, G. F. Graham, E. G. Gross, A. Krongrad, J. L. Lesher, H. K. Park, B. B. Sanders, C. L. Smith and J. R. Taylor, *JAMA J. Am. Med. Assoc.*, 1996, **276**, 1957–1963.
44. R. F. Burk and K. E. Hill, *J. Nutr.*, 1994, **124**, 1891–1897.
45. R. F. Burk and P. E. Gregory, *Arch. Biochem. Biophys.*, 1982, **213**, 73–80.
46. D. Behne, C. Hammel, H. Pfeifer, D. Rothlein, H. Gessner and A. Kyriakopoulos, in 3rd International Symposium on Speciation of Trace Elements in Biological, Environmental and Toxicological Sciences, Port Douglas, Australia, *Analyst*, 1998, **173**, 871–873.
47. K. T. Suzuki and M. Itoh, *J. Chromatogr. B*, 1997, **692**, 15–22.
48. Y. Kobayashi, Y. Ogra and K. T. Suzuki, *J. Chromatogr. B*, 2001, **760**, 73–81.
49. K. T. Suzuki, *J. Health Sci.*, 2005, **51**, 107–114.
50. L. Rottmann and K. G. Heumann, in 2nd Regensburg Symposium on Mass Spectrometric Methods in Element Trace Analysis, Regensburg, Germany, Oct 06-08, 1993. Springer Verlag, New York, USA, 1994.
51. P. Rodriguez-Gonzalez, J. M. Marchante-Gayon, J. I. G. Alonso and A. Sanz-Medel, *Spectrochim. Acta Part B*, 2005, **60**, 151–207.
52. L. Rottmann and K. G. Heumann, *Fresenius' J. Anal. Chem.*, 1994, **350**, 221–227.
53. K. G. Heumann, S. M. Gallus, G. Radlinger and J. Vogl, *Spectrochim. Acta Part B*, 1998, **53**, 273–287.
54. D. Schaumlöffel and R. Lobinski, *Int. J. Mass Spectrom.*, 2005, **242**, 217–223.
55. S. A. Gerber, J. Rush, O. Stemman, M. W. Kirschner and S. P. Gygi, *Proc. Natl. Acad. Sci. U. S. A.*, 2003, **100**, 6940–6945.
56. J. P. Snell, Stewart II, R. E. Sturgeon and W. Frech, *J. Anal. At. Spectrom.*, 2000, **15**, 1540–1545.

57. J. Qvarnstrom, L. Lambertsson, S. Havarinasab, P. Hultman and W. Frech, *Anal. Chem.*, 2003, **75**, 4120–4124.
58. J. R. Encinar, J. Alonso and A. Sanz-Medel, *J. Anal. At. Spectrom.*, 2000, **15**, 1233–1239.
59. L. Yang, Z. Mester and R. E. Sturgeon, *Anal. Chem.*, 2002, **74**, 2968–2976.
60. M. Monperrus, R. C. R. Martin-Doimeadios, J. Scancar, D. Amouroux and O. F. X. Donard, *Anal. Chem.*, 2003, **75**, 4095–4102.
61. J. R. Encinar, D. Schaumlöffel, Y. Ogra and R. Lobinski, *Anal. Chem.*, 2004, **76**, 6635–6642.
62. C. F. Harrington, D. S. Vidler, M. J. Watts and J. F. Hall, *Anal. Chem.*, 2005, **77**, 4034–4041.
63. M. E. D. Busto, M. Montes-Bayon and A. Sanz-Medel, *Anal. Chem.*, 2006, **78**, 8218–8226.
64. M. Hoppler, L. Meile and T. Walczyk, *Anal. Bioanal. Chem.*, 2008, **390**, 53–59.
65. C. L. Deitrich, A. Raab, B. Pioselli, J. E. Thomas-Oates and J. Feldmann, *Anal. Chem.*, 2007, **79**, 8381–8390.
66. A. Prange and D. Schaumlöffel, *Anal. Bioanal. Chem.*, 2002, **373**, 441–453.
67. H. G. Infante, K. Van Campenhout, R. Blust and F. C. Adams, *J. Chromatogr. A*, 2006, **1121**, 184–190.
68. A. Prange, D. Schaumlöffel, P. Brätter, A. N. Richarz and C. Wolf, *Fresenius J. Anal. Chem.*, 2001, **371**, 764–774.
69. C. S. Muniz, J. M. M. Gayon, J. I. G. Alonso and A. Sanz-Medel, *J. Anal. At. Spectrom.*, 2001, **16**, 587–592.
70. M. Wind, A. Wegener, A. Eisenmenger, R. Kellner and W. D. Lehmann, *Angew. Chem. Int. Ed.*, 2003, **42**, 3425–3427.
71. S. P. Gygi, B. Rist, S. A. Gerber, F. Turecek, M. H. Gelb and R. Aebersold, *Nat. Biotechnol.*, 1999, **17**, 994–999.
72. J. Bettmer, N. Jakubowski and A. Prange, *Anal. Bioanal. Chem.*, 2006, **386**, 7–11.
73. T. Prohaska, C. Latkoczy and G. Stingeder, *J. Anal. At. Spectrom.*, 1999, **14**, 1501–1504.
74. D. Schaumlöffel, A. Prange, G. Marx, K. G. Heumann and P. Bratter, *Anal. Bioanal. Chem.*, 2002, **372**, 155–163.
75. A. A. Menegario, M. F. Gine, J. A. Bendassolli, A. C. S. Bellato and P. C. O. Trivelin, *J. Anal. At. Spectrom.*, 1998, **13**, 1065–1067.
76. B. Divjak and W. Goessler, *J. Chromatogr. A*, 1999, **844**, 161–169.
77. M. Wang, W. Y. Feng, W. W. Lu, B. Li, B. Wang, M. Zhu, Y. Wang, H. Yuan, Y. Zhao and Z. F. Chai, *Anal. Chem.*, 2007, **79**, 9128–9134.
78. D. Schaumlöffel, P. Giusti, H. Preud'Homme, J. Szpunar and R. Lobinski, *Anal. Chem.*, 2007, **79**, 2859–2868.
79. H. Zhang, W. Yan and R. Aebersold, *Curr. Opin. Chem. Biol.*, 2004, **8**, 66–75.
80. A. H. P. America and J. H. G. Cordewener, *Proteomics*, 2008, **8**, 731–749.
81. S. Julka and F. Regnier, *J. Proteome Res.*, 2004, **3**, 350–363.

82. W. A. Tao and R. Aebersold, *Curr. Opin. Biotechnol.*, 2003, **14**, 110–118.
83. S. E. Ong and M. Mann, *Nat. Chem. Biol.*, 2005, **1**, 252–262.
84. S. P. Gygi, B. Rist, T. J. Griffin, J. Eng and R. Aebersold, *J. Proteome Res.*, 2002, **1**, 47–54.
85. T. P. J. Dunkley, P. Dupree, R. B. Watson and K. S. Lilley, *Biochem. Soc. Trans.*, 2004, **32**, 520–523.
86. R. J. Zhang, C. S. Sioma, R. A. Thompson, L. Xiong and F. E. Regnier, *Anal. Chem.*, 2002, **74**, 3662–3669.
87. A. Leitner and W. Lindner, *J. Chromatogr. B*, 2004, **813**, 1–26.
88. K. C. Hansen, G. Schmitt-Ulms, R. J. Chalkley, J. Hirsch, M. A. Baldwin and A. L. Burlingame, *Mol. Cell. Proteomics*, 2003, **2**, 299–314.
89. J. V. Olsen, J. R. Andersen, P. A. Nielsen, M. L. Nielsen, D. Figeys, M. Mann and J. R. Wisniewski, *Mol. Cell. Proteomics*, 2004, **3**, 82–92.
90. A. Schmidt, J. Kellermann and F. Lottspeich, *Proteomics*, 2005, **5**, 4–15.
91. P. L. Ross, Y. L. N. Huang, J. N. Marchese, B. Williamson, K. Parker, S. Hattan, N. Khainovski, S. Pillai, S. Dey, S. Daniels, S. Purkayastha, P. Juhasz, S. Martin, M. Bartlet-Jones, F. He, A. Jacobson and D. J. Pappin, *Mol. Cell. Proteomics*, 2004, **3**, 1154–1169.
92. L. DeSouza, G. Diehl, M. J. Rodrigues, J. Z. Guo, A. D. Romaschin, T. J. Colgan and K. W. M. Siu, *J. Proteome Res.*, 2005, **4**, 377–386.
93. R. J. Beynon and J. M. Pratt, *Mol. Cell. Proteomics*, 2005, **4**, 857–872.
94. Y. Oda, K. Huang, F. R. Cross, D. Cowburn and B. T. Chait, *Proc. Natl. Acad. Sci. U. S. A.*, 1999, **96**, 6591–6596.
95. S. E. Ong, B. Blagoev, I. Kratchmarova, D. B. Kristensen, H. Steen, A. Pandey and M. Mann, *Mol. Cell. Proteomics*, 2002, **1**, 376–386.
96. R. J. Beynon, M. K. Doherty, J. M. Pratt and S. J. Gaskell, *Nat. Methods*, 2005, **2**, 587–589.
97. J. M. Pratt, D. M. Simpson, M. K. Doherty, J. Rivers, S. J. Gaskell and R. J. Beynon, *Nat. Protocols*, 2006, **1**, 1029–1043.
98. P. A. Whetstone, N. G. Butlin, T. M. Corneillie and C. F. Meares, *Bioconjugate Chem.*, 2004, **15**, 3–6.
99. H. L. Liu, Y. J. Zhang, J. L. Wang, D. Wang, C. X. Zhou, Y. Cai and X. H. Qian, *Anal. Chem.*, 2006, **78**, 6614–6621.
100. R. Ahrends, S. Pieper, A. Kuhn, H. Weisshoff, M. Hamester, T. Lindemann, C. Scheler, K. Lehmann, K. Taubner and M. W. Linscheid, *Mol. Cell. Proteomics*, 2007, **6**, 1907–1916.
101. P. Patel, P. Jones, R. Handy, C. Harrington, P. Marshall and E. H. Evans, *Anal. Bioanal. Chem.*, 2008, **390**, 61–65.
102. J. X. Li, H. Steen and S. P. Gygi, *Mol. Cell. Proteomics*, 2003, **2**, 1198–1204.
103. C. A. Gartner, J. E. Elias, C. E. Bakalarski and S. P. Gygi, *J. Proteome Res.*, 2007, **6**, 1482–1491.
104. C. Pasquarello, J. C. Sanchez, D. F. Hochstrasser and G. L. Corthals, *Rapid Commun. Mass Spectrom.*, 2004, **18**, 117–127.
105. A. Thompson, J. Schafer, K. Kuhn, S. Kienle, J. Schwarz, G. Schmidt, T. Neumann and C. Hamon, *Anal. Chem.*, 2003, **75**, 1895–1904.

106. Y. H. Lee, H. Han, S. B. Chang and S. W. Lee, *Rapid Commun. Mass Spectrom.*, 2004, **18**, 3019–3027.
107. S. Wiese, K. A. Reidegeld, H. E. Meyer and B. Warscheid, *Proteomics*, 2007, **7**, 340–350.
108. F. L. Brancia, H. Montgomery, K. Tanaka and S. Kumashiro, *Anal. Chem.*, 2004, **76**, 2748–2755.
109. D. R. Goodlett, A. Keller, J. D. Watts, R. Newitt, E. C. Yi, S. Purvine, J. K. Eng, P. von Haller, R. Aebersold and E. Kolker, *Rapid Commun. Mass Spectrom.*, 2001, **15**, 1214–1221.
110. A. Panchaud, E. Guillaume, M. Affolter, F. Robert, P. Moreillon and M. Kussmann, *Rapid Commun. Mass Spectrom.*, 2006, **20**, 1585–1594.

CHAPTER 5
Mössbauer Spectroscopy

YANG QIU[a] AND CHUNYING CHEN[a,b,*]

[a] CAS, Key Laboratory for Biological, Effects of Nanomaterials & Nanosafety, National Center for Nanoscience and Technology of China & Institute of High Energy Physics, Chinese Academy of Sciences, Beijing 100190, China; [b] CAS Key Laboratory of Nuclear Analytical Techniques, Institute of High Energy Physics, Chinese Academy of Sciences, Beijing 100049, China

5.1 Introduction and Fundamentals

5.1.1 The Mössbauer Effect

The *Mössbauer effect*, also known as the *recoilless nucleus resonance absorption of gamma rays*, was first discovered and explained by Rudolf L. Mössbauer in 1957,[1] while he was working on his doctoral thesis at Heidelberg. In the following years, more thorough understanding concerning principles, methods, and applications has been achieved by a myriad of research.

Before elaborating the principle of the Mössbauer effect, we would prefer to present a brief summary of the background concepts relevant to nucleus resonance phenomena. Due to the wave–particle duality of nucleus, the precise value of the energy level of an excited state of mean lifetime τ is unmeasurable. Instead the energy level spreads over a certain energy range of width ΔE, which correlates with the measure uncertainty in time Δt ($\Delta t \approx \tau$) via the Heisenberg uncertainty relation in the form of the conjugate variables energy and time

$$\Delta E \Delta t \geq \hbar \tag{5.1}$$

---

*Corresponding author: Email: chenchy@nanoctr.cn; Tel: +86-10-82545560; Fax: +86-10-62656765.

Mössbauer Spectroscopy

$\hbar = h/2\pi$, the constant $\hbar$ is called "h-bar", while h is Plank's constant; therefore, the profile of a certain energy level other than a ground state energy level (since the life time τ of ground state is at the infinite scale and thus the uncertainty of its energy ΔE is zero) on a graph would be observed to be a distribution within a finite range instead of a sharp line. Moreover, while the energy transit from nuclei and emit gamma quanta coupled with the transition of nuclei, the shape of the gamma-quantum energy has been shown to be a Lorentzian or Breit–Wigner form which could be mathematically given by the Breit–Wigner formula

$$\omega(E) = \frac{\Gamma/2\pi}{(E - E_\gamma^0)^2 + (\Gamma/2)^2} \tag{5.2}$$

where Γ is the full width at half maximum of the probability function $\omega(E)$, and E_γ^0 is the mean value of the emitted gamma quanta.

In a certain system, when an excited nucleus with the energy E_e emits a gamma quantum of energy E_γ, the emitted gamma quantum could be totally absorbed by another nucleus of the same kind at the ground state of energy E_g ($E_\gamma = E_e - E_g$) all the energy E_γ would be used to help the latter one transit to the excited state of energy E_e. This is the so-called *nucleus resonance absorption of gamma rays*. Similarly, the resonance absorption cross-section on graphs could also been mathematically represented by the Breit–Wigner formula of another form:

$$\sigma(E) = \frac{\sigma_0 \Gamma^2}{4(E - E_\gamma^0)^2 + \Gamma^2} \tag{5.3}$$

where σ_0 is the maximum absorption cross-section, which is obtained in the formula

$$\sigma_0 = \frac{\lambda^2}{2\pi} \cdot \frac{2I_e + 1}{2I_g + 1} \cdot \frac{1}{\alpha + 1} \tag{5.4}$$

λ is the wavelength of the gamma ray, I_e and I_g are nucleus spin quantum numbers of the excited and the ground state, respectively, and α is the internal conversion coefficient.[2]

Ideally, the absorption profile would overlap the emission profile; however, due to the wave–particle duality of the gamma quantum, the excited free nucleus which emits the gamma quantum suffers a recoil; hence, the energy of the emitted quanta E_γ becomes

$$E_\gamma = E_0 - E_R = (E_e - E_g) - E_R \tag{5.5}$$

where E_0 is the nuclear transition energy, and E_R is the recoil energy of the nucleus after the emission of gamma quantum, which could be expressed as

$$E_R = E_\gamma^2 / 2Mc^2 \approx E_0^2 / 2Mc^2 \tag{5.6}$$

since E_R is very small compared to E_0 and thus $E_\gamma \approx E_0$.

Due to recoil the position of emission line shifts to smaller energy and, similarly, the absorption line shifts to a larger energy as a result of the same amount of energy lost during absorption, which requires that the energy of the absorbed quanta E_γ should be

$$E_\gamma = E_0 + E_R \qquad (5.7)$$

Consequently, the distance between the emission line and absorption line is $2E_R$, an amount 10^6 larger than the natural line width $\overline{\Gamma}$. Since the lifetime, τ, of the excited state is related to Γ via[3]

$$\Gamma \cdot \tau = \hbar \qquad (5.8)$$

the lifetime broadening of the nuclear transition is impossible to exceed the recoil energy and therefore overlap between the two transition lines as well as the nuclear resonance absorption can never happen for free nuclei in the gaseous or liquid state.

In fact, there are some approaches to employ to shift the emission line so as to increase the probability of its overlapping with absorption lines. One approach is the Doppler effect: when the source (nucleus) is moving at a velocity v_n in the direction of the ray propagation, the gamma quantum of energy, E_γ, receives a Doppler energy E_D,

$$E_\gamma = E_0 - E_R - E_D \qquad (5.9)$$

where E_D could be obtained as follows

$$E_D = \frac{v_n}{c} E_\gamma \qquad (5.10)$$

Nuclei, especially those in gaseous state, move in varied directions; therefore, the mean Doppler broadening of the transition line $\overline{E_D}$ can be evaluated from

$$\overline{E_D} = E_\gamma \overline{(2E_K/Mc^2)^{1/2}} \qquad (5.11)$$

where E_K is the mean kinetic energy of all the nuclei which move at all angles between its moving direction and the direction of wave propagation. Resonance absorption occurs if $\overline{E_D}$ is in the order of E_R or larger as there is a finite probability of overlapping.

In the solid state, another approach could be employed. When attempting to reduce the gamma-ray resonance of solid iridium (Ir) by decreasing the Doppler effect through lowering the temperature in 1957, Rudolf L. Mössbauer accidentally observed an opposite phenomenon in an experiment arranged similarly to the preceding ones[4] to measure the lifetime of the 129 keV state in ^{191}Ir. The seemingly strange results evoked Mössbauer's curiosity, which ultimately

Mössbauer Spectroscopy

led to the discovery of *recoilless nuclear absorption of gamma radiation*, or the *Mössbauer effect*, and a Nobel Prize in Physics awarded to him in 1961. The transition behavior of nuclei in the solid state in that case is quite different from nuclei in the gaseous or liquid state since nuclei are more or less bound to the lattice and thereby there is a certain probability for recoilless absorption.

Actually, the recoil energy E_R of a nucleus in the solid state comprises two parts: the translational energy of the whole lattice E_{tr}, which is the recoil energy transferred to the whole lattice; and the mean lattice vibration energy, $\overline{E_{vib}}$. E_{tr} could be evaluated by the formula (5.6) and, owing to the great mass of the whole lattice M, E_{tr} is usually a considerably small amount. E_R hence is almost the same as $\overline{E_{vib}}$. In the process of emission or absorption of gamma rays in a lattice, if the recoil energy is larger than the characteristic lattice vibration (phonon) energy of some atoms but smaller than the displacement energy, the atoms will remain vibrating in the lattice position and just dissipate the recoil energy by heating surrounding atoms; or if the recoil energy is smaller than the characteristic lattice vibration energy of other atoms, there is, in quantum mechanics, a probability, *f* (*recoil-free fraction*), that no lattice excitation takes place, that is, the *zero-phonon process*. The recoil-free fraction *f* could be attained

$$f = 1 - k^2 \langle x^2 \rangle \tag{5.12}$$

where $\langle x^2 \rangle$ is the mean-square displacement of the nucleus in the x-direction, the direction of the incoming *gamma ray* and k is the propagation vector. Since $k^2 \langle x^2 \rangle \ll 1$,

$$f = \exp\{-k^2 \langle x^2 \rangle\} \tag{5.13}$$

In addition, *f* could be further calculated in Debye's solid model as

$$f(T) = \exp\left\{-\frac{3E_R}{2k_B \theta_D}\left[1 + 4\left(\frac{T}{\theta_D}\right)\int_0^{\theta_D/T} \frac{x}{e^x - 1} dx\right]\right\} \tag{5.14}$$

where k is the Boltzman constant and θ_D is the Debye temperature, which is also recognized as a measure of the strength of the bonds between the atom and the lattice. In this case, *f*(T) is more generally called the *Debye–Waller factor*, or the *Lamber–Mössbauer factor*. Mathematically, the above calculation could be reduced

$$f(T) = \begin{cases} \exp\left[-\frac{E_R}{k_B \theta_D}\left(\frac{3}{2} + \frac{\pi^2 T^2}{\theta_D^2}\right)\right], & T \ll \theta_D \\ \exp\left(-\frac{6E_R T}{k_B \theta_D^2}\right), & T \geq \theta_D \end{cases} \tag{5.15}$$

From above formula, one can see that f increases with decreasing temperature. That is why R. Mössbauer astonishingly observed an increased resonance while originally attempting to reduce the resonance effect by decreasing temperature.

In fact, Mössbauer spectroscopy is implemented as shown in Figure 5.1.[5] The source (S) and the absorber (A) move relative to each other and the gamma rays emitted by the source are registered as a function of the relative velocity (Doppler velocity) by a detector after they have passed the absorber. The plot of the relative transmission versus Doppler velocity is the so-called Mössbauer spectrum. In this plot, the resonance (relative transition) varies as the energy of the emitted gamma rays changes with the altering relative velocity of source, which is closely related to the nuclear structure and dynamics of a certain sample and, therefore, such information can be obtained through conducting Mössbauer spectroscopy.

5.1.2 Hyperfine Structure

A nucleus in a certain environment undergoes lasting interactions with the neighborhood through electric and magnetic fields, which would also perturb nuclear energy levels, and the perturbation, the so-called *nuclear hyperfine interactions*, can change the Hamiltonian of the nucleus:

$$\hat{H}_N = \hat{H}_N^0 + \hat{H}_{Hyp} = \hat{H}_N^0 + \hat{H}_{E0} + \hat{H}_{M1} + \hat{H}_{E2} + \cdots \quad (5.16)$$

where $\hat{H}_{Hyp}$ is the Hamiltonian of the hyperfine interactions and $\hat{H}_{E0}, \hat{H}_{M1}, \hat{H}_{E2}$ are the Hamiltonians of the electric monopole interaction, the magnetic dipole interaction and electric quadruple interaction, respectively.

5.1.2.1 Electric Monopole Interaction

The basic shift of transition line in the Mössbauer spectrum is called the *isomer shift* due to electric monopole interaction. The electric monopole interaction originates from the electrostatic Coulomb interaction between the nucleus and electrons inside the nuclear region and is proportional to the s-electron density at the nucleus. This interaction energy, E_{E0}, which is further defined as electrostatic shift, δE, is attained as

$$E_{E0} = \frac{2}{3} \cdot \pi Z^2 e^2 |\varphi(0)|^2 \cdot \langle r^2 \rangle \equiv \delta E, \quad (5.17)$$

where $\langle r^2 \rangle$ is the mean square nuclear radius, Ze is the nuclear charge and $|\varphi(0)|^2$ is the s-electron density at the nucleus. The energy of excited (e) and ground (g) nuclear state shift to a different extent and hence the resulting

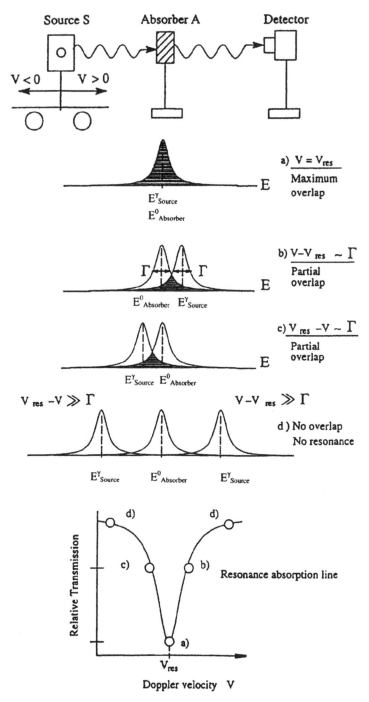

Figure 5.1 Schematic illustration of recording the resonance absorption line in a Mössbauer spectroscopy experiment and relative transmission of gamma quanta as a function of Doppler velocity.[5] © 2000 IOP Publishing Ltd.

nuclear transition shift, δ_N, the electrostatic shift, δE, is obtained as

$$\delta_N = \delta E = (E_e - E_e^0) - (E_g - E_g^0) = \frac{2}{3} \cdot \pi Z^2 e^2 |\varphi(0)|^2 \cdot \left(\langle r^2 \rangle_e - \langle r^2 \rangle_g\right), \quad (5.18)$$

where E_e and E_g are the shifted energy of excited and ground nuclear states and E_e^0 and E_g^0 are the original energy of excited and ground nuclear states for a certain nucleus, $\langle r^2 \rangle_e$, $\langle r^2 \rangle_e$ are the mean square nuclear radius of nucleus in excited and ground state, respectively.

When the source and absorber move relative to each other and one only observes the difference of the electrostatic shifts of the source and absorber, or the isomer shift δ, instead of the shifts of them δ_S and δ_A separately:

$$\delta = \delta_A - \delta_S = \frac{2}{3} \cdot \pi Z^2 e^2 \left(|\varphi(0)_A|^2 - |\varphi(0)_S|^2\right) \cdot \left(\langle r^2 \rangle_e - \langle r^2 \rangle_g\right), \quad (5.19)$$

where δ_A and δ_S are electrostatic shifts of the source and absorber, respectively, and $|\varphi(0)_A|^2$ and $|\varphi(0)_S|^2$ are the s-electron densities at the nuclei of absorber and source, respectively. Furthermore, in the case of a homogeneous charge distribution with a spherical symmetry with a nuclear radius of excited and ground states R_e and R_g, respectively, by allowing $2R = R_e + R_g$ and $\delta R = R_e - R_g$, the calculation of isomer shift is reduced:

$$\delta = \frac{4}{5} \cdot \pi Z e^2 R^2 \frac{\delta R}{R} \left(|\varphi(0)_A|^2 - |\varphi(0)_S|^2\right), \quad (5.20)$$

which is the expression mostly encountered in analyzing Mössbauer spectra; however, while taking consideration the relativistic effect which significantly modifies the wave function, φ, for heavier elements, some problems arise in some detailed studies if formula (5.20) is still employed without modifications; therefore, some corrections have to be made for the formula (5.20). A simple approach is to add a dimensionless factor $S'(Z)$, the value of which is calculated through employing Dirac functions and first-order perturbation theory, into formula (5.20).[2,6] The following equation is then obtained:

$$\delta = \frac{4}{5} \cdot \pi Z e^2 S'(Z) R^2 \frac{\delta R}{R} \left(|\varphi(0)_A|^2 - |\varphi(0)_S|^2\right). \quad (5.21)$$

5.1.2.2 Electric Quadruple Interaction

In nuclei without spherically constant distribution of charges, another type of hyperfine interaction, *electric quadruple interaction*, may occur. In this case, atoms with both an observable nuclear quadruple moment, a measure of the nuclear charge distribution's deviation from spherical symmetry, and a non-zero electric field gradient (EFG) at the nucleus, a measure of the inhomogeneity of electric field.

The electric quadruple moment is a (3×3) second-rank tensor with elements

$$Q_{ij} = \int \rho_n(r)(x_i x_j - \delta_{ij} r^2) dr, \tag{5.22}$$

where ρ_n is the nuclear charge, x_i, x_j are Cartesian coordinates of r, and δ_{ij} is the Kronecker symbol.

The electric field gradient EFG is also a (3×3) second-rank tensor with elements

$$V_{ij} = \frac{\partial^2 V}{\partial x_i \partial x_j} = q(3x_i x_j - r^2 \delta_{ij}) r^{-5}, \tag{5.23}$$

with the requirements

$$\sum_i V_{ii} = 0, \tag{5.24}$$

and

$$V_{ij} = V_{ji}, \tag{5.25}$$

where x_i, x_j are Cartesian coordinates, and δ_{ij} is the Kronecker symbol.

One can simplify the situation in the principal axes system with all the off-diagonal elements being zero. The quadruple moment is hence given as

$$Q = \frac{1}{e} \int \rho_n(r)(3z^2 - r^2) dr, \tag{5.26}$$

if the z-axis is chosen as the axis of preferred orientation, and the EFG, if the principal axes are chosen such that $|V_{zz}| \geq |V_{xx}| \geq |V_{yy}|$, can be specified by two independent parameters: V_{zz}, or eq, and $\eta = \frac{V_{xx} - V_{yy}}{V_{zz}}$, the asymmetry parameter.

To investigate the electric quadruple interaction, the Hamiltonian of electric quadruple interaction can be employed:

$$\hat{H}_{E2} = \frac{eQV_{zz}}{4I(2I-1)} \left[3\hat{I}_z^2 - \hat{I}^2 + \eta(\hat{I}_+^2 + \hat{I}_-^2)/2 \right], \tag{5.27}$$

with $\hat{I}$ being the nuclear spin operator, $\hat{I}_\pm = \hat{I}_x \pm i\hat{I}_y$ being the shift operators, and $\hat{I}_x, \hat{I}_y, \hat{I}_z$ being the operators of the nuclear spin projections onto the principal axes. The interaction does not affect the baricenter of the energy level of a nuclear state but splits the level of nucleus state with spin quantum number $I > 1/2$ (since the quadruple moment in nuclei with $I = 0, 1/2$ is spectroscopically unobservable as described in the formula below) into $(2I+1)/2$ sublevels, i.e. $|I, \pm m_I\rangle$, where $m_I = -I, 1-I, \ldots, -1+I$, I is the nuclear magnetic spin quantum number. Through computing with perturbation theory, the eigenvalue E_{E2} to the Hamiltonian of electric quadruple interaction is obtained as

$$E_{E2} = \frac{eQV_{zz}}{4I(2I-1)}[3m_I^2 - I(I+1)](1+\eta^2/3)^{1/2}. \qquad (5.28)$$

In the frequently met case of ^{57}Fe, one then further obtains the quadruple splitting

$$\Delta E_{E2} = \frac{eQV_{zz}}{2} \qquad (5.29)$$

between the two substates $|3/2, \pm 3/2\rangle$ and $|3/2, \pm 1/2\rangle$. The corresponding distance Δ between the two resonance lines is called *quadruple splitting*, which is closely related to bond properties and molecular and electronic structures.

5.1.2.3 Magnetic Hyperfine Interaction

Besides the electric hyperfine interactions, magnetic hyperfine interaction significantly modifies the profile of nuclear resonance lines as well. Such interaction occurs with the nucleus in the state with spin quantum number $I > 0$ and is described by the Hamiltonian

$$\hat{H}_{M1} = -\hat{\vec{\mu}} \cdot \vec{H} = -g_N \beta_N \hat{\vec{I}} \cdot \vec{H}, \qquad (5.30)$$

where $\vec{\mu}$ is magnetic moment, $\vec{H}$ is effective magnetic field at the nucleus which is a combination of external magnetic field $\vec{B}_0$ and the resulted internal magnetic hyperfine field $\vec{B}_{hf}$ thereof (H_{hf} consists of the field contribution of orbital motion of d-electrons at nucleus; the Fermi contact field, originating from a spin polarization of core shells at the nucleus; the spin-dipolar field, arising from the spin of electron shells as well as a negligible field contributed from lattice), g_N is the nuclear g factor (or nuclear Landé factor), and β_N is the nuclear magneton. In analyzing the experimental results, it is valuable to obtain the eigenvalue E_{M1} to $\hat{H}_{M1}$. This can also be accomplished by computing according to perturbation theory and the eigenvalue E_{M1} is given as

$$E_{M1} = -\frac{\mu H m_I}{I} = -g_N \beta_N H m_I. \qquad (5.31)$$

The interaction, also called the *magnetic dipole interaction* (M1) or the *nuclear Zeeman effect*, is the major cause of $(2I+1)$ fold splitting of the nuclear state with spin quantum number I.

In the case of ^{57}Fe, the magnetic interaction would split the nuclear level of the nucleus with $I = 3/2$ into four sublevels with $|3/2, +3/2\rangle$, $|3/2, +1/2\rangle$, $|3/2, -1/2\rangle$ and $|3/2, -3/2\rangle$, and that with $I = 1/2$ into two sublevels with $|3/2, +$

$1/2\rangle$ and $|3/2, -1/2\rangle$. Since the selection rules for an M1 transition are $\Delta I = 1$ and $\Delta m_I = 0, \pm 1$, the transition line splits into six lines as shown in Figure 5.2.[5]

5.2 Equipment and Experiments

A brief introduction is given of the physical principles, facilities, performance, and data processing for Mössbauer spectroscopy. Due to the extremely narrow line width of Mössbauer gamma rays, the spectroscopy is sufficiently sensitive for probing the interaction between a nucleus and the surrounding electrons, which is essential for studying the chemical environment of atoms. Since the major task of this chapter is to introduce the principle and relevant applications of Mössbauer spectrometery in metalloprotein studies (especially for iron-containing protein) and the methodology and apparatus for studying other compounds may differ slightly, the information discussed in this section is more related to Mössbauer spectrometry of iron compounds.

5.2.1 Equipment for a Mössbauer Spectrometer

The apparatus of a Mössbauer spectrometer is not especially complicated compared with other nuclear-related spectral techniques and fundamentally consists of three parts: a source, an absorber, and a detector.

There are several types of source, e.g. ^{57}Co, ^{151}Sm, ^{151}Gd, and ^{119m}Sn, which are applicable for varied samples, respectively. When probing ^{57}Fe samples, the excited radioactive source ^{57}Co/Rh (^{57}Co in a rhodium matrix) decays via electron capture with emitted gamma rays at 122 and 14.4 keV with additionally imparted Doppler energy since the source moves relative to the absorber.[7] The source under the control of a velocity transducer can move back and forth with constant velocity, sinusoidally, in a symmetric sawtooth, or any other complicated way;[2] therefore, the Doppler energy as well as the energy of gamma quanta can be modified in a controllable manner.

The second part of a Mössbauer spectrometer is the absorber, in which the sample resonantly absorbs the source emitted gamma rays. Part of the gamma rays are transmitted through the absorber while others are absorbed and excite the sample to re-emit gamma rays, X-rays, or electrons.

There hence exist two modes of arrangement for the third part of the spectrometer, the detector: the transmission geometry and the scattering geometry. The detector of the former counts the number of unaffected transmitted gamma rays whereas that of the latter counts that of re-emitted gamma rays, X-rays or electrons. Since the count rate, I(v), is affected by many factors as shown in the formula below[7]

$$I(v) = I_0(1 - f_S) + I_{nr} + I_s + I_c$$
$$+ f_S I_0 \int_{-\infty}^{\infty} S(v - E) \cdot \exp[-10^{-3}\sigma(E)f_A c l N_A] dE, \quad (5.32)$$

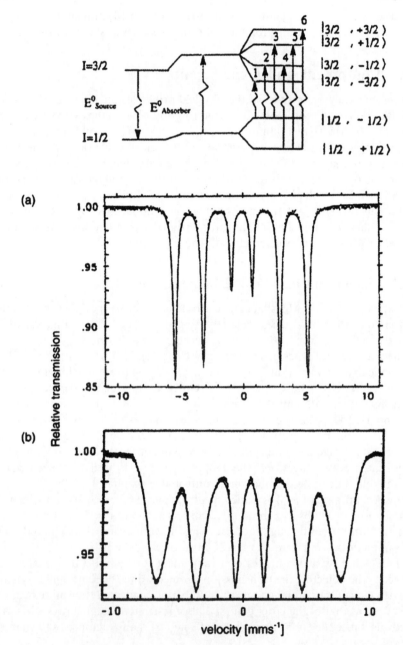

Figure 5.2 The influence of magnetic interaction with respect to the excited nuclear state of ^{57}Fe. (From Schünemann and Winkler.[5]) The influence of magnetic interaction with respect to the excited nuclear state of ^{57}Fe. The latter splits into four sublevels. The selection rule for magnetic dipole (M1) transitions ($\Delta I = 1$ and $\Delta M_I = 0, \pm 1$) yields the observed six-line pattern. (a) The Mössbauer spectrum of a metallic iron foil. (b) The Mössbauer spectrum of the iron carrier protein ferritin. © 2000 IOP Publishing Ltd.

where I_0 is the original count rate of source, I_{nr}, I_s, and I_c are the count rate of the non-resonant background, source background without attenuation of sample, and the cosmic background, respectively, f_S and f_A are the Debye–Waller factors for the source and absorber, $S(v - E)$ is the sour lineshape, $\sigma(E)$ is the cross-section for the absorber, c is the absorber molarity, l is the sample thickness, and N_A is Avogadro's number; therefore, it is of fundamental importance to choose suitable detectors when conducting the experiment in an appropriate manner. Commonly used detectors for gamma rays and X-rays include an argon-filled proportional counter for energies below 20 keV and a solid-state counter for higher energies, e.g. a NaI(Tl) crystal detector, and a Si(Li) crystal detector, while those for emitted electrons are a normal electric or magnetic electron spectrometer and He/CH_4 filled proportional counter alike.[2]

Besides the three major parts, modern Mössbauer spectrometers, to satisfy the requirement of high count rates, also utilize a fast pre-amplifier, a multi-channel analyzer (which is now generally replaced by a computer or microprocessor), a calibration system (which is normally an interferometer) and a stable cryostat (such as simple He baths, He-flow cryostats, and $^3He/^4He$ refrigerators) during plotting of the Mössbauer spectrum.

Recently, due to the needs of outer-space exploration, the miniaturized Mössbauer spectrometer (MIMOS) has been developed by E. Kankeleit and his student G. Klingelhöfer. MIMOS uses the decay of ^{57}Co to ^{57}Fe as a gamma-ray source. The gamma rays bombard the sample and the Doppler shift in energy spectrum is measured.[8] A more recent type of MIMOS II was employed in the Mars exploration rovers, *Spirit* and *Opportunity*, for mineralogical investigations on the Martian surface by NASA in 2002.[9,10] The instrument design of MIMOS II is shown in Figure 5.3.[11] The instrument, which operates in backscattering geometry without sample preparation, mainly consists of a gamma- and X-ray detector system, a collimator, the MB drive, and a control unit. The detector system contains Si PIN-diodes, charge sensitive pre- and filter amplifiers, and a single-channel analyzer. The spectrometer control and data acquisition system generates the velocity reference signal and collects the data of up to five individual detector channels. The data are stored in a dedicated on-board memory system, but can be transferred at any time via a standard serial interface to a computer.[11]

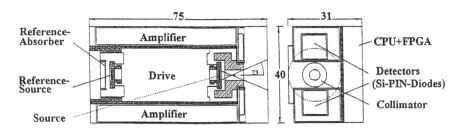

Figure 5.3 Scheme of the MIMOS II instrument design having two detector channels: dimension are in mm.[11] Reprinted with permission from Springer. © 1998 J.C. Baltzer A.G., Scientific Publishing Company.

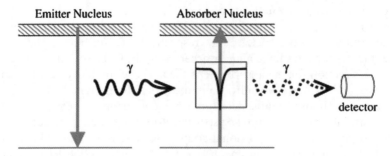

Figure 5.4 Simple Mössbauer spectrum from an identical source and absorber.[12]
© 2009 Royal Society of Chemistry.

5.2.2. The Mössbauer Spectrum

In the course of Mössbauer measurement, the energy of gamma quanta is ordinarily modulated by a mechanical movement of the source relative to the absorber. The spectrum is essentially a plot of Mössbauer transition count rates as a function of velocity of the source relative to the absorber. If no resonance occurs, the spectrum would be a horizontal line with no variations; while resonance occurs, there would be a decrease in the intensity at certain velocity values as shown in Figure 5.4. In interpreting the spectrum the Mössbauer parameters can be obtained, i.e. the isomer shift, the electric quadruple splitting, and the magnetic dipole splitting.

5.2.2.1 Isomer Shift

The isomer shift provides considerable valuable information concerning oxidation states and bond properties of certain atoms as well as the interaction with surrounding ligands thereof since it is a measurement of the difference of both the radius of the nucleus and the electronic charge density near the nucleus between the source and absorber.

Isomer values must be reported with respect to a given reference material since the isomer shifts of a certain sample measured with different source in different conditions would show different values in Mössbauer spectra. In the case of ^{57}Fe Mössbauer spectroscopy, metallic iron, α-Fe and sodium nitroprusside dehydrate (SNP), $Na_2[Fe(CN)_5NO] \cdot 2H_2O$ are common standard reference materials. The sign of the isomer shift values is determined according to conventions: the positive isomer shift indicates source and absorber moving toward each other, i.e. gamma rays of higher energy. Under the condition of a positive isomer shift, from (5.19), we can further conclude that a positive $\delta R/R$ implies a higher density of absorber than source and vice versa.

Experimentally, isomer shift is presented in the dimensions of velocity rather than in energy dimensions. Useful diagrams of its ranges as a function of oxidation states of atoms is introduced to help analyze the Mössbauer spectrum

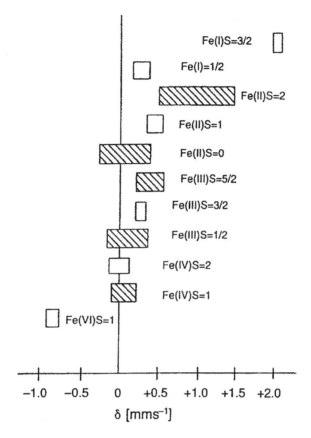

Figure 5.5 Ranges of isomer shifts observed in various oxidation and spin states of iron (given relative to α-iron at 300 K).[5] Shadowing indicates the most frequently met configurations. © 2000 IOP Publishing Ltd.

of certain compounds. One typical diagram is that of iron compounds, as shown in Figure 5.5.[5] It can help characterize the oxidation and spin states of iron atoms in unknown compounds; however, a more quantitative correlation of electron densities and isomer shifts should be achieved based upon molecular orbital calculations, which is too complicated to elaborate as for the purpose of this chapter. Another diagram is presented by L. R. Walker as shown in Figure 5.6.,[13] which is also called Walker diagram for ^{57}Fe, from which one can get to know the isomer shift is affected both by the s-electron density and d configuration.[13] In the study by Lees and Flinn, an empirical function is given to predict the isomer shift on the basis of the degree of occupation of the 5s and 5p electron orbitals while using a Mg$_2$Sn source:[14]

$$\delta = -2.36 + 3.01 n_s - 0.20 n_s^2 - 0.17 n_s n_p \ (\text{mm} \cdot \text{s}^{-1}) \tag{5.33}$$

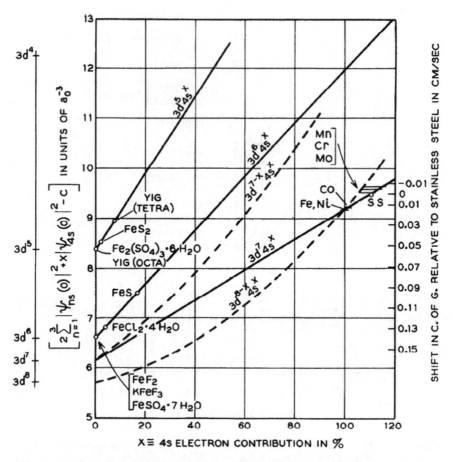

Figure 5.6 A possible interpretation of the ^{57}Fe Mössbauer isomer shifts in various solids.[13] The total s-electron density is potted as a function of the percentage of 4s character for various d-electron configurations. © 1961 The American Physical Society.

5.2.2.2 Electric Quadrupole Splitting

Electric quadruple splitting is related to the electric quadruple moment, the inherited property of a certain type of atom, and the EFG, a measurement of the electric field surrounding an atom. This parameter concerns the electronic structure of the nucleus and the chemical environment and is hence quite useful for determining the fingerprint for valence and co-ordination while correlated with isomer shift, i.e. local distortions, local defects, spin reorientations, and crystal field asymmetry.[15]

5.2.2.3 Magnetic Dipole Splitting

Magnetic dipole splitting provides information concerning spin interaction processes. Information can be derived with reference to the electron configuration, magnetic relaxation, magnetic moment, spin value, magnetic transition

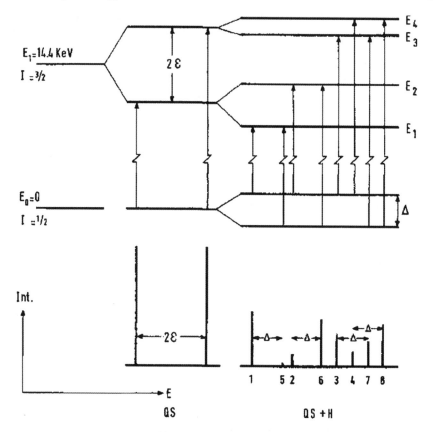

Figure 5.7 Energy levels of ^{57}Fe for combined quadrupole and magnetic interaction.[16] $\Delta = \mu_g |H|$ and $E_i = \alpha_i + 1/2\Delta - IS$ where the α_i are the transition energies. © 1975 North-Holland Publishing Company.

temperature, and spin flop processes.[15] For instance, when the Mössbauer spectrum of an iron compound is investigated, the intensity ratio for the split sextet is important for identifying the orientations of the magnetic moment directions per site.[15]

A typical hyperfine splitting scheme of ^{57}Fe is shown in Figure 5.7,[16] in which the $I = 3/2$ level splits into two due to quadruple interaction and the $|3/2, \pm 3/2\rangle$, $|3/2, \pm 1/2\rangle$, and $|1/2, \pm 1/2\rangle$ levels further split into six levels as a result of magnetic interaction. Consequently, the spectrum will contain eight transition lines; however, according to selection rule, $\Delta m = \pm 1$ or 0, only a sextet can be observed in the spectrum.

5.3 Applications for Chemical Speciation and Metalloproteins

This section will introduce the application of Mössbauer spectroscopy for the study of elemental speciation in environmental and biological samples. The protein samples subjected to Mössbauer spectroscopy are generally purified proteins in the solid state or in frozen solution. Some tissue specimens are also used for an *in situ* Mössbauer spectroscopy survey to elucidate the relationship between iron and certain diseases.

Mössbauer spectroscopy is a useful technique to test any changes of the heme iron electronic structure that is related to protein biosynthesis. Since iron plays a critical role in metabolic pathways of both humans and other living organisms, and Mössbauer spectroscopy is superb for providing structural and dynamic information of iron compounds, with high resolution even at room temperature, the main applications of Mössbauer spectroscopy in metalloprotein studies mainly focus on the structural characterization of iron-containing proteins and their qualitative and quantitative changes during pathological processes, or the effect of environmental factors, and their conformational changes and reaction dynamics.[17]

5.3.1 Sample Preparation

As described previously, it is necessary that samples subjected to Mössbauer spectroscopy should be prepared in the solid state as the Mössbauer effect is insufficient for measurement of other states. The iron-containing proteins should also be purified in the solid state or in frozen solutions.[18] To prepare the frozen solutions, the commonly employed approach is a rapid freezing method, first developed by Kerler *et al.*[19] and Hazony and Bukshpan,[20] which is useful for investigating the behaviors of reactive intermediates in kinetic studies, though it may sometimes change the chemical structure of solutions, i.e. the coordination, chemical bonding conditions, and the oxidation states in solutions.[21–24] Nevertheless, recent progress concerning rapid freeze–quench have largely circumvented this problem.

In addition to the preparation of a solid sample, one should also notice that the content of iron as well as the natural abundance of its Mössbauer isotope ^{57}Fe is manifestly low. Thus, the intensity of Mössbauer resonance may have difficulty in meeting the requirement since the resonance absorption cross-section is relatively low. More often, samples containing iron have to be enriched with the isotope ^{57}Fe.[25] If the samples are purified from bacteria or cells *in vitro*, ^{57}Fe can be added to culture media following extracting Fe from the media, or if the samples are purified from animals, e.g. rats or rabbits, they can be injected with ^{57}Fe-enriched ferric citrate solution after withdrawing the feed of ^{56}Fe in the diet.

5.3.2 Some Examples of Mössbauer Spectroscopy Applied to Metalloprotein Studies

5.3.2.1 Heme Proteins

Heme proteins are the first among those studied by Mössbauer spectroscopy.[26] Varied types of heme proteins play essential roles in metabolism: hemoglobin in oxygen transport, myoglobin in oxygen storage, cytochrome in electron transport, and peroxidases and catalase in biocatalysis; nevertheless, all heme proteins possess a common prosthetic group, the heme, in which an iron ion is in the central position of a porphyrin.[27] This prosthetic heme is the oxidation and reduction center, determining the redox reaction efficiency in the body's oxygen-related metabolism. Lang[28] has carried out intensive studies concerning the structure and dynamics of heme based on Mössbauer spectroscopy, which is a useful technique to test any changes of the heme iron electronic structure that is related to protein biosynthesis.

Hemoglobin (Hb) transports oxygen in the red blood cells of vertebrates. Oxygen combines with the heme at the sixth coordination site of the central iron; therefore, one needs to know the oxidation state of iron in order to well understand how oxygen interacts with hemoglobin. Lang reported the low spin Fe(II) exhibiting a large, heavily temperature-dependent quadruple splitting, which was far greater than expected and gave an assumption that when O_2 is oriented parallel with the heme plane, one π^* orbital of O_2 forms a molecular orbital filled with a probability of one-half.[28] Ofer et al. investigated human red cells to analyze thalassemia, a blood disease, with the Mössbauer spectroscopy. It has been identified that the resonance spectrum of different types of thalassemia red cells more or less differed from that of normal cells.[29] Trautwein et al. theoretically proposed an electronic structure for deoxygenated hemoglobin, corresponding well with the experimental data collected through Mössbauer spectroscopy. The iron atom is shown to be penta-coordinated and projected out of the heme plane.[30]

Myoglobin (MB) is a single-chain globular protein of 153 amino acids. Myoglobin is also one of the most studied heme proteins widely found in muscle tissues containing a heme (iron-containing porphyrin) prosthetic group in the center around which the remaining apoprotein folds. Although its physiological role has not been determined conclusively, the identification of its structure by Kendrew et al.[31] in 1958 led to the Nobel Prize in chemistry, together with Max Perutz, in 1968 for Kendrew's contribution to the development of X-ray crystallography. Lang studied the Mössbauer spectrum of Fe(III) myoglobin in an external magnetic field and reached a conclusion that the iron is high-spin Fe(III).[28] Frauenfelder calculated the energy barriers in the activation energy spectrum of the myoglobin–carbon monoxide binding process on the basis of the Mössbauer spectroscopy data.[32] MB generally serves as a model in the study of the protein structure dynamics. A typical spectrum of MB crystal often displays a superposition of the well known narrow Mössbauer absorption line and a broader one which is representative for a nucleus performing diffusive motions (see Figure 5.8), and this could be explained by diffusive motions in limited space.[33] In more recent research, Parak and Achterhold studied the protein dynamics in the timescale between 100 ns and 100 ps with Mössbauer spectroscopy. The spectrum as shown in Figure 5.9. could be fitted in a two-state model in which the molecule can be either trapped in a "rigid" conformational substrate or it performs strongly overdamped harmonic Brownian oscillations in a flexible state.[34]

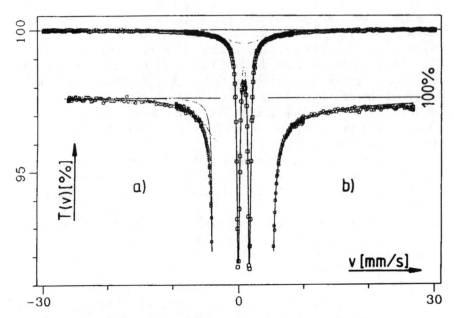

Figure 5.8 Mössbauer spectrum of deoxymyoglobin crystals at 235 K fitted with a doublet of narrow and broad Lorentzians.[33] (a) Amplification of this fit; (b) a least squares fit assuming a distribution of conformational substates. Reprinted with permission. © 1988 J.C. Baltzer A.G., Scientific Publishing Company.

Cytochrome is critical for life in a vast range of organisms since it plays an essential role in electron transport. Electron transport is an elementary part of oxidative phosphorylation that occurs in mitochondria or chloroplasts, and which is the basic process for energy generation. Cooke investigated the Mössbauer spectra of equine heart cytochrome c and calculated the molecular orbitals, concluding that the vacancy in the t_{2g} orbitals is the cause of the quadruple splitting of oxygenated cytochrome c.[35] G. Lang et al. reported that two hemes of cytochrome c oxidase in bovine heart are bound to the native protein in different ways: one of the components corresponds to low-spin Fe(III) oxidized cytochrome c and the other to high-spin Fe(III) oxidized cytochrome c peroxidase while the spectrum of reduced cytochrome c oxidase only contains the low-spin Fe(II) component corresponding to cytochrome c.[36]

Peroxidases are another family of heme-containing enzymes and are generally considered to function as defenders against pathogens. Most of them employ hydrogen peroxide (H_2O_2) as the electron acceptor to catalyze oxidative reactions while some exceptions also exist. In a study of horseradish peroxidase (HRP) and chloroperoxidase (CPO), two of the most thoroughly studied peroxidases, Schulz et al. analyzed tensors of the coupling between a porphyrin π-cation radical and the ferryl unit from the data of Mössbauer spectra and concluded that the coupling is weak in HRP but strong and antiparallel in CPO.[37,38] Additional studies by Antony et al. even showed that the type and

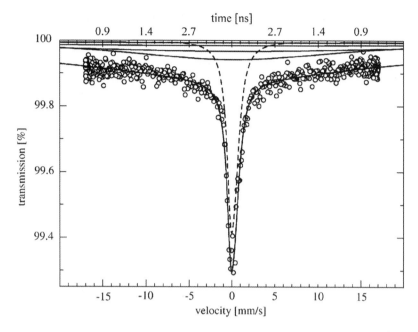

Figure 5.9 Mössbauer absorption spectrum of metmyoglobin at T = 295 K.[34] Thick line: fit of the Brownian oscillator model to the data. Dashed line: elastic nuclear resonance absorption. Thin lines: broad lines due to overdamped harmonic oscillations. Characteristic times given by the width of these broad Lorentzians are 0.87, 0.44, 0.29 and 0.22 ns. © 2005 Elsevier Ltd.

binding orientation of the proximal axial ligand may significantly define the type of magnetic coupling between the ferryl and the π-cation radical.[39]

Catalase, just as the meaning of its name, catalyzes the decomposition of hydrogen peroxide to water and oxygen in almost all living species. The research of this field was first reported by Karger, in whose study lyophilized catalase was measured by a Mössbauer spectrometer.[40] Maeda *et al.* concluded that the four hemeatins in bacterial catalase interact differently in pairs, coinciding with the different Fe–Fe interactomic distances in a research upon the Mössbauer spectrum of catalase in the range 2.1–195 K.[41]

5.3.2.2 Iron–Sulfur Proteins

Another type of frequently studied metalloproteins is the iron–sulfur (Fe–S) protein. It widely exists in all sorts of bacteria, plants, and animals, and plays a part in a wide range of biochemical processes. The best known function of iron–sulfur protein is its role in the redox process of electron transfer. The active center is an iron–sulfur cluster, through which several proteins link together with the coordination between the iron atoms in the cluster motif and sulfur atoms of cysteines in protein motifs. There are several primary types of the Fe–S clusters, i.e. [Fe$_2$S$_2$] cluster, [Fe$_4$S$_4$] cluster, [Fe$_3$S$_4$] cluster, and other more complex clusters.

There are several major types of iron–sulfur proteins, i.e. the rubredoxin (Rb), ferredoxin, Rieske protein, and the high-potential iron-protein (HiPIP).

Rubredoxin (Rb) acts as an electron carrier widely found in many sulfur-metabolizing organisms and is generally classified as iron–sulfur protein. Four cysteine sulfur atoms coordinate to a single iron atom in the iron–sulfur cluster. The Mössbauer data of the rubredoxin of *Clostridium pasteurianum* by Philipps *et al.* showed sextet lines corresponding to a high-spin Fe(III) in the oxidized state and a high-spin Fe(II) in the reduced state.[42]

Ferredoxins mediate electron transfer in many sorts of redox reactions in plant species and some protozoa as well. A number of interests have been put upon them due to their critical functions in photophosphorylation reactions of photosynthesis. Rao *et al.* believed that oxidized ferredoxin contains two high-spin Fe(III) (S = 5/2) atoms and the reduced one contains one high-spin Fe(III) atom and one high-spin Fe(II) atom as an overlap of two kind of magnetic splitting observed in the Mössbauer spectrum of reduced ferredoxin under a strong external magnetic field.[43] Miao *et al.* investigated Yah1p, an [Fe$_2$S$_2$]-containing ferredoxin located in the matrix of *Saccharomyces cerevisiae* mitochondria and functioning in the synthesis of Fe–S clusters and heme α-prosthetic groups. They performed Mössbauer spectroscopy upon mitochondria samples from Gal-YAH1 cells as shown in Figure 5.10.[44] The signal of [Fe$_4$S$_4$] cluster ($\delta = 0.45$ mm s^{-1}, $\Delta E_Q = 1.15$ mm s^{-1}) increased and that of high-spin Fe^{2+} ($\delta = 1.35$ mm s^{-1}, $\Delta E_Q = 3.3$ mm s^{-1}) decreased in the Yah1p-deplete/O$_2$ mitochondria as compared with the Yah1p-Replete/O$_2$ mitochondria; however, only the high-spin Fe^{2+} signal could be observed in the Yah1p-deplete/Ar mitochondria. If the Yah1p-deplete/O$_2$ mitochondria are studied in an 8.0 T parallel field, the spectrum is quite different. It is different from the spectrum of high-spin Fe^{3+} although it appears similar to it and the author suggested it could be attributed to a heterogeneous distribution of ferric nanoparticles.[44]

Rieske protein is essential in the respiratory electron transport in mitochondria, chloroplasts and some bacteria. Its structure is quite different from other proteins containing [Fe$_2$S$_2$] in so far as the iron atom in Rieske protein is coordinated to two cysteines and two histidines. Münck *et al.* investigated the EFG tensor of the ferrous site and found a fast relaxation between the two Zeeman levels of the S = 1/2 Kramers doublet leading to a diminished internal hyperfine field and a quasidiamagnetic spectrum in a filed of 4 T at 200 K, which is an indication of a magnetic splitting due to external field.[45]

HiPIP are a varied series of high redox potential [Fe$_4$S$_4$] containing iron–sulfur proteins which transport electrons in anaerobic metabolisms. Its naming is based on the high redox potential contrary to the negative redox potential of ferredoxin. Middleton *et al.* analyzed the Mössbauer spectra of the [Fe$_4$S$_4$]$^{3+}$ cluster and concluded that oxidized PiPIP from *Chromatium* consists of a Fe$^{2.5+}$Fe$^{2.5+}$ pair with $S_{12} = 9/2$ and a Fe^{3+}Fe^{3+} pair, whose S_{34} value may be 5.[46]

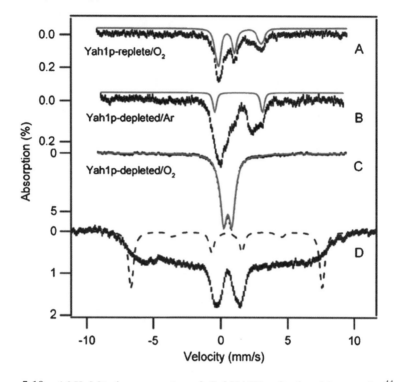

Figure 5.10 4.2 K Mössbauer spectra of Gal-YAH1 mitochondria samples.[44] (A) Yahp1-replete/O_2 mitochondria. The solid grey line outlines the sum of two major doublets, representing 70% of the iron. (B) Yah1p-depleted/Ar mitochondria. The doublet drawn above the data outlines a high-spin ferrous species with $\Delta E_Q = 3.55$ mm s^{-1} and $\delta = 1.33$ mm s^{-1}. (C) Yah1p-depleted/O_2 mitochondria. The solid line is a quadrupole doublet with $\Delta E_Q = 0.62$ mm s^{-1} and $\delta = 0.52$ mm s^{-1}. (D) Yah1p-replete/O_2 mitochondria of (C) studied in an 8.0 T parallel field. If the material of (C) represented magnetically isolated Fe^{3+} ions, a spectrum as outlined by the dashed black line would be expected; for the simulation we used $A/g_n/\beta = -21$ T for the ^{57}Fe magnetic hyperfine coupling constant (this A-value is appropriate for ferric phosphate). © 2008 American Chemical Society.

Layer et al. employed UV–visible absorption spectroscopy and Mössbauer spectroscopy to characterize *Escherichia coli* CYaY as an iron donor for the assembly of [Fe_2S_2] cluster in the scaffold IscU.[47] In this study, the author provided iron in the form of CyaY-Fe^{3+}. At time intervals, formation of iron–sulfur clusters on IscU was monitored by UV–visible absorption spectroscopy from the increase of characteristic absorption bands in the 300–700 nm region (Figure 5.11).[47] The absorption bands at 320, 410, and 456 nm and the shoulder at 510 nm present in the UV–visible spectrum at the end of the reaction are characteristic for [Fe_2S_2] clusters assembled in IscU. The author also provided iron in the form of CyaY-$^{57}Fe^{3+}$, and recorded the Mössbauer spectrum

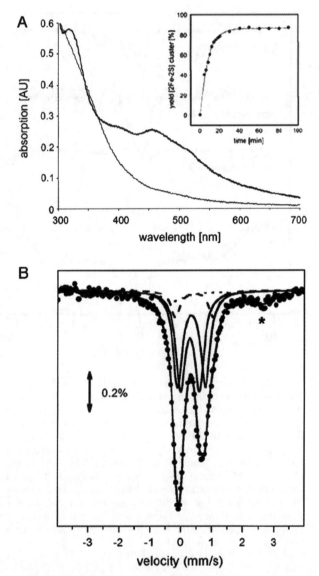

Figure 5.11 Reconstitution of [2Fe–2S] IscU with CyaY-Fe^{3+}, IscS, and cysteine.[47] (A) UV–visible analysis. ApoIscU (50 μM) was incubated with 1 μM IscS, 2 mM cysteine, and $CyaYFe^{3+}$ (150 mM iron) in 0.1 M Tris-HCl, pH 8, 50 mM KCl. The reaction was followed by UV–visible absorption spectroscopy. *Thin line*, initial spectrum; *bold line*, spectrum after 2 h of reaction. *Inset*, a plot of the yield (%) of [Fe–S] formation as a function of time was fitted to a rate equation for a first order process. (B) Zero field Mössbauer spectrum at 4.2 K of IscU reconstituted with CyaY-Fe^{3+}. The *continuous lines* are theoretical simulations assuming four different doublets as described in the text. *Solid lines*, doublets A and B; *dashed line*, doublet C. The *asterisk* indicates the high energy line of a doublet with parameters $d = 1.26$ mm s^{-1}, $\Delta E_Q = 2.80$ mm s^{-1} (*dotted line*) attributed to a high spin ferrous impurity. © 2006 The American Society for Biochemistry and Molecular Biology, Inc.

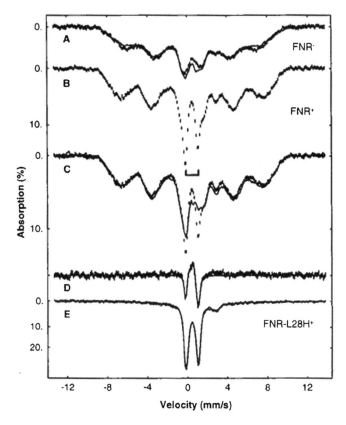

Figure 5.12 Mössbauer spectra of *E. coli* cells recorded at 1.5 K (A) and 4.2 K (E).[48] (A) FNR-cells, anaerobic (dash marks) and after exposure to air (solid line). (B) Anaerobic FNR+ cells. Bracket marks 4Fe-FNR. (C) Anaerobic FNR+ cells from B (dash marks) and FNR+ cells exposed for 15 min to air (solid line). (D) Difference spectrum of the anaerobic FNR+ sample minus the spectrum of the air-exposed FNR+ sample, by using the spectra shown in C. The solid line in D is a theoretical difference spectrum assuming 4Fe-FNR and 2Fe-FNR in 1:1 cluster ratio. (E) Whole-cell spectra of FNR-L28H mutant, anaerobic (dash marks), and exposed to air for 20 min (solid line). Reprinted with permission from PNAS. © 1998 The National Academy of Sciences.

(Figure 5.12).[47] The spectrum exhibits a strong asymmetric doublet at 0–1.0 mm s^{-1} and a rather broad weak peak at $\sim$2.5 mm s^{-1}. The latter is attributed to a high-spin ferrous impurity accounting for 10% of total iron. The strong asymmetric doublet was simulated assuming three doublets: A, B, and C. A and B account for the $[Fe_2S_2]^{2+}$ IscU protein with: $\delta = 0.30$ mm s^{-1}, $\Delta E_Q = 0.60$ mm s^{-1} for A, and $\delta = 0.34$ mm s^{-1} and $\Delta E_Q = 0.88$ mm s^{-1} for B. Doublet C was proposed to arise from $[Fe_4S_4]^{2+}$ (S = 0) species with $\delta = 0.44$ mm s^{-1}, $\Delta E_Q = 1.10$ mm s^{-1}.[47]

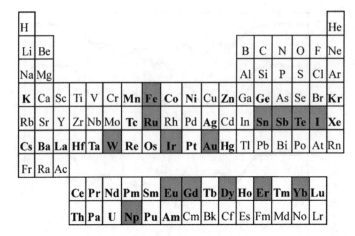

Figure 5.13 Elements in the periodic table with Mössbauer isotopes (in bold font). Those shaded in grey are commonly used elements.[17] © 2007 Royal Society of Chemistry.

In a study investigating $[Fe_4S_4]^{2+}$ containing transcription factor FNR (fumarate nitrate reduction) in *Escherichia coli*, Popescu et al. used ^{57}Fe Mössbauer spectroscopy and showed the $[Fe_4S_4]^{2+}$ cluster in FNR could converted to $[Fe_2S_2]^{2+}$ form *in vivo* on exposure to O_2.[48] As shown in Figure 5.13,[48] the FNR + *E. coli* cells showed reduced 4Fe-FNR signal when exposed for 15 min to air, and such a phenomenon was not observed in the FNR cells and oxygen-stable FNR-L28H mutant cells.

5.3.2.3 Other Types of Iron-containing Proteins

Besides heme and iron–sulfur proteins, other types of metalloproteins could also been investigated by Mössbauer spectroscopy. For instance, ferritin, transferirin, ovotransferrin, lactoferrin, siderochromes, and hemerythrin.

Ferritin is a globular protein complex broadly existing in prokaryotes and eukaryotes with a function to collect free iron in organisms since free iron may catalyze the formation of undesired reactive oxygen species and thus cause harm to organisms. It has been recognized as an intracellular iron storage protein and is called apoferritin while no iron is bound to it. Boas and Window observed an isomer shift consistent with octahedral coordination while the observed effective magnetic field was an indication of tetrahedral coordination.[49]

Transferrin literally means a delivery protein. It is found in plasma and acts as a vehicle to intelligently adjust the distribution of iron in the body with the help of its pH-sensitive affinity. Similarly, the transferring without iron is called apotransferrin in the same way. There are two binding sites for this protein and intensive studies in this respect have been done. Early studies based on ESR

Mössbauer Spectroscopy

and Fe activity measurement showed they are different, while later works based on Mössbauer spectroscopy showed opposite results.[27]

Ovotransferrin, or conalbumin, is found in egg white and functions similarly to transferrin. Aisen *et al.* observed that the two iron-binding sites are different in a Mössbauer spectrum at 4.2 K in an external magnetic field of 550×10^3 A m^{-1} (550 Oe).[50]

Lactoferrin is quite similar to transferring and has been first purified from milk of mammals. Ladriere *et al.* detected distinct paramagnetic hyperfine structure. The result showing that it was not possible to discriminate between the Mössbauer spectrum of lactoferrin and that of transferrin at 77 K indicated that the two binding sites are equivalent.[51]

5.3.3 Mössbauer Spectroscopy for the Study of Elemental Speciation

Besides the application on iron-containing proteins, Mössbauer spectroscopy has also been applied to specify the composition of samples. Since the resonance absorption of a certain element occurs independently, this approach is quantitatively applicable even in the presence of other elements or other isotopes. Meanwhile, the Mössbauer parameters can change with the chemical environment of the nucleus; therefore, useful qualitative information can be obtained by Mössbauer measurement. Currently, about 90 isotopes with 110 transitions have been applied in Mössbauer spectroscopy as shown in Figure 5.13.[17]

5.3.3.1 Quantitative Speciation

Similar to other quantitative analytical approaches, the area under the spectral lines is proportional to the number of Mössbauer atoms detected. Practically, in one approach, a calibration graph of the area value of a series of external analogous samples with known concentrations would be constructed and compared with that of the unknown samples. In another approach a graph of the area value of unknown sample added with analogous samples with a series of known concentration can be drawn for calibration. Both approaches are common methods employed in instrumental analysis.

Quantitative analysis could be used to measure the concentration of the element component in a sample as well as to identify the oxidation state and phase. In a study by Korecz *et al.*, the Mössbauer spectrum is used to follow, quantitatively, the course of incorporation of iron into the corundum lattice. The composition of the iron aluminium oxide formed was also found to be $FeAl_5O_{18}$, which means an aluminium oxide polymer with an iron atom in every five aluminium atoms.[52]

Bonchev *et al.* studied the abdomen of 50 ants by smearing them with an absorbent of tin oxide. These authors then analyzed the line width caused by the movement of the living ants and calculated the rate of motion in the ant nest.[53]

5.3.3.2 Qualitative Speciation

Mössbauer atoms under certain conditions can show certain types of profile in observed spectra. In specifying some unknown samples, the spectra can be compared with those of known samples, and, hence, the components of the unknown samples might be deduced.

Conventionally, a new substance would be added as a standard into a sample. If the substance is identical with a constituent of the sample, an increase of intensity without any other changes can be observed; while if the substance matches nothing in the sample, new lines would appear in spectrum. This approach sometimes may be frustrated when the spectra of two different substances are similar. To eliminate such problems, an external magnetic field would be added to complicate the spectrum, in order to reduce the possible interference caused by the spectra of different substances.

5.4 Other Techniques for Hyperfine Interaction Studies

There are some other nuclear-related techniques, like nuclear resonance spectroscopy (NMR), electron paramagnetic resonance (EPR), and resonance Raman spectroscopy for studying the structures of metalloproteins. This section will briefly introduce their application in metalloproteins from different aspects.

5.4.1 Nuclear Resonance Spectroscopy

Nuclear resonance spectroscopy is based on the nuclear resonance effect, which can occur only in nuclei containing odd numbers of protons and/or neutrons in so far as such nuclei have intrinsic magnetic moments. In other words, NMR is theoretically applicable for nuclei with the *nuclear spin number* I larger than 0; however, practically only nuclei with $I = 1/2$ are those most used for NMR study since those with $I > 1/2$ possess electric quadrupole moments which broaden the line width and reduce the resolution of the spectra.

As a result of the Zeeman effect, the degenerated energy levels with magnetic spin number $m = \pm 1/2$ (or $I = 1/2$) split into two separate levels with different energy in the presence of a static magnetic field B_0. The energy difference between the two levels is

$$\Delta E = \hbar \gamma B_0, \tag{5.34}$$

where $\hbar$, h-bar, is the Plank's reduced constant, γ is the gyromagnetic ratio (the ratio of a nucleus's magnetic dipole moment to its angular momentum). When electromagnetic radiation with the frequency

$$\nu = \frac{\Delta E}{h} = \frac{\gamma B_c}{2\pi}. \tag{5.35}$$

is applied to the nuclei, NMR resonance occurs and absorption can be observed. Since the energy gap between the two levels is perturbed by the population of surrounding electrons, i.e. the matched frequency of magnetic radiation shifts due to the shielding effect; therefore, a spectroscopy is available in so far as the correlation between the information of the chemical environment of a nucleus and the characteristic resonance frequency can be obtained. This frequency shift due to chemical environment changes is the so-called *chemical shift* δ, the basic parameter for NMR spectroscopy. In experiments, one further phenomenon should also be noted. The corresponding lines of a certain sample may even split into more lines due to spin–spin coupling, which results from interactions between a measured nucleus and surrounding nuclei. A constant, called the *coupling constant* J, is consequently introduced to denote the strength of coupling.

Among applications of NMR spectroscopy in metalloprotein studies, the situation in studying the paramagnetic systems is quite different from that in studying diamagnetic systems. One initial challenge comes from the broadening due to paramagnetism, which renders the signal-to-noise ratio of paramagnetic system spectrum lower then expected. Banci *et al.* proposed some solutions and solved the three-dimensional structure in solution of the paramagnetic high-potential iron–sulfur protein I from *Ectothiorhodospira halophila* through NMR.[54] Since then, intensive studies concerning the structure of paramagnetic metalloproteins have been done and the structure can now be successfully determined by NMR coupled with the help of theoretical calculation.

The parameters in NMR and Mössbauer spectroscopy are related. Polam *et al.* investigated the valence electron cloud asymmetry of the ^{57}Fe(II) complex of tetramesitylporphyrin and octaethylporphyrin by NMR and Mössbauer spectroscopy. A strong correlation between the chemical shift and electric quadrupole splitting was observed.[55] There is a growing awareness that both shielding and the EFG at the iron nucleus are local phenomena and should require a detailed understanding of the remaining structure in recent years and more and more studies with reference to computations (especially those employing density function theory) on NMR data to studies the structure of iron-containing metalloproteins have been published.[56]

Besides direct probing, another approach utilized to study metalloproteins is the use of nuclear resonance of metal nuclei to probe the metal binding sites in proteins and many such investigations were conducted upon Zn(II), Ca (II), and Cu(I) metalloproteins.[57]

Mao *et al.* studied the NMR spectroscopic shifts of ^{1}H, ^{13}C, ^{15}N, and ^{19}F in paramagnetic metalloprotein and metalloporphyrin systems. Using DFT methods, they obtained very good correlations between experimental and theoretical results.[58]

5.4.2 Electron Paramagnetic Resonance

Electron paramagnetic resonance (EPR) is a phenomenon analogous to nuclear magnetic resonance. It occurs only when there are some unpaired electrons, i.e.

the spin quantum number $s = 1/2$ is achieved. Due to the Zeeman effect, when an external magnetic field B_0 is imposed the energy level $s = 1/2$ splits into two with $m_s = +1/2$ and $m_s = -1/2$ and the energy separation between the two levels is

$$\Delta E = g_e \mu_B B_0, \qquad (5.36)$$

where g_e is the electron's *g-factor* (or *Landé g-factor*), and μ_B is the Bohr magneton. When absorption or emission of electromagnetic radiation of frequency

$$\nu = \frac{\Delta E}{h} = \frac{g_e \mu_B B_c}{h}. \qquad (5.37)$$

occurs, the unpaired electron moves between the two energy levels. According to Maxwell–Boltzmann distribution, the number of atoms in the lower state is much larger than that in the higher state and hence it is absorption that is commonly monitored and converted into a spectrum. The measurement is often conducted by altering the magnetic field while keeping the frequency of electromagnetic radiation constant. In a spectrum, the line intensity, width, g-factor, and hyperfine coupling are crucial information which can be utilized to identify free radicals and paramagnetic centers as well as to study the kinetic process coupled with them.

By noting that metal centers in metalloproteins can be prepared in a redox state in which their ground state is paramagnetic, data provided by EPR have therefore played a critical role in the elucidation of structure and function of these centers.[59] The application of EPR in studying metalloproteins can be traced back to 1959 when Bray *et al.* first studied the presence of metal centers in biological systems at low temperature.[60] Since that time, a very large number of EPR study methods have been developed, e.g. pulsed EPR, and high-field high-frequency EPR. Since the orientation of paramagnetic centers with respect to the applied field, B_0, can change the resonance lines, one may sometimes wish to prepare a single crystal sample which is rarely possible or convenient. Actually, a common method is to prepare a frozen solution with randomly oriented paramagnetic centers, which display typical features at some peculiar B values. The values can be correlated to certain values of the local g tensor. In the interpretation of the EPR spectra, the spectral features must be understood in terms of molecular information. The theoretical calculation by quantum chemistry is thus indispensable. Nevertheless, it should be kept in mind that ordinary approaches of quantum chemistry are not applicable for EPR calculations because the transition between energy levels is quite small and many small interactions must be taken into account. As a result, EPR spectra should be dealt with by a so-called phenomenological spin-Hamiltonian which only contains adjustable parameters fitted to experiments.[61] This is especially crucial for studying metalloproteins, the interpretation of whose EPR spectrum is challenging due to the complexity of the surrounding environment and its interactions.

Practically, EPR spectroscopy can be used with Mössbauer spectroscopy to investigate the reaction mechanisms of metalloproteins. Garcia-Serres *et al.* employed both EPR spectroscopy and Mössbauer spectroscopy to characterize the electronic states and reaction pathways of reactive oxygen intermediates generated by 77 K radiolytic cryoreduction and subsequent annealing of oxyheme oxygenase and oxy-myoglobin. Their study has directly demonstrated the idea that ferric-hydroperoxo hemes are indeed the precursors of the reactive ferryl intermediates.[62]

ENDOR (electron nuclear double resonance) spectroscopy is a special EPR technique that is very useful for learning about the structure of paramagnetic molecules, as well as providing information about the distances and orientations of atoms surrounding paramagnetic centers. Gurbiel *et al.* have employed continuous wave electron nuclear double resonance (CW ENDOR) to study the geometry of [2Fe–2S] Rieske-type clusters and rationalize some of the properties of these novel centers. They have concluded that all the geometry of the cluster is essentially the same for all Rieske and Rieske-type proteins.[63]

5.4.3 Resonance Raman Spectroscopy

Resonance Raman (RR) spectroscopy, as with conventional Raman spectroscopy, is a spectroscopic technique helpful for studying characteristic vibrational, rotational, and other low-frequency modes of matter. However, this method differs from conventional Raman spectroscopy in the excitation wavelength, which is tuned to be near an electronic transition and the intensity of vibration associated with the transition increases tremendously and the remaining Raman signals are overwhelmed. As this approach can be used to study some certain vibrational modes without interruption of other signals, it is especially useful for determining the structure or components of complex samples, e.g. biomolecules. Moreover, the detection limits is about 10^6 times larger than conventional Raman spectroscopy and, hence, substances of much lower concentrations in complex systems can be studied.

Metalloproteins constitute a class of special proteins containing coordination complexes in which there are intensive electronic absorption bands in the near UV or visible range, while in common proteins only deep UV absorption is available. Consequently, it is theoretically possible to attain the information of a complex exclusive of that of the other parts of the proteins when selectively exciting the metalloproteins. In exploring the resonance Raman spectroscopy of an iron complex, the charge transfer transition and π–π^* transition for resonance Raman studies is utilized. Similar to Mössbauer spectroscopy, RR spectroscopy is also used for exploring the process by which ligands (e.g. CO, NO, O_2) bind to heme proteins. Recently, Ibrahim *et al.* investigated the vibrational modes of proteins, porphyrin rings and

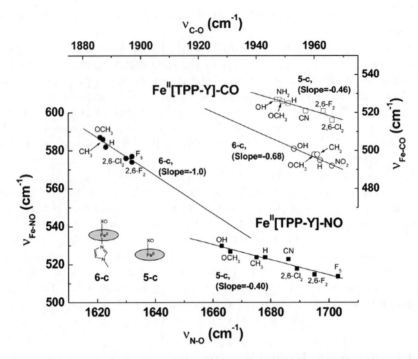

Figure 5.14 FeXO backbonding correlations for 5-c and 6-c (N-MeIm axial ligand) NO and CO adducts of Fe(II)TPP-Y with the indicated phenyl substituents, Y, in organic solvents (CH_2Cl_2, DMF, Bz).[64] The main difference between the CO and NO plots is that the 6-c line lies below the 5-c line for CO but above the 5-c line for NO. Adding an axial ligand depresses the vFeC frequency but elevates the vFeN frequency. The reason for that is mainly thought to be that the Fe-N-O angle is bent, the Fe-N stretching coordinate mixes with the Fe-N-O bending coordinate, and this mixing increases when an axial ligand is bound. © 2006 American Chemical Society.

heme-bound ligands in RR spectra and observed differential sensing of proteins influenced by NO and CO vibrations in the heme adduct. Based on DFT calculation as shown in Figure 5.14[64] they proposed the deviation results from closing of the Fe-N-O angle due to a shift in the valence isomer equilibrium towards the Fe(III)(NO$^-$) form, an effect absent in CO adducts. This study may give an insight into how the heme sensor proteins regulate the avid binding of CO, NO, and O_2 to heme.[64]

When exploring dynamic structures of metalloproteins which may be challenging for conventional methods, i.e. X-ray crystallography, theoretical calculation, or EPR, RR combined with mutagenesis may be appropriate for the investigation of changes in properties which occur as a result of changes in protein sequences. In this regard, a brief review of heme peroxidase is now available.[65]

5.5 Limitations and Conclusions

Mössbauer spectroscopy has the advantages of high resolution, great anti-interference abilities, non-destructiveness, simple facility and manipulation, and no requirement for specific purification and crystallization of samples. As a result, it is quite suitable for the preliminary investigation of bulk or surface analysis.

However, although about 90 isotopes are Mössbauer active isotopes, only a few can be utilized for Mössbauer spectroscopic studies, and these are mostly restricted to iron compounds, tin compounds, ruthenium compounds, and antimony compounds.

Some problems may arise from the limitation as to the state the sample. Since the Mössbauer effect can only take place in the solid state, measurements can only be carried out on samples in the solid state, which renders sample preparation inconvenient and prevents some unfrozen samples from being measured.

A further restriction is the detection limit. A reliable Mössbauer spectrum for studying a certain component in a mixture can be obtained only when its concentration is no less than 5–10% of the overall content of Mössbauer atoms of interest.[27] Another limitation is the sensitivity. The correlation between the concentration of the Mössbauer atoms and the area under the spectral lines is ambiguous in some cases.[27] When Mössbauer spectroscopy is used to study the analytical or conformational information of a certain sample, the above limitations should be taken into account and undesired factors which may affect results should be ruled out. More specifically, Mössbauer spectroscopy is more convenient in cases where either a complex system needs to be roughly studied in a short period, or the sample needs to be retained without destruction throughout analysis, or where other approaches are too costly for analysis.

Acknowledgments

The authors acknowledge the financial support by the Ministry of Science and Technology of China as the National Basic Research Programs (2006CB705603 and 2010CB934004), the Natural Science Foundation of China (10975040), the NSFC/RGC Joint Research Scheme (20931160430) and the Knowledge Innovation Program of the Chinese Academy of Sciences.

References

1. R. Mössbauer, *Zeitschrift f Physik A Hadrons and Nuclei*, 1958, **151**, 124–143.
2. P. Gütlich, R. Link and A. Trautwein, *Mössbauer Spectroscopy and Transition Metal Chemistry: Fundamentals and Application*, Springer-Verlag, Berlin, 1978.
3. V. Weisskopf and E. Wigner, *Zeitschrift f Physik A Hadrons and Nuclei*, 1930, **63**, 54–73.

4. K. G. Malmfors, *Ark Fys*, 1953, **6**, 49–56.
5. V. Schünemann and H. Winkler, *Rep. Prog. Phys.*, 2000, **63**, 263–353.
6. D. A. Shirley, *Rev. Mod. Phys.*, 1964, **36**, 339–351.
7. W. Dunham, L. Harding and R. Sands, *Eur. J. Biochem.*, 1993, **214**, 1–8.
8. G. Klingelhöfer, P. Held, R. Teucher, F. Schlichting, J. Foh and E. Kankeleit, *Hyperfine Interact.*, 1995, **95**, 305–339.
9. G. Klingelhöfer, B. Bernhardt, J. Foh, U. Bonnes, D. Rodionov, P. A. De Souza, C. Schroder, R. Gellert, S. Kane, P. Gutlich and E. Kankeleit, *Hyperfine Interact.*, 2002, **144**, 371–379.
10. R. Rieder, R. Gellert, J. Bruckner, G. Klingelhöfer, G. Dreibus, A. Yen and S. W. Squyres, *J. Geophys. Res. Planet*, 2003, **108**, 8066–8079.
11. G. Klingelhöfer, *Hyperfine Interact.*, 1998, **113**, 369–374.
12. *Introduction to Mössbauer Spectroscopy: Part I*, Royal Society of Chemistry, London, 2009, http://www.rsc.org/Membership/Networking/InterestGroups/MossbauerSpect/Intropart1.asp:.
13. L. Walker, G. Wertheim and V. Jaccarino, *Phys. Rev. Lett.*, 1961, **6**, 98–101.
14. J. K. Lees and P. A. Flinn, *J. Chem. Phys.*, 1968, **48**, 882.
15. P. Carretta and A. Lascialfari, *NMR-MRI, USR and Mössbauer Spectroscopies in Molecular Magnets*, Springer Verlag, Berlin, 2007.
16. J. van Dongen Torman, R. Jagannathan and J. M. Trooster, *Hyperfine Interact.*, 1975, **1**, 135–144.
17. Y. Gao, C. Chen and Z. Chai, *J. Anal. At. Spectrom.*, 2007, **22**, 856–866.
18. A. Kamnev, L. Antonyuk, L. Kulikov and Y. Perfiliev, *BioMetals*, 2004, **17**, 457–466.
19. W. Kerler, W. Neuwirth, E. Fluck, P. Kuhn and B. Zimmermann, *Zeitschrift f Physik A Hadrons and Nuclei*, 1963, **173**, 321–346.
20. Y. Hazony and S. Bukshpan, *Rev. Mod. Phys.*, 1964, **36**, 360.
21. A. Vértes, K. Burger and I. Dézsi, in *Proceedings of the Fifth Hungarian Complex Chemistry Colloquium, Siófok, 1969*.
22. A. Vértes, in *Proceedings of the Chemical Society Meeting, Southampton, 1969*.
23. A. Vértes, *Acta Chim. Hung.*, 1970, **63**, 9.
24. K. Burger and A. Vértes, in *Proceedings of the 2nd Conference on Coordination Chemistry, Smolenice–Bratislava*.
25. A. Upadhyay, A. Hooper and M. Hendrich, *J. Am. Chem. Soc.*, 2006, **128**, 4330–4337.
26. P. G. Renstein and J. B. Swan, in *Proceedings of the International Biophysics Congress*, 1961, p. 147. (This item is an oral talk in a meeting, more information can be obtained in http://www.mtakpa.hu/kpa/kereso/slist.php?inited = 1&co_on = 1&ty_on = 1&la_on = &if_on = 1&st_on = 1 &url_on = 1&cite_type = 2&orderby = -7&Scientific = &top10 = &lang = 1 &location = kpa&debug = &stn = 1&AuthorID = 2028540&DocumentID = #).
27. A. Vértes, L. Korecz and K. Burger, in *Mossbauer Spectroscopy*, Elsevier, Amsterdam, 1979, pp. 344–386.

28. G. Lang, *Mossbauer Spectroscopy and its Applications,* International Atomic Energy Agency, Vienna, 1972.
29. S. Ofer, S. Cohen, E. Bauminger and E. Rachmilewitz, *J. Physique Colloques,* 1976, **37**, C6, 199–202.
30. A. Trautwein, R. Zimmermann and F. Harris, *Theoretical Chemistry Accounts: Theory, Computation, and Modeling, Theor. Chim. Acta,* 1975, **37**, 89–104.
31. J. C. Kendrew, G. Bodo, H. M. Dintzis, R. G. Parrish, H. Wyckoff and D. C. Phillips, *Nature,* 1958, **181**, 662–666.
32. R. H. Austin, K. Beeson, L. Eisenstein, H. Frauenfelder and I. C. Gunsalus, *Phys. Rev. Lett.,* 1975, **34**, 845.
33. F. Parak, J. Heidemeier and G. U. Nienhaus, *Hyperfine Interact.,* 1988, **40**, 147–157.
34. F. G. Parak and K. Achterhold, *J. Phys. Chem. Solids,* 2005, **66**, 2257–2262.
35. R. Cooke, *Enzyme Studies with the Mössbauer Effect,* New England Nuclear Corporation, Boston, 1968.
36. G. Lang, S. J. Lippard and S. Rosen, *Biochim. Biophys. Acta,* 1974, **336**, 6–14.
37. C. E. Schulz, P. W. Devaney, H. Winkler, P. G. Debrunner, N. Doan, R. Chiang, R. Rutter and L. P. Hager, *FEBS Lett.,* 1979, **103**, 102–105.
38. C. Schulz, R. Rutter, J. Sage, P. Debrunner and L. Hager, *Biochemistry,* 1984, **23**, 4743–4754.
39. J. Antony, M. Grodzicki and A. Trautwein, *J. Phys. Chem. A,* 1997, **101**, 2692–2701.
40. W. Karger, *Berich Bunsen Gesell,* 1964, **68**, 793.
41. Y. Maeda, A. Trautwein, U. Gonser, K. Yoshida, K. Kikuchi-Torii, T. Homma and Y. Ogura, *Biochim. Biophys. Acta,* 1973, **303**, 230–236.
42. W. Phillips, M. Poe, J. Weiher, C. McDonald and W. Lovenberg, *Nature,* 1970, **227**, 574–577.
43. K. Rao, R. Cammack, D. Hall and C. Johnson, *Biochem. J.,* 1971, **122**, 257.
44. R. Miao, M. Martinho, J. G. Morales, H. Kim, E. A. Ellis, R. Lill, M. P. Hendrich, E. Munck and P. A. Lindahl, *Biochemistry,* 2008, **47**, 9888–9899.
45. E. Munck, J. C. M. Tsibris, I. C. Gunsalus and P. G. Debrunne, *Biochemistry,* 1972, **11**, 855–863.
46. P. Middleton, D. Dickson, C. Johnson and J. Rush, *Eur. J. Biochem.,* 1980, **104**, 289–296.
47. G. Layer, S. Ollagnier-de Choudens, Y. Sanakis and M. Fontecave, *J. Biol. Chem.,* 2006, **281**, 16256–16263.
48. C. V. Popescu, D. M. Bates, H. Beinert, E. Munck and P. J. Kiley, *Proc. Natl. Acad. Sci. U. S. A.,* 1998, **95**, 13431–13435.
49. J. Boas and B. Window, *Aust. J. Phys,* 1966, **19**, 573.
50. P. Aisen, G. Lang and R. Woodworth, *J. Biol. Chem.,* 1973, **248**, 649–653.
51. J. Ladriere, R. Coussement and B. Theuwissen, *J. Physique,* 1974, **35**, C6, 351–353.

52. L. Korecz, I. Kurucz, E. Pappmoln, K. Burger, E. Pungor and G. Menczel, *Talanta*, 1972, **19**, 1599–1604.
53. T. Bonchev, I. Vassilev, T. Sapundzh and M. Evtimov, *Nature*, 1968, **217**, 96–98.
54. L. Banci, I. Bertini, L. Eltis, I. Felli, D. Kastrau, C. Luchinat, M. Piccioli, R. Pierattelli and M. Smith, *Eur. J. Biochem.*, 1994, **225**, 715–725.
55. J. R. Polam, J. L. Wright, K. A. Christensen, F. A. Walker, H. Flint, H. Winkler, M. Grodzicki and A. X. Trautwein, *J. Am. Chem. Soc.*, 1996, **118**, 5272–5276.
56. N. Godbout, R. Havlin, R. Salzmann, P. Debrunner and E. Oldfield, *J. Phys. Chem. A*, 1998, **102**, 2342–2350.
57. G. D. Pountney and I. Armitage, *Biochem. Cell Biol.*, 1998, **76**, 223–234.
58. J. H. Mao, Y. Zhang and E. Oldfield, *J. Am. Chem. Soc.*, 2002, **124**, 13911–13920.
59. C. More, V. Belle, M. Asso, A. Fournel, G. Roger, B. Guigliarelli and P. Bertrand, *Biospectroscopy*, 1999, **5**, S3–S18.
60. R. Bray, B. Malmström and T. Vänngard, *Biochem. J.*, 1959, **73**, 193–197.
61. F. Neese, *Curr. Opin. Chem. Biol.*, 2003, **7**, 125–135.
62. R. Garcia-Serres, R. M. Davydov, T. Matsui, M. Ikeda-Saito, B. M. Hoffman and B. H. Huynh, *J. Am. Chem. Soc.*, 2007, **129**, 1402–1412.
63. R. J. Gurbiel, P. E. Doan, G. T. Gassner, T. J. Macke, D. A. Case, T. Ohnishi, J. A. Fee, D. P. Ballou and B. M. Hoffman, *Biochemistry*, 1996, **35**, 7834–7845.
64. M. Ibrahim, C. Xu and T. Spiro, *J. Am. Chem. Soc.*, 2006, **128**, 16834–16845.
65. G. Smulevich, A. Feis and B. D. Howes, *Acc. Chem. Res.*, 2005, **38**, 433–440.

CHAPTER 6
X-ray Absorption Spectroscopy

YU-FENG LI[a] AND CHUNYING CHEN[a,b,*]

[a] CAS Key Laboratory of Nuclear Analytical Techniques, Institute of High Energy Physics, Chinese Academy of Sciences, Beijing 100049, China
[b] CAS, Key Laboratory for Biological, Effects of Nanomaterials & Nanosafety, National Center for Nanoscience and Technology of China & Institute of High Energy Physics, Chinese Academy of Sciences, Beijing 100190, China

6.1 Introduction

X-ray techniques have played a crucial role in finding the answers for many questions related to metallomics and metalloproteomics. The photons in the X-ray wavelength range are a form of electromagnetic radiation produced by the ejection of an inner orbital electron and subsequent transition of atomic orbital electrons from states of high to low energy. When a beam of X-ray photons falls onto a given specimen three basic phenomena may occur; namely, absorption, scatter, or fluorescence. The three basic phenomena form the basis of three important X-ray analytical methods: the absorption technique, which is the basis of X-ray absorption spectrometry; the scattering effect, which is the basis of X-ray diffraction; and the fluorescence effect, which is the basis of X-ray fluorescence spectrometry (XRF).[1] Detailed information on the application of XRF in metallomics and metalloproteomics is described in Chapter 3.

Nowadays, protein crystallography is perhaps the most powerful tool in the field of metallomics, especially metalloproteomics, for analysis of protein structures using X-ray diffraction techniques. At many synchrotron radiation facilities all over the world, dedicated beamlines for this purpose are already

---

*Corresponding author: Email: chenchy@nanoctr.cn and chenchy@ihep.ac.cn; Tel: +86-10-82545560; Fax: +86-10-62656765.

Nuclear Analytical Techniques for Metallomics and Metalloproteomics
Edited by Chunying Chen, Zhifang Chai and Yuxi Gao
© Royal Society of Chemistry 2010
Published by the Royal Society of Chemistry, www.rsc.org

operational. Protein crystallography can determine the macromolecular three-dimensional structure at a resolution of 0.15–2 nm; however, the requirement for single crystals will severely limit its application to numerous biological samples. Readers can refer to Chapters 7 and 8 for more information.

This chapter will focus on X-ray absorption spectroscopy (XAS) including X-ray absorption near edge structure (XANES) and extended X-ray absorption fine structure (EXAFS). This powerful technique for probing the local structure around almost any specific element in the periodic table (except the lightest) gives information on the number and chemical identities of near neighbors and the average interatomic distances up to 0.1–1 pm without the requirement for preparation of crystalline samples. XANES, the low kinetic energy range (5–150 eV), can provide information about vacant orbitals, electronic configuration, and site symmetry of the absorbing atom. The absolute position of the edge contains information about the oxidation state of the absorbing atom. In the near edge region, multiple scattering events dominate. Theoretical multiple scattering calculations are compared with experimental XANES spectra in order to determine the geometrical arrangement of the atoms surrounding the absorbing atom. EXAFS, 150–1000 eV above the absorption edge, can provide accurate distances and identities of the nearby atoms. For example, Hg in human hair and blood samples from long-term mercury-exposed populations have been studied by EXAFS and structural information, such as bond distances and coordination numbers of Hg, has been obtained.[2] Further, EXAFS can give a refinement of the structure derived from X-ray crystallography since EXAFS has a higher spatial resolution than X-ray crystallography, especially for local structures.[3–5] Apart from the information contained in this chapter, readers are expected to refer to a number of excellent reviews (and references therein) dealing with the principles of XAS and its applications in metallomics and metalloproteomics.[6–10]

6.2 Basics of X-Ray Absorption Spectroscopy

6.2.1 X-ray Absorption and Fluorescence

XAS is the measurement of the X-ray absorption coefficient of a material. X-rays of a narrow energy resolution are shone on the sample and the incident and transmitted X-ray intensity is recorded as the incident X-ray energy is incremented. When a beam of monochromatic X-rays passes through matter, it loses its intensity due to interaction with the atoms in the material. The intensity decreases exponentially with incident distance if the material is homogeneous, and after transmission, according to Beer's law, the intensity is:

$$I = I_0 e^{-\mu t} \tag{6.1}$$

here I_0 is the X-ray intensity incident on a sample, μ is the absorption coefficient, t is the sample thickness, and I is the intensity transmitted through the sample.

At most X-ray energies, the absorption coefficient μ is a smooth function of energy, with a value that depends on the sample density ρ, the atomic number Z, atomic mass A, and the X-ray energy E roughly as:

$$\mu \approx \frac{\rho Z^4}{A E^3} \quad (6.2)$$

The strong dependence of μ on both Z and E is a fundamental property of X-rays, and is the key to interpreting why X-ray absorption is so useful for medical and other imaging techniques, including X-ray computed tomography. Due to the Z^4 dependence, the absorption coefficient μ for O, Ca, Fe, and Pb differs greatly – spanning several orders of magnitude – so that good contrast between different materials can be achieved for nearly any sample thickness and concentrations by adjusting the X-ray energy.

The absorption coefficient μ generally decreases with increasing X-ray energy; however, a sudden increase, known as the absorption edge, occurs when the X-ray energy is sufficient to overcome the binding energy of a core electron in an element (Figure 6.1). The sudden increase is explained by the fact that at the corresponding energy position the photon energy is sufficient to ionize a more tightly bound core electron. The absorption edge structure often consists of discrete absorption bands superimposed on the steeply rising continuum absorption edge. These discrete absorption bands are caused by transitions of core electrons to discrete bound valence levels. The absorption edges

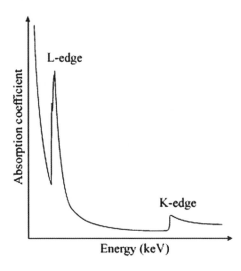

Figure 6.1 The absorption coefficient decreases smoothly with higher energy, except for special photon energies. When the X-ray energy is sufficient to overcome the binding energy of a core electron in an element, the absorption coefficient will increase sharply.

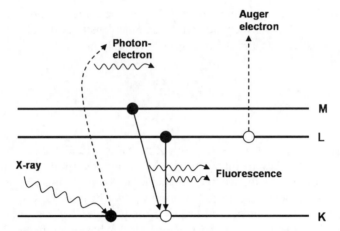

Figure 6.2 X-ray fluorescence and the Auger effect of the decay of the excited state after X-ray absorption.

that are of most interest are the K edge (1s–3p), followed by the three L edges: L_1 edge (2s–5p), L_2 edge ($2p_{1/2}$–$5d_{3/2}$) and L_3 edge ($2p_{3/2}$–$5d_{5/2}$).[11] These edges are element specific and shift to higher energies when the atomic number increases. Since the core levels depend on the element and its chemical environment, they also show chemical specificity. X-ray absorption spectroscopy is therefore not only an element specific technique, but it is also sensitive to the immediate environment of the absorbing atom.

The atom is in an excited state after an absorption event, with one of the core electron levels left empty (a so-called core hole), and a photoelectron is emitted. The excited state will eventually decay typically within a few femtoseconds after the absorption event. Two main processes occur for the decay of the excited atomic state following an X-ray absorption event (Figure 6.2). The first is X-ray fluorescence, in which a higher energy core-level electron fills the deeper core hole. The fluorescence energies emitted in this way are characteristic of the atom, and can be used to identify the atoms in a system and quantify their concentrations by the fluorescence intensity. For example, an L shell electron dropping into the K level gives the Kα fluorescence line.

The second process for de-excitation of the core hole is the Auger effect, in which an electron drops from a higher electron level and a second electron is emitted into the continuum (and possibly even out of the sample). In the hard X-ray regime (>2 keV), X-ray fluorescence is more likely to occur than Auger emission, but for lower energy X-ray absorption, Auger processes dominate. Both processes are proportional to the rate of absorption, therefore, either of the two processes can be used to measure the absorption coefficient μ; however, fluorescence is somewhat more commonly used.

Typical X-ray absorption spectra generally have three regions (Figure 6.3), including the pre-edge, X-ray absorption near-edge structure (XANES) or near edge X-ray absorption fine structure (NEXAFS), including the first 50–150 eV

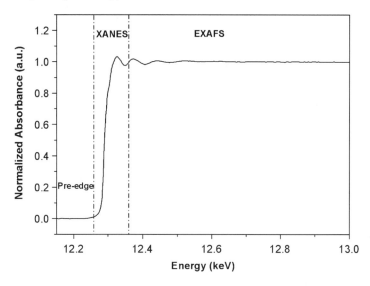

Figure 6.3 Mercury L$_{III}$-edge spectrum showing pre-edge (below 12 250 eV), X-ray absorption near-edge structure (XANES) (12 250–12 350 eV) and extended X-ray absorption fine structure (EXAFS) (12 350–13 000 eV) regions.

after the absorption edge and extended X-ray absorption fine structure (EXAFS), which can be detected in suitable samples till 1000 eV and more behind the edge, respectively.[12] XANES is also known as NEXAFS when applied to surface and molecular science.

In the pre-edge region the incident photon energy is below the ionization threshold. The edge inflection point can be used to estimate approximately the core binding energy E_0. Not only the position of an edge but also the shape and position of the peaks in the pre-edge region depend on the oxidation state, the geometry of the coordination sphere and the character of the binding between the neighboring atoms and the absorbing atom. By analyzing the edge and the pre-edge features, information about the electronic structure and the surrounding of the photoabsorbing atom can be obtained. In the near edge region-XANES, the incident X-ray energy E is higher than that of the absorption edge E_0, and the excited electron leaves the absorbing atom as a photoelectron with relatively low kinetic energy: $E_K = E - E_0$. Multiple scattering of the photoelectron then occurs within the first and second shell of the surrounding atoms and the absorption spectrum may display a complex structure known as "shape resonances". In practice, the pre-edge, the edge, and the near edge region are called XANES. XANES spectra can be used to determine the average oxidation state of the element in the sample. The XANES spectra are also sensitive to the coordination environment of the absorbing atom in the sample. Fingerprinting methods have been used to match the XANES spectra of an unknown sample to those of known "standards". Linear combination fitting of several different standard spectra can give an estimate to the amount of each of the known standard spectra within an unknown sample.

In the EXAFS region, the photoelectron then obtains high kinetic energy and single back-scattering from the nearest neighboring atoms dominates, providing information about their number and identity and also the distance from the absorbing atom.

6.2.2 Samples and Sample Preparation

Samples for XAS can be prepared in a way that preserves their chemistry; no crystallization is necessary since no long-range order is required, and there is no need to purify a single compound. Samples may include solid, gaseous, or liquid forms. Solid samples are usually placed in spacers consisting, for example, of an aluminium plate with a central rectangular opening, covered with Kapton or Mylar tape, which does not absorb X-rays and is resilient to cold. Liquid samples can be placed in cells, again bound with cold-tolerant materials being transparent to X-rays. For metallomics and metalloproteomics studies, samples are generally frozen for analysis, and are usually kept cold in a cryostat, although spectra can also be collected at room temperature if necessary. Low temperatures help keep the sample fresh, and minimize the formation of free radicals by the high energy of the X-ray beam, which can cause photoreduction of the sample. Very low temperatures (on the order of 10 K with liquid He cooling) are generally used for EXAFS experiments to minimize molecular movement and variation in bond distance.

6.2.3 XAS Measurement

XAS requires a very good measurement of absorption coefficient $\mu(E)$. To measure the energy dependence of $\mu(E)$ in the X-ray region, first of all, a continuous and tunable X-ray source is needed. The X-ray source typically used is a synchrotron, which provides a full range of X-ray wavelengths, and a monochromator made from silicon that uses Bragg diffraction to select a particular energy.

The absorption coefficient $\mu(E)$ is most commonly measured either in transmission or fluorescence mode according to the following formulas. Transmission mode:

$$\mu(E) = \log(I_0/I) \tag{6.3}$$

X-ray fluorescence (or Auger emission) mode:

$$\mu(E) \propto I_f/I_0 \tag{6.4}$$

where I_f is the monitored intensity of a fluorescence line (or, again, electron emission) associated with the absorption process.

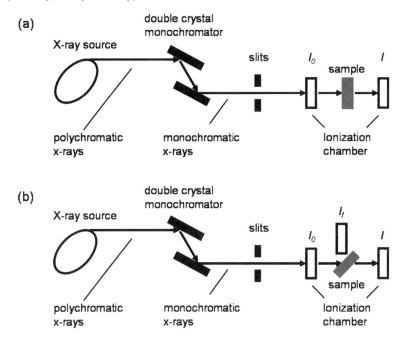

Figure 6.4 Schematic setup for X-ray absorption measurements: (a) in transmission mode and (b) in fluorescence mode.

In the transmission mode, the basic idea is to measure the intensities of the monochromatic photon beam before (I_0) and after (I) it has passed through a sample of interest according to the formula (6.3). These intensities are measured usually in ionization chambers, which allow the measurement of the ion current induced by the interaction of the monochromatic beam with the gas filing the ionization chamber as shown in Figure 6.4a.

For concentrated samples (that is, samples in which the element of interest is a major component), XAS should be measured in transmission mode in suitable thickness. Typically, the sample thickness t is adjusted so that $\mu t \approx 2.5$ above the absorption edge and/or the edge step $\Delta\mu(E)t \approx 1$. For many solid metal-oxides, t is often tens of microns. For dilute solutions, the sample thickness is typically in the millimeter range. Besides the requirement of the right thickness for transmission measurements, the sample must be uniform, and free of pinholes. For a powder, the grain size cannot be much bigger than an absorption length. If these conditions can be met (which can be a challenge at times), a transmission measurement will be very easy to perform and give excellent data. This method is usually appropriate for model compounds, and elements with concentrations $>10\%$.

A typical setup for measurements in fluorescence mode is shown in Figure 6.4b. In this mode, an ionization chamber is also used to monitor the incident intensity of the incoming monochromatic photon beam, but now the fluorescence emitting from the sample at a given angle is measured. The most common types of detector used for fluorescence measurement are Lytle chamber setups and different solid state detectors.

The fluorescence mode is preferred for thick samples or lower concentrations (down to the mg kg^{-1} level and even lower level depending on instrumentation and parameters of a given beamline). However, in this detection mode a high concentration of the target atom in a sample leads to a distortion of the obtained spectra. This is due to the fact that the energy of the fluorescence photon matches exactly a resonant absorption band of the target atom, which is called self-absorption, leading to a significant reduction of fluorescence observed by the detector, and consequently a faulty determination of the absorption coefficient.

Another possibility for measurement of the absorption coefficient $\mu(E)$ is to measure the electrons created in the absorption process, like total electron yield (TEY), Auger electron yield (AEY) and fluorescence yield (FY) mode to measure the XAS. The escape depth for electrons is generally much less than a micron, making these measurements much more surface-sensitive than X-ray fluorescence measurements. Although neither pinholes nor self-absorption can occur in this detection mode, for non-conducting samples, the accumulation of charge on the sample leads to a distortion of the spectra. Intense mixing of the sample material with graphite is a common approach to remedy this, but this method may lead to chemical or phase-transformations, which can also distort the spectra.

In XAS measurement, multiple spectra are generally collected to improve signal-to-noise ratios, from a minimum of two spectra for concentrated samples to many hours, for example, for dilute biological samples. As beam intensity increases, the signal-to-noise ratio improves and less measurement time is required. Since other elements in the sample can cause scattering of X-rays, filters are sometimes used to remove this scatter.

6.2.4 Data Analysis

After X-ray absorption measurement, the energy dependence of $\mu(E)$ can be obtained by plotting as shown in Figure 6.3. It should be noted that no matter whether $\mu(E)$ is measured in transmission or fluorescence (or electron emission), data reduction and analysis are essentially the same.

Since the EXAFS region of the XAS spectrum is dominated by single scattering processes, it is much easier for the theoretical understanding of the observed structures and this has been achieved by Sayers et al.[13] They showed that the Fourier transform of an EXAFS spectrum should peak at the distances corresponding to the radial distance of the excited atom to the neighboring coordination shells.

Interpretation of EXAFS data is normally based on the EXAFS equation:

$$\chi(k) = \sum_j \frac{N_j S_0^2 f_j(k) e^{-2R_j/\lambda(k)} e^{-2k^2\sigma_j^2}}{kR_j^2} \sin[2kR_j + \delta_j(k)] \quad (6.5)$$

Here R is the distance between the absorber and neighboring atom, N is the coordination number of neighboring atom, σ^2 is the mean-square disorder of neighbor distance, S_0^2 is the amplitude reduction factor, f(k) is the scattering amplitude, $\delta(k)$ is the phase-shift, and $\lambda(k)$ is the mean free path. The photoelectron wavevector k is inversely proportional to the De Broglie wavelength λ. The EXAFS $\chi(k)$ for an unknown sample can be fit to theoretical or known functions to give the best fit for N, R, E_0 and σ^2. Scattering amplitude f(k) and phase-shift $\delta(k)$ are generally obtained from theory such as FEFF[14] or taken from model compounds. More details of the EXAFS theory and the process of data analysis are extensively discussed in the literature.[12,15]

The aim of data analysis in the EXAFS region is to determine the structure information such as distances (R), coordination numbers (N), disorder parameters (σ^2) and types of atoms in the various coordination shells for the "unknown" sample. To achieve this, a series of data reduction of the raw data is necessary. The standard sequence of data reduction consists of: normalization of the data to unit edge step; interpolation to k-space; Fourier transformation to R-space; isolation of the amplitude and phase of individual coordination shell windowing and inverse transformation; and analysis of the amplitudes and phase using the EXAFS equation by nonlinear least squares fitting, the cumulant expansion/ratio method, and beat analysis.[12] Nowadays, several commercial or freely available programs can be applied for the EXAFS data analysis, e.g. WinXAS,[16] IFEFFIT,[17] Horae,[18] EXAFSPAK,[19] XFIT,[20] and USTCXAFS.[21]

An example for the above procedure is illustrated on the basis of data of manganese(II)-activated aminopeptidase P (Mn-AMPP).[22] The XAS data were first normalized as shown in Figure 6.5. From the XANES of Mn-AMPP at the Mn K edge shown in Figure 6.5, the oxidation state of the Mn atoms is +2.

The EXAFS $\chi(k)$ was isolated, k^3-weighted, and was Fourier transformed to R-space. As it is a general feature of the amplitude of the EXAFS oscillations to decrease drastically with increasing k, it is common to introduce a k-dependent weighting factor; in this case it is k^3. This function contains information on the scattering of all coordination information on the scattering of all coordination shells. This problem can be overcome by performing a Fourier transformation, which separates contributions from the various shells in R-space. It is then possible to isolate the contributions of different coordination shells by performing a windowed back-transformation and extracting the information on this coordination shell by fitting a model of the environment to the experimental data (Figure 6.6). This process yields the results that Mn in Mn-AMPP is coordinated predominantly by O donor atoms at an average Mn–ligand distance of 0.215 nm. The Mn–Mn interaction at 0.35 nm is not detected in the EXAFS.

The XANES signals are much larger than the EXAFS, but the interpretation of XANES is complicated by the fact that there is no simple analytic (or even physical) description of XANES. The main difficulty is that the EXAFS equation breaks down at low k, due to the 1/k term and the increase in the mean-free-path at very low k. Still, there is much chemical information from the XANES region, notably formal valence (very difficult to experimentally determine in a nondestructive way) and coordination environment. For

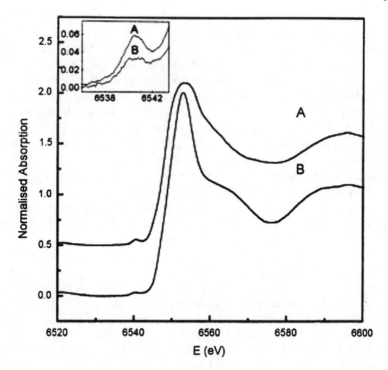

Figure 6.5 Mn K-edge X-ray absorption spectra at 77 K. (A) Mn-AMPP; (B) $[Mn(H_2O)_6]^{2+}$. Inset: magnifications of the 1s–3d feature.[22] © 1998 SBIC.

example, selenium with different oxidation states has shifted position of the absorption edge,[23] as demonstrated in Figure 6.7. Therefore, from a set of references with known formal oxidation state, it is possible to derive the formal oxidation state of the unknown sample. Besides, even at the same oxidation state, as with SeO_2 and Na_2SeO_4 in Figure 6.7, the difference of local electronic environment also results in changes of XANES spectra, which makes XANES fingerprinting feasible. These approaches of assigning formal valence state are based on edge features and as a fingerprinting technique it makes XANES somewhat easier than EXAFS to be crudely interpreted, even if a complete physical understanding of all spectral features is not available.

6.3 Application of XAS in Metallomics and Metalloproteomics

6.3.1 Fingerprints Studies and Quantitative Speciation by XANES

As mentioned above, it is possible to extract information about the electronic and geometric environment of absorber atom based on the comparison

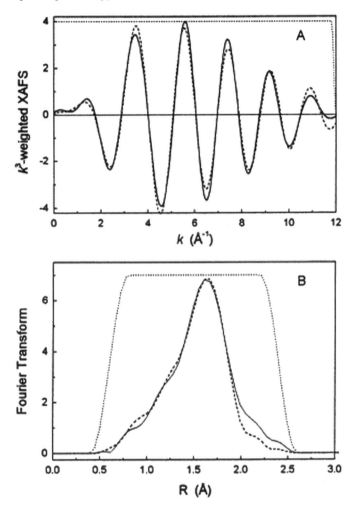

Figure 6.6 The observed and calculated EXAFS of Mn-AMPP (A), and their Fourier transforms (B). The observed EXAFS and its transform are drawn as solid lines, the calculated EXAFS and its transform as dashed lines and the filter applied to each function as a dotted line.[22] © 1998 SBIC.

between a spectrum obtained from an unknown sample and the spectra of known references. An example using fingerprints of XANES is the study of mercury forms in fish. By comparison of the Hg L_{III}-edge spectra of swordfish with other fish living in the different parts of the food chain, with spectra of model compounds, Harris et al.[24] found the same type of mercury compound in both fish, and the spectra matched well with methylmercury cysteine as seen in Figure 6.8.

In the study of mercury and selenium interaction, an examination of the energies and shapes of Se K-edge XANES and Hg L_{III} XANES of plasma from

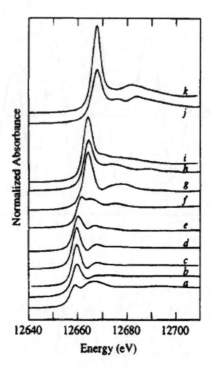

Figure 6.7 Se K-edge XAS spectra for selected selenium standards: (a) FeSe; (b) elemental selenium (gray hexagonal); (c) elemental selenium (red monoclinic); (d) SeS$_2$; (e) selenocystine (R-Se-Se-R); (f) selenomethionine (RSe-H); (g) SeO$_2$; (h) Na$_2$SeO$_3$ (i) SeO$_3^{2-}$(aq); (j) Na$_2$SeO$_4$ (k) Se$_2$O$_4^{2-}$(aq).[23] © 1995 American Chemical Society.

rabbits exposed to HgCl$_2$ and Na$_2$SeO$_4$ revealed that mercury formed complexes with the structures similar to the synthesized Hg-Se-S model compounds formed by reaction of sodium selenite with mercuric chloride and glutathione.[25] In another study of mercury and cadmium accumulated in the livers and kidneys of seal, albatross, dolphin, and squid,[26] Hg L$_{II}$ XANES and Se K-edge XANES from these animals were compared with HgO, HgSe, and m-HgS (metacinnabar) for Hg XANES and with Se, HgSe, Na$_2$SeO$_3$, and Na$_2$SeO$_4$ for Se XANES. A visual comparison of the Hg L$_{II}$ XANES indicated that Hg ligation in the liver samples was closest to that in HgSe and HgS. Comparisons of Se XANES indicated that the liver samples contained Se^{2-}, most likely as HgSe. From the Cd K-edge XANES, it was found that Cd was bound mainly to sulfur and that the Cd-O bond was observed in the tissues of seal.

In general, the fingerprints techniques by XANES are widely used in metallomics and metalloproteomics studies. More studies are exemplified as Se metabolism of the purple bacterium *Rhodobacter sphaeroides*,[27] reduction of As in Indian mustard,[28] Fe oxidation states in tissues,[29] Mn in mitochondria isolated from brain, liver, and heart,[30] Cu and Zn intake by *Bradybaena similaris* (land snail),[31] Ni speciation in hyperaccumulator and non-accumulator *Thlaspi*

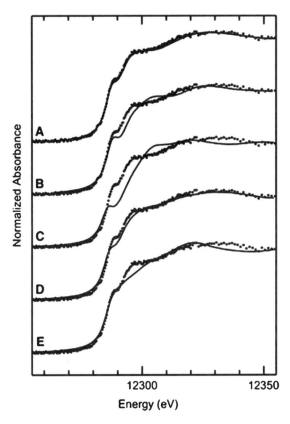

Figure 6.8 Comparison of the Hg L_{III} near-edge spectra of swordfish with selected solution spectra. The points show the spectrum of the fish; the solid lines show the spectra of aqueous solutions of standard species, as follows: (A) $CH_3HgS(Cys)$, (B) CH_3HgCl, (C) Hg^{2+} ($Hg(NO_3)_2$ solution), and (D) $Hg(SR)_2$, (E) $[Hg(SR)_4]^{2-}$ (from Harris et al.[24] © 2003 The American Association for the Advancement of Science.)

species,[32] and Cu and Zn in copper–zinc superoxide dismutase isoforms.[33] By using softer X-rays, XANES has also been applied to lighter atoms such as S, P, Na, K, Mg, and Ca.[34–36] Details can be seen from the literatures and references therein.

Besides the straightforward fingerprints studies, XANES can also be applied to quantitative speciation. This is because XAS is a local probe technique, which implies no long-range order in the sample is required. Therefore, if the absorbing atoms are present in the sample at two different sites, the XANES spectrum obtained from this material can be represented by the weighted addition of the spectra of suitable reference samples. For many systems, XANES analysis based on linear combinations of known spectra from "model compounds" is sufficient to estimate ratios of different species. More sophisticated linear algorithms, such as principle component analysis and factor analysis, can also be applied to XANES spectra.[37]

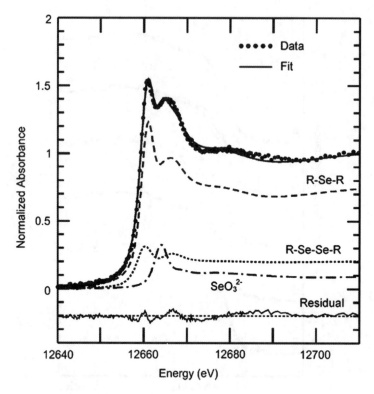

Figure 6.9 Least-squares fit of the spectrum of mayflies (*Epeorus* sp.) to the sum of spectra of standard selenium species. The figure shows the fit, and the individual components, scaled according to their contributions to the fit: SeO_3^{2-} (selenite), R-Se-R (selenides), and R-Se-Se-R (diselenides).[38] © 2007 American Chemical Society.

An example using XANES for quantitative speciation is the study of selenium species in stream insects by Andrahennadi et al.[38] In their study, selenomethionine, trimethylselenonium, and selenocystine were used to represent the selenium environments of R-Se-R, R_3Se^+, and R-Se-Se-R, respectively. When spectra of samples from the stream sites were fitted to the sum of spectra of the standard species (Figure 6.9), organic selenides (R-Se-R), modeled as selenomethionine, were found to be the most abundant among all selenium compounds tested, ranging from 36 to 98%. Together with organic diselenides (R-Se-Se-R), these organic forms account for more than 85% of selenium in all nymphal and larval insects measured.

In our study of mercury species in human hair and blood samples from people exposed to mercury,[2] XANES speciation was applied and the results are shown in Table 6.1. Least-squares fitting of the X-ray absorption near-edge spectra found that inorganic mercury was the major mercury species in hair samples (92%), while both inorganic and methyl mercury were about 50% of total mercury in RBC and serum samples, which was in agreement with the

Table 6.1 Least-square fitting analysis of the XANES spectra for Hg species in hair, RBC and serum from people living in Wanshan.[2] © 2007 Elsevier Inc.

	CH_3Hg-SR (%)	Hg-SR (%)	$^a CH_3Hg^+$ (%)	$^a Hg^{2+}$ (%)
Hair	8.43	91.74	7.08	92.92
RBC	47.21	52.47	48.31	51.69
Serum	41.63	58.31	43.52	56.48

^{a}Results were obtained by acidic extraction and ICP-MS determination.

data obtained by acidic extraction, fractionation of Hg^{2+} and CH_3Hg^+ and quantification by ICP-MS.

Conventional speciation techniques require sample pretreatments such as digestion that may alter the chemical species of the element of interest. As shown above, quantitative speciation by XANES offers an excellent method for elucidating the chemical species of an element in biological tissues and in intact organisms with minimal sample preparation. The detection limits are generally in tens to hundreds of $\mu g\, g^{-1}$ level, but they are dependent on both the incident X-ray intensities and specific atoms. For example, the detection limit for selenium speciation may reach $0.2\, \mu g\, g^{-1}$.[39]

6.3.2 Fingerprints and Structural Information by EXAFS

EXAFS fingerprinting proceeds in much the same way as XANES fingerprinting, but using the oscillations in the EXAFS region for fitting, rather than at the edge. An example is the EXAFS fingerprint study on arsenic species and transformation in arsenic hyperaccumulator.[40] Figure 6.10 shows that the arsenic in the plant is mainly coordinated with oxygen, except some arsenic coordinated with S as As-GSH in root. The complexation of arsenic with GSH might not be the predominant detoxification mechanism in Cretan brake. Although some arsenic in root exists as As^{5+} in Na_2HAsO_4 treatments, most of arsenic in plants is present as As^{3+}-O in both treatments.

Besides fingerprinting, the widespread array of applications of EXAFS is in structure–function studies of metallomics and metalloproteomics. One of the main applications of EXAFS in the case of metalloproteins is the investigation of accurate bond lengths and active site geometries of metals. This field of research has been pioneered by Shulman et al.,[41] who investigated iron–sulfur distances in the nonheme iron–sulfur protein rubredoxin obtained from *Peptococcus aerogenes*. The data obtained in this study revealed that all four FeS (cysteine) distances are mainly equal. This did not correspond with previous investigations of the similar protein from *Clostidium pasteurianum* using single-crystal XRD, which supposed that one FeS distance is significantly shorter than the average distance of the other three distances. The atomic environment around the iron site in rubredoxin of *C. pasteurianum* was also studied using EXAFS yielding an analogous result to that obtained by Shulman et al.[42]

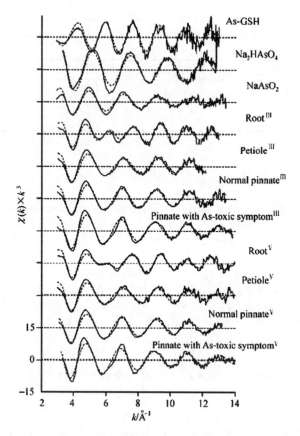

Figure 6.10 Arsenic K-edge EXAFS of Cretan brake samples and model compounds. Solid lines show the data and the dashed lines show the best fits.[40] © 2004 Science in China Press.

Later, improved single-crystal XRD also led to a symmetric structure which is completely in accordance with the EXAFS data.

In our study on the structural information of Hg in human hair and blood samples,[2] EXAFS fitting found Hg is coordinated by three sulfur atoms in hair and four sulfur atoms in both RBC and serum, as shown in Table 6.2.

EXAFS has become a standard method for probing the metal site structure in different metalloproteins and metalloenzymes and also in the study of metallodrugs.[43,44] It has been applied to many metalloproteins that have been purified and isolated containing Fe,[45] Mo,[46,47] Mn,[22] Cu,[48] Co,[49] Zn,[50] Ni,[51] Cd,[52] and others. For example, nickel EXAFS of carbon monoxide dehydrogenase (CODH) from *Clostridium thermoaceticum* strain DSM 521 were collected and the Fourier filtered data was best fit to sulfur at 0.216 nm.[51] A reasonable fit to oxygen/nitrogen could only be obtained by setting the ΔE_0 to 40 and the Debye–Waller term to 5×10^{-5} nm. A reasonable Ni-S fit was also obtained by assuming two Ni-S bond lengths at 0.222 and 0.211 nm, respectively. Addition of a Ni-M (M=Fe, Ni, Zn) term at 0.325 nm, also improved the

Table 6.2 EXAFS curve-fitting analysis of the mercury spectra in hair, RBC and serum samples from people living in Wanshan.[2] © 2007 Elsevier Inc.

Sample	Interaction	N	R (nm)	σ^2 (nm^2)	Relative error
Hair	Hg-S	3.10	0.248 ± 0.002	0.00013 ± 0.00002	0.083
	S-Hg	4.05	0.236 ± 0.004	0.00025 ± 0.00004	0.078
RBC	Hg-S	4.09	0.251 ± 0.003	0.00012 ± 0.00001	0.053
HHB-Hg	Hg-S	4.07	0.248 ± 0.005	0.00022 ± 0.00002	0.049
Serum	Hg-S	4.08	0.238 ± 0.002	0.00010 ± 0.00004	0.058
HSA-Hg	Hg-S	4.10	0.241 ± 0.002	0.00012 ± 0.00003	0.049

N, coordination number; R, interatomic distance, σ^2, the Debye–Waller factor; relative error is defined as $[\Sigma k^6(\chi_{exp} - \chi_{cal})^2/\Sigma k^6 \chi_{exp}^2]^{1/2}$.

fit although a high ΔE_0 (19 eV) was required to obtain a reasonable fit. Similar behavior was observed with the nickel model compound where a Ni–Ni distance could be best fit by setting the ΔE_0 to 21 eV. The nickel in the model compound is a square planar, and the environment of nickel in CODH may be square planar as well. This would leave open axial coordination sites for CO, methyl, or acetyl CoA.

Alzheimer's disease (AD) is a progressive neurodegenerative disorder with the characteristic presence of amyloid plaques composed primarily of amyloid β peptide (Aβ, up to 42 amino acids). Aβ is normally soluble and found in all biological fluids, and the structural transition from its native state to a β-sheet aggregation is accompanied by the concurrent gain of neurotoxic function. Due to the growing evidence of involvement by transition metals, e.g. Cu, Zn, and Fe, in the elevated level of amyloid deposits in AD-affected brains, accurate details of copper binding site in Aβ may be critical to the etiology of AD.[53] Recent XAS analysis (see Figure 6.11) of Aβ(1–40)-Cu^{2+} in a 1:1 ratio suggests that Cu^{2+} is bound to three nitrogens from three histidines (His6, His13, and His14) at 0.185–0.194 nm, and two oxygens (one from Tyr10 at 0.200 nm, and the other belonging to either a water molecule or one other amino acid at 0.191 nm).[54]

6.3.3 Micro-XAS in Metallomics and Metalloproteomics

In the metallomics and metalloproteomics studies, spatial analysis of absorbing atom of interest is also fascinating for elucidating the uptake and biotransformation of different species of elements. An example studied the transformation of arsenic (Na$_2$HAsO$_4$ or NaAsO$_2$) by the arsenic hyperaccumulator, Cretan brake (*Pteris cretica* L. var *nervosa* Thunb).[40] It was found that As^{5+} tended to be reduced to As^{3+} after it was taken up into the root, and arsenic was kept as As^{3+} when it was transported to the above-ground tissues like petioles and pinnas. However, this example shows the spatial analysis at bulk scale; therefore, this kind of XAS techniques can be called bulk-XAS. Using a microscopic mode of XAS, it is possible to obtain the information provided by bulk XAS at a spatial resolution of only a few micrometers or even

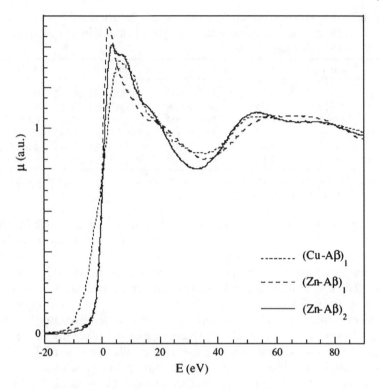

Figure 6.11 Absorption coefficient in the X-ray absorption near-edge region of the spectrum of: (Cu–Ab)$_1$, Ab-peptide complexed with Cu^{2+} in solution (dotted line); (Zn–Ab)1, Ab-peptide complexed with Zn^{2+} in solution (broken line); (Zn–Ab)$_2$, Ab-peptide complexed with Zn^{2+} in the re-dissolved pellet (continuous line). To ease comparison, spectra are shifted in energy (by about 680 eV) in order for the K edges of Cu^{2+} and Zn^{2+} to be located at almost the same energy. The energy scale is given with reference to the edge energy of the absorbers. The latter is empirically taken as the energy where the first derivative of the absorption spectrum attains its maximum.[54] © 2006 EBSA.

smaller.[55,56] The high sensitivity for trace species in combination with a good spatial resolution permits the investigation of the lateral distribution of various species of the absorbing atom of interest.

A common application of micro-XAS is to provide speciation at specific locations of interest combining with elemental mapping techniques like micro-X-ray fluorescence (micro-XRF) spectroscopy.[57] The study of Mn oxidation states in wheat rhizospheres infested with the fungus *Gaeumannomyces* is such a case.[58] The take-all disease, caused by *Gaeumannomyces graminis* var. *tritici*, is one of the world's most damaging root diseases of wheat. It has been hypothesized that the fungus reduces the host's defense mechanism prior to invasion by catalyzing the oxidation of soluble Mn^{2+} to insoluble Mn^{4+} on the rhizoplane and in the soil surrounding the root. By using micro-XAS, a direct

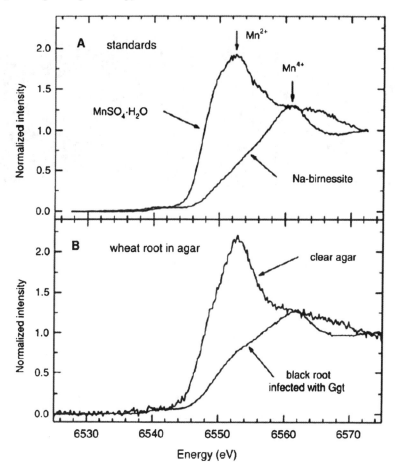

Figure 6.12 Mn K micro-XAS spectral of: (A) standard mixtures containing of Mn^{2+} or Mn^{4+}; and (B) a section of wheat root infected with *Gaeumannomyces graminis* var. *tritici* (Ggt) compared with an adjacent area of clear agar.[58] © 1995 American Phytopathological Society.

test of this hypothesis has been accomplished to obtain information about the spatial distribution of Mn oxidation states in and around live wheat roots growing in agar infected with *G. graminis* var. *tritici*. As shown in Figure 6.12, Mn in clear agar occurred only as Mn^{2+}, whereas Mn around dark roots infected with *G. graminis* var. *tritici* was predominately present as Mn^{4+}. The distribution of Mn oxidation states clearly showed the presence of Mn^{4+}-containing precipitates in the interior of a root infected with *G. graminis* var. *tritici*.

Another interesting study is the determination of the distribution of Cr valence states in human cells after *in vitro* exposure to soluble or particulate chromium compounds.[59] The chromium biological activity depends strongly upon its oxidation state and solubility. Several hexavalent chromium, Cr(VI),

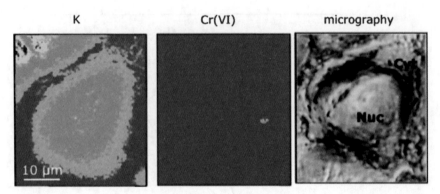

Figure 6.13 Synchrotron radiation microprobe mapping of potassium and chromium oxidation states in a single human epithelial cell exposed to particulate chromate, and corresponding optical microscopy view. Beam spatial resolution (V×H): $0.5 \times 1\,\mu m^2$; color scale in counts per pixel. Cr(VI) distribution shows a perinuclear localization.[59] © 2005 American Chemical Society.

compounds are known to be human carcinogens, particularly for the respiratory tract, whereas trivalent chromium, Cr(III), compounds are considered as nontoxic and noncarcinogenic agents. Cr(VI) as an oxyanion form can penetrate the cell via the sulfate transport system while Cr(III) cannot enter the cell. In Figure 6.13, freeze-dried cultured human ovarian cells were mapped with a resolution as low as 0.5 μm. It was shown that soluble Cr(VI) compounds are fully reduced to Cr(III) in cells. Cr(III) is homogeneously distributed within the cell volume and therefore present within the nucleus. In the case of low solubility particulate chromate compounds, Cr(VI) can coexist in the cell environment, as particles in the perinuclear region, together with intracellular and intranuclear Cr(III). Chemical distribution maps also suggest that intracellular Cr(III) originates from extracellular dissolution and reduction of lead chromate rather than from intracellular engulfed particles. More examples can be seen in the *in vivo* studies of thallium speciation and compartmentation in *Iberis intermedia*,[60] selenium in the epithelial layer on the turtles shell and the bone tissue.[61]

Micro-XAS can also be applied in combination with chemically specific imaging as done by Pickering *et al.*,[62] who reported the spatial distribution and chemical species of selenium in *Astragalus bisulcatus* (two-grooved poison or milk vetch), a plant capable of accumulating up to 0.65% of its shoot dry biomass as Se in its natural habitat. By selectively tuning incident X-ray energies close to the Se K-absorption edge, quantitative and 100-μm resolution images of the spatial distribution, concentration, and chemical form of Se in intact root and shoot tissues were collected. Plants exposed to $5\,\mu mol\,L^{-1}$ selenate for 28 days contained predominantly selenate in the mature leaf tissue at a concentration of 0.3–$0.6\,mmol\,L^{-1}$, whereas the young leaves and the roots contained organoselenium almost exclusively, indicating that the ability to

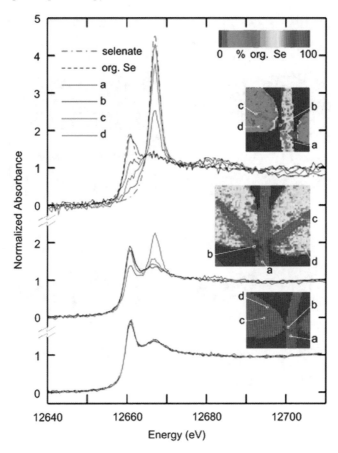

Figure 6.14 Spatially resolved X-ray absorption near-edge spectra recorded by using a $100 \times 100\,\mu m^2$ beam from selected portions of mature (top), intermediate (middle), and young (bottom) leaves. The spectra of selenate and selenomethionine (org. Se) are also included. The insets show maps of the percentage of total Se as organoselenium (balance selenate) and indicate the spatial location for the spectra. The percentages of selenate in each leaf specimen as determined by near-edge fitting for (a) stem, (b) petiole, (c) mid-vein, and (d) leaf blade, respectively, are: mature 36, 0, 69, 88; intermediate 3, 0, 7, 28; young 3, 0, 1, 1.[62] © 2000 The National Academy of Sciences.

biotransform selenate is either inducible or developmentally specific. While the concentration of organoselenium in the majority of the root tissue was much lower than that of the youngest leaves (0.2–0.3 compared with 3–4 mmol L^{-1}), isolated areas on the extremities of the roots contained concentrations of organoselenium of an order of magnitude greater than the rest of the root. These imaging results were corroborated by spatially resolved X-ray absorption near-edge spectra collected from selected $100 \times 100\,\mu m^2$ regions of the same tissues (Figure 6.14).

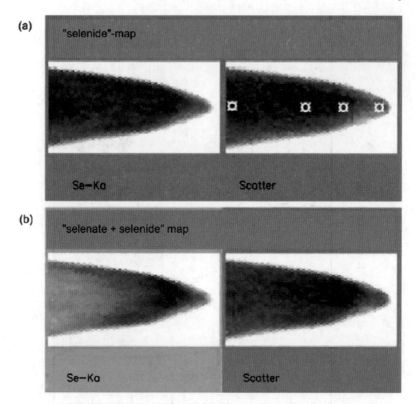

Figure 6.15 Chemical state maps of a leaf tip of onion grown in a selenate solution [Se(VI)]. Size of the map: 4 mm × 2 mm. (a) Map recorded at 0 eV relative energy (position of the inflection point of the Se-methylselenocysteine or selenomethionine reference spectra); (b) map recorded at the 7 eV relative energy (position of the maximum of the selenate reference spectrum).[63]
© 2006 American Chemical Society.

Another similar chemical state mapping were conducted in onion leaf by Bulska and co-workers, who found the local accumulation of Se(IV) in onon leaf tip (see Figure 6.15).[63]

The above-mentioned works using micro-XAS to investigate virgin sample materials have been performed almost exclusively on the tissues cooled down to 10 K. An *in vivo* study of the distribution of a different Se-containing species in living plant tissue was performed by Bulska *et al.*[63] Microscopic X-ray absorption near-edge structure spectroscopy (micro-XANES) and confocal micro-XRF analysis were used for the *in vivo* determination of the distribution of total selenium and for the local speciation of selenium in roots and leaves of onion. Selected *Allium cepa* L. plants were grown hydroponically in a standard medium containing inorganic selenium compounds (selenite or selenate). It was found that the ratio of inorganic to amino acid selenium compounds differs in various subparts of the plant. Detailed *in vivo* investigation of the distribution of various selenium species in virtual cross sections of root tips and green leaf showed that

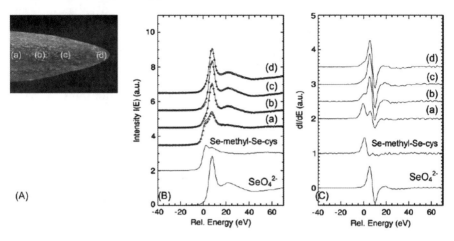

Figure 6.16 (A) Optical micrograph of a leaf of a Se(VI)-exposed onion, indicating the different sampling positions and (B) the corresponding micro-XANES spectra. Solid line, experimental data; symbols, result of fitting a linear combination of SeO_4^{2-} and Se-methylselenocysteine reference spectra to the experimental data. (C) First derivatives of the XANES spectra shown in B.[63] © 2006 American Chemical Society.

the selenium transport takes place via different mechanisms, depending on the nature of the selenium compounds originally taken up (cf. Figure 6.16).

Micro-XAS with softer X-rays are most suitable for the study of lighter atoms like C, N, O, and S. By using micro-XANES to examine the content and distribution of DNA and protein in mature sperm cells, Zhang et al.[64] found that the total nuclear protein to DNA ratio is similar in the sperm of many eutherian mammals, which indicated that the total protamine content of sperm chromatin had to be constant among mammalian species, independent of the extent of expression of the protamine 2 gene. In an attempt to analyze the spatial distributions of the constitutent elemetns in a mammalian cell by micro-XANES, Ito et al.[65] found marked absorption changes at the absorption edges of carbon, nitrogen, and oxygen, while minor but significant changes for iron and calcium were also observed, particularly in the cytoplasmic areas.

6.4 Combination of XAS and other Techniques in Metallomics and Metalloproteomics

As just mentioned, XAS can give information like oxidation state, neighboring atoms, bond lengths, and coordination numbers in metallomics and metalloproteomics with minimum sample preparation. Further, XAS can present its full advantages through combination with other techniques in metallomics and metalloproteomics. Therefore, in this part, we show the combined application of XAS with separation techniques like size-exclusion chromatography (SEC) and gel electrophoresis (GE), the combination of XAS with XRF to give both spatial and structural information, and the combination of XAS with other structural determination techniques like protein crystallography (PX), computational chemistry, nuclear magnetic resonance (NMR), neutron scattering (NS), circular dichroism spectroscopy (CD), Raman spectroscopy, and electron spin resonance (ESR) for studying structure–function relationships in many important biochemical compounds.

6.4.1 Combination of XAS with Separation Techniques in Metallomics and Metalloproteomics

Separation techniques can be performed through extraction or electrophoresis; for example, in the study of the antagonism between inorganic mercury and selenium in mammals, the synthesized Hg-Se-S species was purified by size-exclusion chromatography (Sephadex G-25) and was used as model compounds in XAS study.[25]

Electrophoresis-based separation techniques have been widely used in proteomics studies. The most commonly used electrophoresis-based technique is gel electrophoresis (GE), which is a procedure for separating a mixture of molecules through a stationary material (gel) in an electrical field. The GE, especially two-dimensional (2D) polyacrylamide gel electrophoresis (PAGE), is a powerful tool for protein separation.[66]

Recently, Chevreux and co-workers[33] have proposed the use of XAS for the analysis of non-crystalline metalloproteins directly on IEF-PAGE (isoelectrofocusing PAGE). This method allows the direct analysis of the metal structural environment in its native state as the gel electrophoresis is performed under non-denaturing conditions. The oxidation state of copper and zinc in bovine and human copper zinc superoxide dismutase (CuZnSOD) was determined by XANES on IEF gels (Figure 6.17). The results obtained are in good agreement with XAS analysis of bovine CuZnSOD crystals,[67] and indicate the existence of mixed Cu(I)/Cu(II) oxidation states. In addition, the use of IEF enables the separation of isoelectric point isoforms of CuZnSOD, which could not be resolved in protein crystals, and indicates that in a minor isoform copper was present as Cu(II), while in the major isoform as a mixed redox state Cu(I)/Cu(II). Overall, the combination of IEF and XANES offers unique capabilities for the characterization of metalloproteins because metal oxidation state can be obtained in proteins using non-denaturing electrophoresis conditions, thus preserving their native state. In addition, IEF separation enables the speciation of protein isoforms with distinct isoelectric points.

6.4.2 Combination of XAS with XRF in Metallomics and Metalloproteomics

In section 6.3.3, we mentioned that micro-XAS can be combined with micro-XRF to provide oxidation information at a specific location at a spatial

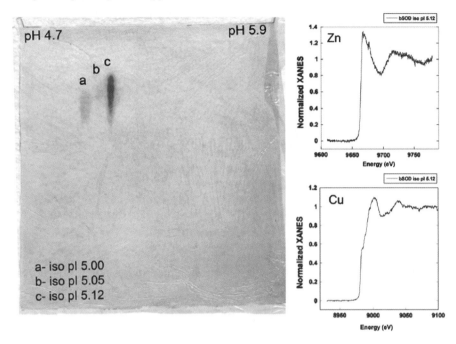

Figure 6.17 Left: 2D gel electrophoresis of commercial bovine CuZnSOD run under non-denaturing conditions showing three p*I* isoforms at 5.00, 5.0, and 5.12. XANES spectra were recorded on an unstained gel prepared simultaneously to the stained gel presented here. Right: zinc and copper absorption K-edge spectra of bovine CuZnSOD isoform at p*I* = 5.12.[33] © 2008 The Royal Society of Chemistry.

resolution of only a few micrometers. However, the concentrations of the element of interest in sliced samples for micro-XAS analysis are generally lower than that in bulk samples, which will hinder the application of micro-XAS in sliced samples. Therefore, a more common application of these two X-ray techniques is based on bulk XAS and micro-XRF. For example, Howe and co-workers[68] studied the localization and speciation of chromium in subterranean clover using bulk XANES, micro-XRF, and EPR spectroscopy. They found that the uptake, translocation, and form of Cr in the plant were dependent on the form and concentration of supplied Cr. Using XRF mapping, they found that at low Cr(VI) treatment concentrations, Cr in the leaves was observed predominately around the leaf margins, while at higher concentrations Cr was accumulated at leaf veins. By comparison of the XANES spectra of fresh tissue from subclover plants grown with 0.04 and 1.6 mmol Cr(VI) L^{-1} with those of the 5% (w/w) standards of Cr(VI) and Cr(III), Chromium was found predominately in the +3 oxidation state, regardless of the Cr source supplied to the plant, though at high Cr(VI) treatment concentrations, Cr(VI) and Cr(V) were also observed. At low Cr(VI) concentrations, the plant effectively reduced the toxic Cr(VI) to less toxic Cr(III), which was observed both as a Cr(III) hydroxide phase at the roots and as a Cr(III)–organic complex in the roots and shoots.

Another example using bulk-XAS and micro-XRF is the investigations of the intracellular metabolism of arsenic performed on human hepatoma cells HepG2 following their exposure to high doses of arsenite (1 mM) or arsenate (20 mM).[69] Microprobe XRF elemental mapping of thin-sectioned cells showed As accumulation in the euchromatin region of the cell nucleus (following arsenite exposure) synonymous with As targeting of DNA or proteins involved in DNA transcription (Figure 6.18). XAS analysis of bulk arsenite-treated cells, however, showed the predominance of an As tris-sulfur species, providing increased credence to As interactions with nuclear proteins as a key factor in As-induced toxicity.

6.4.3 Combination of XAS with Protein Crystallography in Metallomics and Metalloproteomics

In structure characterization of metalloproteins, protein crystallography (PX) is the most powerful tool for the determination of macromolecular 3D structure at a resolution of 0.15–2 nm. However, few studies at a higher resolution can be found up to now owing to (1) the limited diffraction power of most crystals; (2) photoreduction, or radiation damage, of the protein during data collection; and (3) micro-heterogeneity of samples.[70] This is not sufficient for illuminating the structural basis related to activities of the metal centers in functional metalloproteins, where the structural changes to the metal coordination during redox or substrate-binding reactions happen generally at ≤0.01 nm. On the other hand, XAS provides an alternative tool for the local structure around certain atoms at a resolution of 0.1–1 pm, depending on the circumstances.[3,71] Recently, an even higher resolution of one femtometer was detected by differential X-ray absorption spectroscopy at the ID24 beamline of the European Synchrotron Radiation Facility (ESRF).[72]

Currently, the combination of crystallographic information and XAS, specifically EXAFS, is a powerful approach for studying the structure–function relationship of proteins, particularly when subtle structural changes are associated with a chemical reaction. Cheung *et al.*[4] made direct use of three-dimensional information from crystal structures of azurin in the analysis of EXAFS data for the oxidized and reduced form of the protein. Subtle structural changes (<0.01 nm) take place at the Cu site during a single-electron redox process. This approach is likely to be of most interest in cases where crystallographic information is available for some state of a protein but not for others. Under such conditions, EXAFS data on different states of the protein may be solved by the combined approach to predict structural changes and examine how these intermediated states perturb the metal at the catalytic site.

Corbett and co-workers[73] compared both the structural and Mo-localized electronic features of the iron–molybdenum co-factor (FeMoco) in isolated MoFe protein and in the ADP·AlF$_4^-$ stabilized complex of the MoFe protein with the Fe protein. The local metal structure of the iron–molybdenum cofactor of nitrogenase in isolated MoFe protein has been determined by XAS.

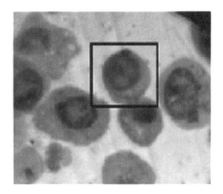

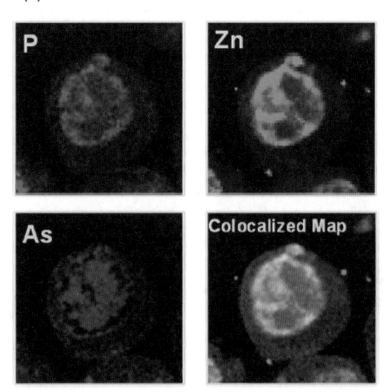

Figure 6.18 (A) Correlative light micrograph of a toluidine-blue stained thin section of HepG2 cells that have been exposed to arsenite (1 mM, 4 h), and (B) the corresponding microprobe SR-XRF maps of P, Zn, As, and the colocalization of the elements.[69] © 2008 American Chemical Society.

The Mo K- and Mo L-edge XAS edge and K-edge EXAFS analysis provided a detailed comparison of isolated MoFe protein in solution with that in a single crystal and with the $ADP \cdot AlF_4^-$ stabilized nitrogenase complex in solution. The results indicated a lack of perturbation of either the Mo electronic environment or local structure upon nitrogenase complex formation facilitated by $ADP \cdot AlF_4^-$ in agreement with previous findings, but at a more detailed resolution. The double-edge XAS studies on the isolated and complexed MoFe proteins highlight the utility of XAS in studying the nitrogenase system and, more importantly, provide a baseline for future studies on nitrogenase intermediates.

XAS can also provide additional information for PX. An example is described for formate dehydrogenaseH (FDH_H) of *Escherichia coli* containing molybdenum and selenocysteine at its active site.[74] The crystal structure was reported indicating an unusual des-oxo molybdenum site with four MoS ligands at 0.24 nm, one MoOH (or $MoOH_2$) at 0.21 nm and a coordinated selenocysteine with MoSe of 0.26 nm.[75] A des-oxo molybdenum site with four MoS (0.235 nm), one MoO (0.210 nm) and one MoSe ligand (0.262 nm) has been confirmed by subsequent EXAFS analysis at the Mo K edge (Figure 6.19).[76] Selenium EXAFS at the Se K edge is also in good accordance with that of molybdenum and, furthermore, indicates the presence of SeS ligation at 0.219 nm, which had been missed in the crystal structure.[76] The crystallographic analysis suggested a close Se–S contact of 0.29 nm with one of the MoS ligands. EXAFS, however, indicated that the active site of the formate dehydrogenase H of *E. coli* contained a novel seleno-sulfide ligand to molybdenum, whereas sulfur was likely originated from one MoS ligand.[74,76]

Another aspect that brings a significant link between XAS and PX is in the determination of oxidation states at metal centers. Therefore, it was recommended by Hasnain and Strange[5] that all metalloprotein crystallography data

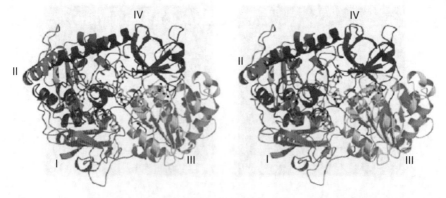

Figure 6.19 Stereo view of the overall fold of formate dehydrogenase H (FDH_H). Domains I, II, III, and IV are shown. The MGD co-factors are shown as ball-and-stick models with MGD^{801} on the left and MGD^{802} on the right. Mo is shown as a ball in the center, and the Fe_4S_4 cluster is shown as a ball-and stick model in domain I.[75] (Reprinted with permission from Science.)

collections should be accompanied by XAS measurements, so that oxidation states of the protein to which the crystallographic structure corresponds can be ascertained. Pohl et al.[77] presented a model for the activation and DNA binding of ferric uptake regulator (Fur) from *Pseudomonas aeruginosa* by the combined approach of the crystal structure analysis of the Fur protein in complex with Zn^{2+} at a resolution of 0.18 nm, and characterization of the distinct zinc and iron binding sites in solution with XAS and micro-PIXE. More work on the combined application of XAS and PX can be found in reviews and references therein.[78]

6.4.4 Combination of XAS with Computational Chemistry in Metallomics and Metalloproteomics

For XAS, the identification of the coordination geometry cannot always be determined with high precision to identify the type, number, and geometry of all the attached ligands. This problem arises especially when the metal binding site is complicated, which is the case for metalloproteins with multiple metals in either one site or a number of isolated sites within the same molecule, because the EXAFS and XANES data are an average of all metal–ligand bonds. Unambiguous confirmation of the structure of the metal binding site in the metalloprotein has proven difficult without PX diffraction data. And in some cases, crystallization has proven to be difficult or impossible so that PX data cannot be obtained. Thus, new ways of improving the accuracy and precision of the determination of the coordination shell of the bound metals are needed for the structural elucidation of these metal-binding biomolecules. Computational techniques offer a possible route to this goal by combining information from a number of experimental sources.[79]

In metalloproteomics, there has been progress in recent years in developing theoretical methods to study protein structures locally at metal-binding sites and at locations of catalytic activity.[80,81] The use of quantum chemical calculations can supplement experimental data of PX and can also be used to interpret the structures, e.g. to decide the protonation state of metal-bound ligands. By providing subatomic resolution information that is local to the metal environment, XAS is a valuable ally of computational chemistry.

Geometry-sensitive, metal-edge XAS data from biologically bound metals can be directly compared with simulated XAS data calculated for model compounds of the metal binding sites. For multiple metals in biological molecules, the XAS data can be calculated for the environment proposed for each metal site individually and, as well, different ligand combinations can be tested in the coordination sphere. Through the comparison between the measured XAS data from chemically synthesized analogue molecules and biomolecules, and the calculated XAS data from theoretically generated models, the coordination environment that best matches the metal binding site in the biomolecule can be identified.

The XAS data, especially the EXAFS data can be used directly by the computational calculation to predict protein structures. The recent applications combining XAS with computational chemistry included the study of the unusually high redox potential of copper in rusticyanin (Figure 6.20),[82] the geometric and electronic properties of the red copper site in nitrocyanin[83] and the properties of an Fe–Fe hydrogenase active site in four protonation states involved with hydride binding.[84] Polarised single-crystal XANES and time-dependent density functional theory (DFT) were used to examine the origins of electronic transitions of high-valent Mn relevant to the catalytic cycle of PSII.[85] XAS has also been used with molecular dynamics (MD) to identify high-affinity Mn^{2+}-binding sites on the extracellular region of bacteriorhodopsin near the retinal pocket.[86] Sulfur K-edge XAS was used with DFT to probe ligand–metal bond covalency and the electronic structure and reactivity of metal–sulfur sites in proteins.[87] In these types of studies, XAS provides the accurate metrical data that can be used to both gauge the success of the theoretical treatments and develop combined XAS molecular mechanics (MM)/quantum mechanics (QM) refinement methods.[88]

6.4.5 Combination of XAS with Neutron Scattering in Metallomics and Metalloproteomics

Neutron scattering (NS) possesses three advantages: (1) it is easier to sense light atoms, such as hydrogen, in the presence of heavier ones; (2) neighboring elements in the periodic table generally have substantially different scattering cross-sections and, thus, can often be distinguished; and (3) the nuclear dependence of scattering allows differentiation of isotopes of the same element with substantially different scattering lengths for neutrons. Additionally, because of the electrical neutrality of the neutron, the interaction of a neutron with the nucleus in samples is weak, so that the neutron serves as a non-destructive probe to characterize protein structure and properties.[89]

The three-dimensional structure determinations of biological macromolecules such as proteins and nucleic acids by X-ray crystallography have improved our understanding of many of the mysteries involved in biological processes. At the same time, these results have clearly reinforced the commonly held belief that H atoms and water molecules around proteins and nucleic acids play a very important role in many physiological functions. However, since it is very hard to identify H atoms accurately in protein molecules by X-ray diffraction alone, a detailed discussion of protonation and hydration sites can only be based upon speculations so far. In contrast, it is very well known that neutron diffraction provides an experimental method of directly locating H atoms (Figure 6.21).[90] More details about neutron scattering can be seen in Chapter 7. Therefore, the combination of these two diffraction techniques can give complementary information of the three-dimensional protein structure. Besides, we have just mentioned XAS can give supplemental information to PX, the combination of NS, PX, and XAS can give the structural information

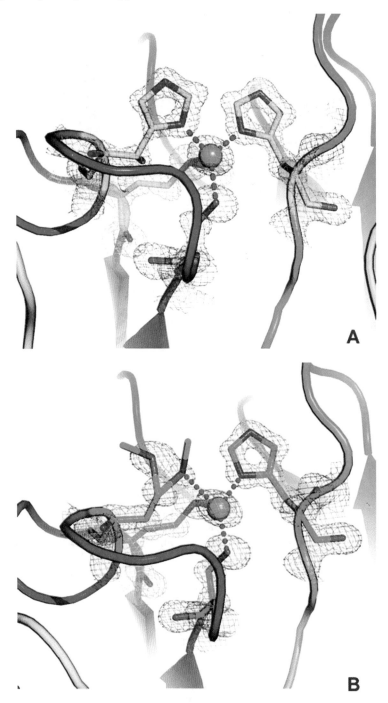

Figure 6.20 Active site electron density for (A) native and (B) H143M rusticyanin.[82] © 2006 American Chemical Society.

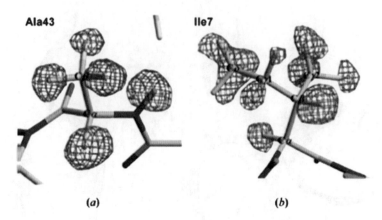

Figure 6.21 $|Fo| - |Fc|$ omit nuclear-density map of the H atoms around the residues (a) Ala43 and (b) Ile7 of wild-type rubredoxin.[90] © 2008 International Union of Crystallography. Printed in Singapore.

including both the locating of H atoms and metal atoms. The combination of NS and XAS has been proposed in the studies of SRNiO$_3$ perovskites,[91] bulk GeS$_2$ and GeSe$_2$ glasses,[92] monoclinic zirconia[93] and water,[94] however, to our best knowledge, no report has been found on the combined application of both NS and XAS in metallomics and metalloproteomics yet, but this should be a promising research direction.

6.4.6 Combination of XAS with Circular Dichroism in Metallomics and Metalloproteomics

Circular dichroism (CD) is the measurement of the difference in absorption of left- and right-circularly polarized light as it passes through an optically active or chiral sample.[95] Optical activity arises due to the influence of neighboring groups on the electronic structure of a chromophore involved in an electronic transition.[96] CD has found its principal application in studies of protein secondary structures and in detecting conformational changes in proteins, as the spectra are sensitive to small alterations in polypeptide backbone structures.[97]

The combination application of XAS and CD can give both structural information and conformation changes in proteins. For example, Lu et al.[98,99] studied the interaction of metal ions with metallothioneins using CD. Jiang et al.[100] and Gui et al.[101] from the same research group further characterized the structures by a combination of XAS with CD. In their studies, the structural parameters for [M$_4$(SPh)$_{10}$]$^{2-}$ (M) Cd^{2+}, Zn^{2+}), M$_7$-MT (M) Cd^{2+}, Zn^{2+}), M$_{12}$-MT (M) Cu$^+$, Ag$^+$), and Ag$_{17}$-MT were obtained by XAS. CD spectroscopy is an essential technique for identifying metal–MT species with well-defined tertiary structures. They found that during titration of apo-MT 1 with Ag(I) at low pH and 50°C, Ag_{12}-MT 1 was characterized by the CD bands at 263 nm (−, very strong) and 314 nm (+, strong), and Ag_{17}-MT 1 was identified by the CD bands at 302 nm (+, very strong) and 370 (−, weak). The difference in the CD features of these two silver metallothioneins was attributed to changes in the tertiary structures that arise from changes in the fraction of bridging and terminal sulfur atoms, rather than from changes in coordination geometry around the Ag(I) (Figure 6.22).

Another example on the combined application of XAS and CD is the study of the effects of MgATP or MgADP binding to the *Azotobacter vinelandii* nitrogenase Fe protein on the properties of the [4Fe–4S] cluster.[102] It was found that MgADP or MgATP binding to the oxidized nitrogenase Fe protein resulted in distinctly different CD spectra, suggesting distinct changes in the environment of the [4Fe–4S] cluster, which suggested that MgADP or MgATP

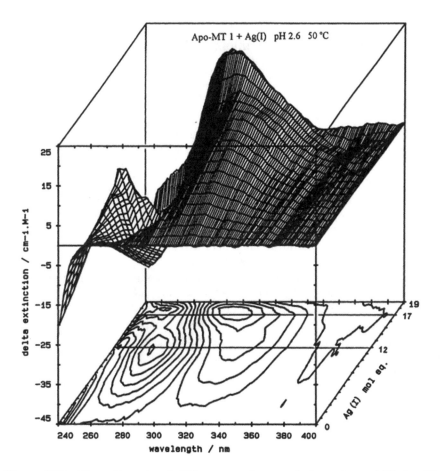

Figure 6.22 Three-dimensional CD spectra for a solution of apo-MT 1 with Ag(I):MT from 0 to 19 recorded at pH 2.7 and 50°C. The grid lines added to the contour diagram are drawn for Ag(I) molar ratios of 12 and 17, respectively.[101] © 1996 American Chemical Society.

binding to the nitrogenase Fe protein induced different conformational changes. The CD spectrum of a [2Fe–2S] form of the nitrogenase Fe protein was also investigated to address the possibility that the MgATP- or MgADP-induced changes in the CD spectrum of the native enzyme were the result of a partial conversion from a [4Fe–4S] cluster to a [2Fe–2S] cluster. No evidence was found for a contribution of a [2Fe–2S] cluster to the CD spectrum of oxidized Fe protein in the absence or presence of nucleotides. Fe K-edge XAS spectra of the oxidized Fe protein revealed no changes in the structure of the [4Fe–4S] cluster upon MgATP binding to the Fe protein. The present results reveal that MgATP or MgADP binding to the oxidized state of the Fe protein results in different conformational changes in the environment around the [4Fe–4S] cluster.

6.4.7 Combination of XAS with Nuclear Magnetic Resonance in Metallomics and Metalloproteomics

Nuclear magnetic resonance (NMR) is one of the techniques chosen for protein structure determination and it is intensively used in structural studies.[103] The NMR procedure for solving protein structures relies on obtaining of constraints on different structural parameters, like interproton distances, dihedral angles, and interatomic vectors, from a variety of NMR experiments.[104] The constraints provide upper limits to the values of the above parameters, which can be used in distance–geometry calculations that eventually determine the protein structure. When dealing with metalloproteins, however, no classical constraints can be obtained on the metal center and NMR is not able to provide its structure. The capability of EXAFS to provide the structure of a metal atom bound to a protein is then perfectly suited to complement the process of the structure determination, as demonstrated by the study of copper transporters in bacteria done by Banci and colleagues[105] who exploited this synergism between NMR and XAS for a rapid and complete elucidation of the structure of proteins involved in copper trafficking in bacteria.

The solution structure of the metal-free form of CopC (copper transport proteins from *Pseudomonas syringae*) has been obtained by NMR methods[106] and the protein shows a Greek-key-barrel fold similar to that adopted by blue copper proteins like plastocyanin, which are involved in electron-transfer processes and host a type-I copper site.[107] In contrast, CopC was shown to be able to bind copper (II) in a type-II binding site probably identified in a His2GluAsp site. A second possible copper(I) binding site is constituted by a Met-rich region present in CopC.[106] In order to quickly identify the nature of the copper binding sites in CopC, an X-ray absorption study of this protein was applied. The two EXAFS spectra indicates that the Cu(I) and Cu(II) ions are bound to different binding sites present in the protein. The data analysis shows that Cu(II) is bound to two histidine residues (0.199 nm) and two oxygen ligands within a short distance (0.197 nm), and that the coordination is completed by one/two further light (O/N) ligands, most probably from water molecules at 0.283 nm. In the

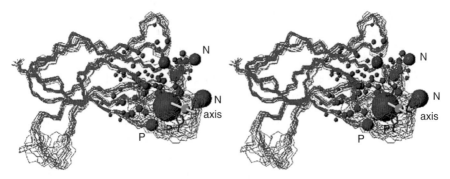

Figure 6.23 Stereoview of the Cu(II)-CopC structure family calculated using NMR constraints and information derived from EXAFS data. Pseudocontact shifts are depicted on the protein frame as spheres. Examples of positive (P) and negative (N) spheres are shown. The radius of each sphere is proportional to the absolute value of the pseudocontact shift. The principal axis of the magnetic susceptibility tensor is also shown (axis).[108] © 2003 American Chemical Society.

other sample the Cu (I) ion is bound to three S atoms at 0.230 nm and one histidine at 0.195 nm. By applying this information to the structure of the apo-CopC, the location of the binding sites in the protein becomes evident since only two sites are present in CopC which fulfil the requirements obtained from the EXAFS analysis. Furthermore, the XAS experiment has shown in a very simple and elegant way that the same copper ion migrates from one site to the other depending on the copper oxidation state. This finding discloses a new functional aspect of CopC that may act as a redox-driven switch in copper transport.[106] The application of the constraints obtained from EXAFS has also provided valuable help in the structure determination of the paramagnetic Cu(II)-bound form of CopC (Figure 6.23).[108]

6.4.8 Combination of XAS with Raman Spectroscopy in Metallomics and Metalloproteomics

Raman spectroscopy is a well-established technique which provides information on the local symmetry and the local atomic structure.[109,110] Being sensitive to local bonds as well as to molecular units, its results appear complementary to those extracted from an XAS experiment, because both give very precise information about the local structure of a sample, both are not restricted to crystalline materials, and in both cases the volumes of the material probed are similar. The Raman spectra of many organic or inorganic compounds have been gathered in several databases, and this can provide strong guidelines for XAS analysis. Similarly, XAS analysis is grounded on the study of model compounds which have a well-defined signature. The key point is to have information at the same time from exactly the same location in the sample, this area being as small as possible for a precise description of the system.

Therefore, combining the two techniques could be a very powerful method, structurally. The experimental setup has recently been developed by Briois *et al.* and has been applied to materials on a macro scale[111,112] or a micro scale.[113]

Tang and co-workers[114] studied methyl-coenzyme M reductase (MCR) to characterize the Ni coordination and oxidation states of the catalytically important MCR_{ox1} and MCR_{red1} forms of the enzyme and to compare these states with the inactive Ni(II) forms of MCR and with the Ni(I), Ni(II), and Ni(III) states of co-factor F_{430} in solution. The Raman data indicated that co-factor F_{430} underwent a significant conformational change when it bound to MCR (Figure 6.24). Furthermore, the vibrational characteristics of the ox1 state and red1 states were significantly different, especially in hydrocorphin ring modes with appreciable CN stretching character. Analyses of the XANES and EXAFS data revealed that both the ox1 and red1 forms were best described as

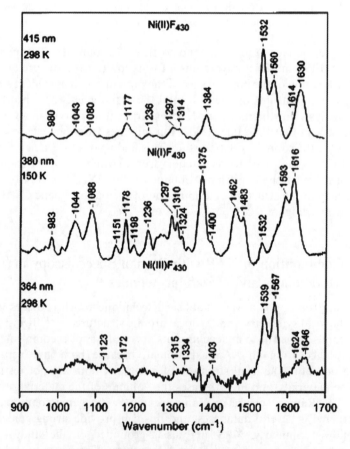

Figure 6.24 High-frequency RR spectra of Ni(II), Ni(I), and Ni(III) co-factor F430. The excitation wavelength and temperature are indicated for each spectrum.[114] © 2002 American Chemical Society.

hexacoordinate and that the main difference between ox1 and red1 was the absence of an axial thiolate ligand in the red1 state.

6.4.9 Combination of XAS with Electron Spin Resonance in Metallomics and Metalloproteomics

Electron spin resonance (ESR), also called electron paramagnetic resonance (EPR), detects unpaired electrons by observing changes in an electron's properties when a magnetic field is applied to a sample. One of the most valuable aspects of EPR spectroscopy is identification of the molecules on which unpaired electrons reside.[115] ESR and XAS are complementary techniques. The former gives geometrical information about paramagnetic metallic cations such as copper(II), whereas the latter is generally used to describe the local structure including bond distance, coordination number, and type of near-neighbors surrounding a specific element.

The association of copper to riboflavin binding protein (RBP) from egg white has been studied by ESR and XAS.[116] EPR results confirm that the copper binding site is a well-ordered type II site. The XAS studies, performed on copper loaded RBP, were used to examine the bound metal's oxidation state and metal–ligand coordination structure. A substantial pre-edge feature at ca. 8978 eV and the minor feature at 8984 eV, indicative of $1s \rightarrow 3d$ and $1s \rightarrow 4p$ electronic transitions typically seen for Cu(II) and Cu(I), respectively, suggest that while the bulk of the copper bound to RBP is Cu(II) a substantial distribution of Cu(I) is also bound to RBP (see Figure 6.25). EXAFS simulations, used to characterize the copper nearest-neighbour ligand structure, indicate an average coordination environment constructed by three oxygen/nitrogen based ligands at an average bond distance of 0.196 nm.

Since photoreduction of copper is common in XAS, EPR spectroscopy was used to quantify the relative amounts of copper(I) and copper(II) bound to the protein by comparison to copper sulfate standard samples. Spin quantification shows that the protein, as loaded aerobically with copper(II) in the absence of external reducing agents contains between 30 and 50% copper(I) in various samples tested, suggesting that while some of the signals for copper(I) seen in the XANES data might be attributed to photoreduction, some reduction of the copper(II) occurs as a result of the binding interaction.

Nitrogenase is a multicomponent metalloenzyme that catalyzes the conversion of atmospheric dinitrogen to ammonia. For decades, it has been generally believed that the [8Fe–7S] P-cluster of nitrogenase component 1 is indispensable for nitrogenase activity. However, Hu et al.[117] identified two catalytically active P-cluster variants by ESR spectroscopic studies, and found that both P-cluster variants resemble [4Fe–4S]-like centers based on XAS experiments.

ESR and XAS have also been used to study the oriented photosystem II (PS II) particles from spinach chloroplasts to determine more details of the structure of the oxygen evolving complex.[118] The nature of halide binding to Mn is

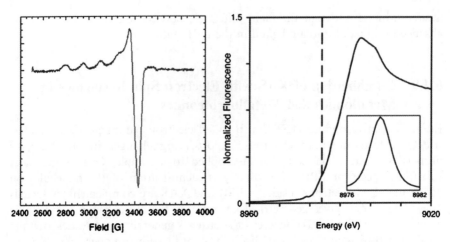

Figure 6.25 Left: continuous wave (CW) EPR of copper-loaded riboflavin binding protein (1.65 mM RBP) in 0.10 mM Tris, pH 7.2. Microwave frequency: 9.7 GHz; microwave power: 2.0 μW; modulation frequency: 10 kHz; modulation amplitude: 10 G; temperature: 10 K. Right: XANES of Cu(II) loaded RBP (top). Vertical dashed line indicates the 1s→4p transition at 8984 eV. Inset shows the expansion of the background-subtracted 1s→3d region of the XANES spectra.[116] © 2008 Bentham Open.

also studied with Cl K-edge and Mn EXAFS of Mn-Cl model compounds, and with Mn EXAFS of oriented PS II in which Br has replaced Cl. A linear relationship was found between the dichroism found in the tyrosine and the mosaic spread. Variation of the sample orientation with respect to the X-ray e-vector yields highly dichroic EXAFS, indicative of an asymmetric tetranuclear cluster.

More work on the combined application of XAS with ESR can be seen in the study of *Panulirus interruptus* hemocyanin in the crystalline state,[119] straw lignin,[120] and the hydrogen binding site in hydrogenase.[121]

Besides the above-mentioned applications of XAS in metallomics and metalloproteomics, further information can be obtained by using more specialized experimental setups. Time-resolved QEXAFS measurements of dynamic processes at the time scale of seconds were made possible by use of focussing crystal optics in combination with a CCD camera or photodiode array.[122–124] For example, the thermal decomposition of di-ammonium hexachloroplatinate (($NH_4)_2[PtCl_6]$) to metallic platinum has been investigated by time-resolved *in situ* XAS measurements at the Cl K edge and Pt L_{III} edge.[124] Spectra at the Cl K edge were recorded by an energy dispersive monochromator (EDM), while the Pt L_{III}-edge spectra were recorded by a double crystal monochromator. A temperature interval between 250°C and 350°C has been chosen. Figure 6.26 shows time-resolved *in situ* Cl K XANES of $(NH_4)_2[PtCl_6]$ taken at the EDM while heating with a constant temperature ramp. Below 300°C no changes in the spectra could be observed but at about 310°C a second pre-edge structure at 2822 eV emerges and at about 2825.5 eV a

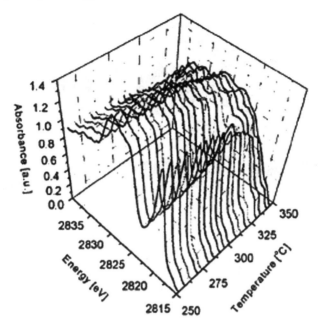

Figure 6.26 Time-resolved *in situ* Cl K-XANES of $(NH_4)_2[PtCl_6]$ taken at the EDM while heating with a constant temperature ramp.[124] © 1999 International Union of Crystallography.

new feature appears. Above 330°C, these spectral indications of an intermediate disappear. Fingerprinting arguments indicate that this intermediate could be cis-platin or some other platinum amine complex.

A combination of time-resolved X-ray absorption spectroscopy, pre-steady state kinetics and computational quantum chemistry is able to study the active site of the zinc ion of bacterial alcohol dehydrogenase during single substrate turnover. A series of alternations in the coordination number and structure of the catalytic zinc ion with concomitant changes in metal–ligand bond distances have been reported.[125] Therefore, probing metalloenzyme reactions by *in situ* XAS is promising for explanations of the changes that occur in critical metal centers during the course of the enzymatic reaction. Specifically for catalytic zinc sites, XAS is the only spectroscopic method that can provide high-resolution structural and electronic information in solution.

6.5 Conclusions and Outlook

Taken together, the application of XAS in metallomics and metalloproteomics and its combination with other techniques can be expressed as in Figure 6.27. XAS, including XANES, EXAFS, and micro-XAS can provide valuable information on the oxidation state, geometry, and nearest coordination shells of absorbing atoms in complex samples. The main advantages are that it can

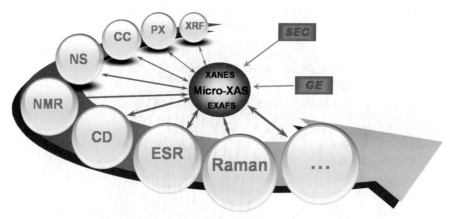

Figure 6.27 The application of XAS in metallomics and metalloproteomics and its combination with other techniques. XRF, X-ray fluorescence spectroscopy; PX, protein crystallography; CC, computational chemistry; NS, neutron scattering; NMR, nuclear magnetic resonance; CD, circular dichroism spectroscopy; ESR, electron spin resonance; SEC, size exclusion chromatography; GE, gel electrophoresis.

determine the percentage of various species in a mixture, and that it can examine the coordination around absorber atoms in spite of the presence of many other types of atoms. By combination with other techniques, XAS shows its greatest capability in metallomics and metalloproteomics.

The disadvantages for XAS are that high concentrations are generally required, although higher X-ray beam intensities are expected to alleviate this obstacle. Speciation via fingerprinting is not absolute; if the proper models are not included, important components can be missed, and some spectra are too similar to be distinguished. This can be alleviated by supplementing XANES with EXAFS and with ligand XANES and EXAFS. EXAFS spectra cannot always distinguish different bonding situations; this can be alleviated by combination with molecular modeling.

The application of XAS in metallomics and metalloproteomics is mainly limited by the availability of beam time at synchrotron radiation sources although it is being improved with the construction of new facilities worldwide. The growing importance of the XAS application is reflected at light sources all over the world by an increasing number of beamlines which are devoted to the study of problems related to metallomics and metalloproteomics. Another general limitation of XAS analysis is that only the average structure can be determined. If the metal of interest is present in multiple environments, such as different oxidation states, the information determined by XAS will represent the mixture of these states[33]. Further, the detection limit of about $100\,\mu g\,g^{-1}$ for XAS may also hinder its application in some metallome and metalloproteome studies.[126]

Despite the wide application of XAS in metallomics and metalloproteomics studies, both theoretical and experimental improvements are still needed. In particular, with the development of proteome, high-throughput (HT) techniques of structural characterization are, therefore, required for metalloproteomics. Scott and co-workers[127] have proposed a HTXAS procedure. They designed a 25-well sample holder with 1.5 mm-diameter holes in a 5×5 arrangement on a 100 wide polycarbonate holder that fits into a liquid helium-flow XAS cryostat. With the beam aperture at 1 mm × 1 mm, the cryostat and sample holder are rastered first to align the wells, then the metal distribution is determined with a multi-channel energy-discriminating solid-state fluorescence detector. The elemental distribution maps detected by XRF are used to target protein samples for further speciation (by XANES) and structural analysis (by EXAFS) with a 30-element intrinsic germanium detector. As a feasibility study, a significant fraction of the B2200 open reading frames (ORFs) in the *Pyrococcus furiosus* (Pf) genome has been individually cloned, tagged, expressed, and purified by high-throughput techniques (Figure 6.28).

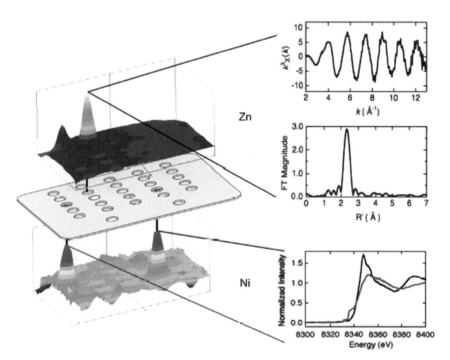

Figure 6.28 Examples of the data collected to test the feasibility of the procedure of HTXAS procedure. Individual gene product samples from the *Pyrococcus furiosus* genome were loaded into a 5×5 spatial array of 3 mL samples (middle left). A single raster scan was used to monitor both Ni (lower left) and Zn (upper left) K emission, identifying the presence of these metals in the indicated wells. Ni XANES speciation (lower right) clearly differentiates the types of Ni binding sites and Zn EXAFS structural analysis (upper right) shows the sensitivity available for determining metal-site structures.[127] © 2005 International Union of Crystallography.

Up to 25 Pf ORF products are loaded (3 mL per well, protein concentration 0.2–1 mM) in a single run. Two Ni-containing proteins and one Zn-containing protein are detected by XRF. Ni XANES speciation clearly differentiates the types of Ni binding sites and Zn EXAFS structural analysis, which shows the sensitivity available for determining metal-site structures.

The development of a new generation of modern synchrotron sources and high efficiency of multi-element detectors will greatly improve the resolution and sensitivity for metallomics and metalloproteomics. Automation of processes for arraying small samples for XAS survey, efficient signal detection, rapid data collection, reduction and analysis for multiple low-volume low-concentration samples must be achieved for HTXAS determination. Recently, a novel Ge pixel array detector (PAD) with 100 segments has been developed for HT and energy-resolution fluorescence XAS. Using a monolithic approach, a nearly perfect packing fraction (>88%), commission rate (99%) and high efficiency over a wide energy range (5–60 keV) were achieved and systematic error related to elastic and inelastic scattering was eliminated, resulting in a significant improvement over a previous Ge multi-element detector. The recent *in situ* XAS study only requires an amount of specimen less than 10^{-3} g.[128]

The theoretical improvement of XAS will also strengthen its application in metallomics and metalloproteomics. As mentioned above, the low-energy part of an XAS spectrum (the XANES region) is of great interest for metallomics and metalloproteomics studies since it is extremely sensitive to the structural details of the absorbing site (overall symmetry, distances and bond angles). However, the quantitative analysis of the full XAS spectrum, including the edge, is a complex many-bodied problem that requires adequate treatment and a time-consuming algorithm in order to calculate the absorbing cross-section in the framework of the full multiple-scattering approach. By using MXAN,[129] the XANES spectra that are related to these configurations are calculated in a reasonable time and applied to a single crystal of the iron protein carbonmonoxy-myoglobin (MbCO) and of its cryogenic photoproduct Mb*CO (Figure 6.29). The extracted local structure of Mb*CO includes an Fe-CO distance of 0.308 (7) nm, with a tilting angle between the heme normal and the Fe-C vector of 37(7)° and a bending angle between the Fe-C vector and the C-O bond of 31.(5)°. More work on the application of MXAN can be found in the study of XANES energy region of *Leptospira interrogans* peptide deformylase (PDF).[130]

In conclusion, with the development of the proteome, high-throughput (HT) techniques of structural characterization are required for metalloproteomics. The development of a new generation of modern synchrotron sources and high efficiency of multi-element detectors will greatly improve the resolution and sensitivity for metallomics and metalloproteomics. Hence, improvements in the state of the art of high-throughput XAS will also facility metallomics and metalloproteomics studies.

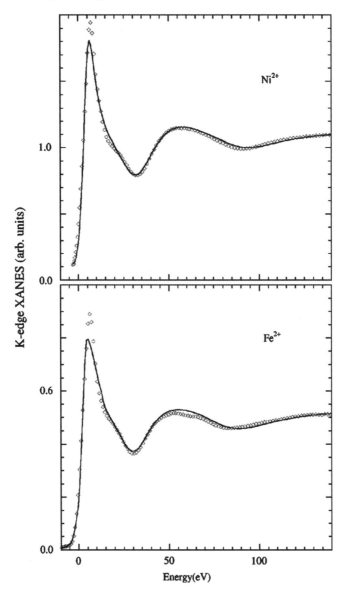

Figure 6.29 Upper frame: experimental (circles) and best-fit calculation (solid line) of the Ni K-edge XANES spectrum of the Ni^{2+} aqua ion. Lower frame: experimental (circles) and best-fit calculation (solid line) of the Fe K-edge XANES spectrum of the Fe^{2+} aqua ion.[129] © 2003 International Union of Crystallography.

Acknowledgments

The authors acknowledge the financial support by the Ministry of Science and Technology of China as the National Basic Research Programs

(2006CB705603), the NSFC/RGC Joint Research Scheme (20931160430) and the Knowledge Innovation Program of the Chinese Academy of Sciences.

References

1. R. Jenkins, in *Encyclopedia of Analytical Chemistry*, ed. R. A. Meyers, John Wiley and Sons Ltd., Chichester, 2000.
2. Y.-F. Li, C. Chen, B. Li, W. Li, L. Qu, Z. Dong, M. Nomura, Y. Gao, J. Zhao, W. Hu, Y. Zhao and Z. Chai, *J. Inorg. Biochem.*, 2008, **102**, 500.
3. V. L. Aksenov, A. Y. Kuzmin, J. Purans and S. I. Tyutyunnikov, *Phys. Part. Nucl.*, 2001, **32**, 1.
4. K. C. Cheung, R. W. Strange and S. S. Hasnain, *Acta Crystallogr.*, 2000, **D56**, 697.
5. S. S. Hasnain and R. W. Strange, *J. Synchrotron Radiat.*, 2003, **10**, 9.
6. Y. Gao, C. Chen and Z. Chai, *J. Anal. At. Spectrom.*, 2007, **22**, 856.
7. S. Hasnain, *J. Synchrotron Radiat.*, 2004, **11**, 7.
8. Y.-F. Li, C. Chen, Y. Qu, Y. Gao, B. Li, Y. Zhao and Z. Chai, *Pure Appl. Chem.*, 2008, **80**, 2577.
9. Y.-F. Li, Y. Gao, C. Chen, B. Li, Y. Zhao and Z. Chai, *Sci. China Ser. B-Chem.*, 2009, **39**, 1.
10. W. Shi and M. Chance, *Cell. Mol. Life Sci.*, 2008, **65**, 3040.
11. J. Stöhr, *NEXAFS Spectroscopy*, Springer, Berlin, 1992.
12. D. C. Koningsberger and R. Prins, *X-ray Absorption: Principles, Applications, Techniques of EXAFS, SEXAFS and XANES*, John Wiley and Sons Inc., New York, 1988.
13. D. E. Sayers, E. A. Stern and F. W. Lytle, *Phys. Rev. Lett.*, 1971, **27**, 1204.
14. A. L. Ankudinov, B. Ravel, J. J. Rehr and S. D. Conradson, *Phys. Rev. B, Condens. Matter Mater. Phys.*, 1998, **58**, 7565.
15. B. K. Teo, *EXAFS: Basic Principles and Data Analysis*, Springer-Verlag, Berlin, 1986.
16. T. Ressler, *J. Synchrotron Radiat.*, 1998, **5**, 118.
17. M. Newville, *J. Synchrotron Radiat.*, 2001, **8**, 322.
18. B. Ravel and M. Newville, *J. Synchrotron Radiat.*, 2005, **12**, 537.
19. G. N. George, S. J. Geroge and I. J. Pickering, *EXAFSPAK* (http://www-ssrl.slac.stanford.edu/~george/exafspak/exafs.htm), Stanford Synchrotron Radiation Laboratory, California, 2001.
20. P. J. Ellis and H. C. Freeman, *J. Synchrotron Radiat.*, 1995, **2**, 190.
21. W. Zhong, B. He, Z. Li and S. Wei, *J. Univ. Sci. Technol. Chin.*, 2001, **31**, 328.
22. L. Zhang, M. J. Crossley, N. E. Dixon, P. J. Ellis, M. L. Fisher, G. F. King, P. E. Lilley, D. MacLachlan, R. J. Pace and H. C. Freeman, *J. Biol. Inorg. Chem.*, 1998, **3**, 470.
23. I. J. Pickering, G. E. Brown and T. K. Tokunaga, *Environ. Sci. Technol.*, 1995, **29**, 2456.

24. H. H. Harris, I. J. Pickering and G. N. George, *Science*, 2003, **301**, 1203.
25. J. Gailer, G. N. George, I. J. Pickering, S. Madden, R. C. Prince, E. Y. Yu, M. B. Denton, H. S. Younis and H. V. Aposhian, *Chem. Res. Toxicol.*, 2000, **13**, 1135.
26. T. Arai, T. Ikemoto, A. Hokura, Y. Terada, T. Kunito, S. Tanabe and I. Nakai, *Environ. Sci. Technol.*, 2004, **38**, 6468.
27. V. Van Fleet-Stalder, T. G. Chasteen, I. J. Pickering, G. N. George and R. C. Prince, *Appl. Environ. Microbiol.*, 2000, **66**, 4849.
28. I. J. Pickering, R. C. Prince, M. J. George, R. D. Smith, G. N. George and D. E. Salt, *Plant Physiol.*, 2000, **122**, 1171.
29. W. M. Kwiatek, M. Galka, A. L. Hanson, C. Paluszkiewicz and T. Cichocki, *J. Alloys Compd.*, 2001, **328**, 276.
30. T. E. Gunter, L. M. Miller, C. E. Gavin, R. Eliseev, J. Salter, L. Buntinas, A. Alexandrov, S. Hammond and K. K. Gunter, *J. Neurochem.*, 2004, **88**, 266.
31. M. Yasoshima, M. Matsuo, A. Kuno and B. Takano, *J. Synchrotron Radiat.*, 2001, **8**, 969.
32. U. Kramer, I. J. Pickering, R. C. Prince, I. Raskin and D. E. Salt, *Plant Physiol.*, 2000, **122**, 1343.
33. S. Chevreux, S. Roudeau, A. Fraysse, A. Carmona, G. Devès, P. L. Solari, T. C. Weng and R. Ortega, *J. Anal. At. Spectrom.*, 2008, **23**, 1117.
34. B. Akabayov, C. J. Doonan, I. J. Pickering, G. N. George and I. Sagi, *J. Synchrotron Radiat.*, 2005, **12**, 392.
35. F. Jalilehvand, *Chem. Soc. Rev.*, 2006, **35**, 1256.
36. K. Kobayashi, K. Hieda, H. Maezawa, Y. Furusawa, M. Suzuki and T. Ito, *Int. J. Radiat. Biol.*, 1991, **59**, 643.
37. T. Ressler, J. Wong, J. Roos and I. L. Smith, *Environ. Sci. Technol.*, 2000, **34**, 950.
38. R. Andrahennadi, M. Wayland and I. J. Pickering, *Environ. Sci. Technol.*, 2007, **41**, 7683.
39. D. Thavarajah, A. Vandenberg, G. N. George and I. J. Pickering, *J. Agric. Food Chem.*, 2007, **55**, 7337.
40. Z. Huang, T. Chen, M. Lei, T. Hu and Q. Huang, *Sci. China Ser. C Life Sci.*, 2004, **47**, 124.
41. R. G. Shulman, P. Eisenberger, W. E. Blumberg and N. A. Stombaugh, *Proc. Natl. Acad. Sci. U. S. A.*, 1975, **72**, 4003.
42. B. Bunker and E. A. Stern, *Biophys. J.*, 1977, **19**, 253.
43. A. Pattanaik, G. Bachowski, J. Laib, D. Lemkuil, C. F. Shaw, D. H. Petering, A. Hitchcock and L. Saryan, *J. Biol. Chem.*, 1992, **267**, 16121.
44. R. C. Elder, K. Ludwig, J. N. Cooper and M. K. Eidsness, *J. Am. Chem. Soc.*, 1985, **107**, 5024.
45. P. J. Stephens, T. V. Morgan, F. Devlin, J. E. Penner-Hahn, K. O. Hodgson, R. A. Scott, C. D. Stout and B. K. Burgess, *Proc. Natl. Acad. Sci. U. S. A.*, 1985, **82**, 5661.
46. S. P. Cramer, J. J. Moura, A. V. Xavier and J. LeGall, *J. Inorg. Biochem.*, 1984, **20**, 275.

47. J. Chen, J. Christiansen, N. Campobasso, J. T. Bolin, R. C. Tittsworth, B. J. Hales, J. J. Rehr and S. P. Cramer, *Angew. Chem. Int. ed. Eng.*, 1993, **32**, 1592.
48. H. Beinert, *Eur. J. Biochem.*, 1997, **245**, 521.
49. O. Y. Gavel, S. A. Bursakov, J. J. Calvete, G. N. George, J. J. G. Moura and I. Moura, *Biochemistry*, 1998, **37**, 16225.
50. L. Jacquamet, D. Aberdam, A. Adrait, J. L. Hazemann, J. M. Latour and I. Michaud-Soret, *Biochemistry*, 1998, **37**, 2564.
51. N. R. Bastian, G. Diekert, E. C. Niederhoffer, B. K. Teo, C. T. Walsh and W. H. Orme-Johnson, *J. Am. Chem. Soc.*, 1988, **110**, 5581.
52. I. J. Pickering, R. C. Prince, G. N. George, W. E. Rauser, W. A. Wickramasinghe, A. A. Watson, C. T. Dameron, I. G. Dance, D. P. Fairlie and D. E. Salt, *Biochim. Biophys. Acta*, 1999, **1429**, 351.
53. D. Strozyk and A. I. Bush, in *Neurodegenerative Diseases and Metal Ions: Metal Ions in Life Sciences,* ed. A. Sigel, H. Sigel and R. Sigel, Wiley, Chichester, 2006, **vol. 1**, p. 1.
54. F. Stellato, G. Menestrina, M. Serra, C. Potrich, R. Tomazzolli, W. Meyer-Klaucke and S. Morante, *Eur. Biophys. J.*, 2006, **35**, 340.
55. C. Jacobsen and J. Kirz, *Nat. Struct. Biol.*, 1998, **5**, 650.
56. T. Bacquart, G. Deves, A. Carmona, R. Tucoulou, S. Bohic and R. Ortega, *Anal. Chem.*, 2007, **79**, 7353.
57. W. Yun, S. T. Pratt, R. M. Miller, Z. Cai, D. B. Hunter, A. G. Jarstfer, K. M. Kemner, B. Lai, H. R. Lee, D. G. Legnini, W. Rodrigues and C. I. Smith, *J. Synchrotron Radiat.*, 1998, **5**, 1390.
58. D. G. Schulze, T. McCay-Buis, S. R. Sutton and D. M. Huber, *Phytopathology*, 1995, **85**, 990.
59. R. Ortega, B. Fayard, M. Salome, G. Deves and J. Susini, *Chem. Res. Toxicol.*, 2005, **18**, 1512.
60. K. G. Scheckel, E. Lombi, S. A. Rock and M. J. McLaughlin, *Environ. Sci. Technol.*, 2004, **38**, 5095.
61. D. B. Hunter, P. M. Bertsch, K. M. Kemner and S. B. Clark, *J. Phys. IV*, 1997, **7**, 767.
62. I. J. Pickering, R. C. Prince, D. E. Salt and G. N. George, *PNAS*, 2000, **97**, 10717.
63. E. Bulska, I. A. Wysocka, M. H. Wierzbicka, K. Proost, K. Janssens and G. Falkenberg, *Anal. Chem.*, 2006, **78**, 7616.
64. X. Zhang, R. Balhorn, J. Mazrimas and J. Kirz, *J. Struct. Biol.*, 1996, **116**, 335.
65. A. Ito, S. K. H. Nakano, T. Matsumura and K. Kinoshita, *J. Microsc.*, 1996, **181**, 54.
66. B. D. Hames, *Gel Electrophoresis of Proteins: A Practical Approach,* Oxford University Press, New York, 1998.
67. M. Hough and S. S. Hasnain, *J. Mol. Biol.*, 1999, **287**, 579.
68. J. A. Howe, R. H. Loeppert, V. J. DeRose, D. B. Hunter and P. M. Bertsch, *Environ. Sci. Technol.*, 2003, **37**, 4091.

69. K. L. Munro, A. Mariana, A. I. Klavins, A. J. Foster, B. Lai, S. Vogt, Z. Cai, H. H. Harris and C. T. Dillon, *Chem. Res. Toxicol.*, 2008, **21**, 1760.
70. I. Ascone, R. Fourme, S. Hasnain and K. Hodgson, *J. Synchrotron Radiat.*, 2005, **12**, 1.
71. G. Dalba, P. Fornasini, R. Grisenti and J. Purans, *Phys. Rev. Lett.*, 1999, **82**, 4240.
72. R. F. Pettifer, O. Mathon, S. Pascarelli, M. D. Cooke and M. R. J. Gibbs, *Nature*, 2005, **435**, 78.
73. M. C. Corbett, F. A. Tezcan, O. Einsle, M. Y. Walton, D. C. Rees, M. J. Latimer, H. B. and K. O. Hodgson, *J. Synchrotron Radiat.* 2005, **12**, 28.
74. G. N. George, B. Hedman and K. O. Hodgson, *Nat. Struct. Biol.*, 1998, **5**, 645.
75. J. C. Boyington, V. N. Gladyshev, S. V. Khangulov, T. C. Stadtman and P. D. Sun, *Science*, 1997, **275**, 1305.
76. G. N. George, C. M. Colangelo, J. Dong, R. A. Scott, S. V. Khangulov, V. N. Gladyshev and T. C. Stadtman, *J. Am. Chem. Soc.*, 1998, **120**, 1267.
77. E. Pohl, J. C. Haller, A. Mijovilovich, W. Meyer-Klaucke, E. Garman and M. L. Vasil, *Mol. Microbiol.*, 2003, **47**, 903.
78. R. W. Strange and S. S. Hasnain, in *Protein–Ligand Interactions: Methods and Applications*, ed. G. U. Nienhaus, Humana Press, New York. 2005, vol. **305**, p. 167.
79. J. Chan, M. E. Merrifield, A. V. Soldatov and M. J. Stillman, *Inorg. Chem.*, 2005, **44**, 4923.
80. U. Ryde, *Dalton Trans.*, 2007, 607.
81. H. M. Senn and W. Thiel, *Curr. Opin. Chem. Biol.*, 2007, **11**, 182.
82. M. L. Barrett, I. Harvey, M. Sundararajan, R. Surendran, J. F. Hall, M. J. Ellis, M. A. Hough, R. W. Strange, I. H. Hillier and S. S. Hasnain, *Biochemistry*, 2006, **45**, 2927.
83. L. Basumallick, R. Sarangi, S. DeBeer George, B. Elmore, A. B. Hooper, B. Hedman, K. O. Hodgson and E. I. Solomon, *J. Am. Chem. Soc.*, 2005, **127**, 3531.
84. S. Loscher, L. Schwartz, M. Stein, S. Ott and M. Haumann, *Inorg. Chem.*, 2007, **46**, 11094.
85. J. Yano, J. Robblee, Y. Pushkar, M. A. Marcus, J. Bendix, J. M. Workman, T. J. Collins, E. I. Solomon, S. DeBeer George and V. K. Yachandra, *J. Am. Chem. Soc.*, 2007, **129**, 12989.
86. F. Sepulcre, A. Cordomi, M. G. Proietti, J. J. Perez, J. Garcia, E. Querol and E. Padros, *Proteins Struct. Funct. Bioinf.*, 2007, **67**, 360.
87. A. Dey, T.-A. Okamura, N. Ueyama, B. Hedman, K. O. Hodgson and E. I. Solomon, *J. Am. Chem. Soc*, 2005, **127**, 12046.
88. Y.-W. Hsiao, Y. Tao, J. E. Shokes, R. A. Scott and U. Ryde, *Phys. Rev. B, Condens. Matter Mater. Phys.*, 2006, **74**, 214101.
89. I. Hazemann, M. T. Dauvergne, M. P. Blakeley, F. Meilleur, M. Haertlein, A. Van Dorsselaer, A. Mitschler, D. A. A. Myles and A. Podjarny, *Acta Cryst. D*, 2005, **61**, 1413.

90. N. Niimura and R. Bau, *Acta Cryst. A*, 2008, **64**, 12.
91. M. Medarde, C. Dallera, M. Grioni, B. Delley, F. Vernay, J. Mesot, M. Sikora, J. A. Alonso and M. J. Martinez-Lope *Physical Review B*, 2009, **80**, 245105.
92. L. F. Gladden, S. R. Elliott, G. N. Greaves, S. Cummings and T. Rayment, *J. Non-Cryst. Solids*, 1985, **77–78**, 1199.
93. M. Winterer, R. Delaplane and R. McGreevy, *J. Appl. Cryst.*, 2002, **35**, 434.
94. M. C. Bellissent-Funel, L. Bosio, A. Hallbrucker, E. Mayer and R. Sridi-Dorbez, *J. Chem. Phys.*, 1992, **97**, 1282.
95. S. Beychok, *Science*, 1966, **154**, 1288.
96. B. Wallace, *J. Synchrotron Radiat.*, 2000, **7**, 289.
97. J. W. Curtis Johnson, *Proteins Struct. Funct. Genet.*, 1990, **7**, 205.
98. W. Cai and M. J. Stillman, *J. Am. Chem. Soc.*, 1988, **110**, 7872.
99. W. Lu, A. J. Zelazowski and M. J. Stillman, *Inorg. Chem.*, 1993, **32**, 919.
100. D. T. Jiang, S. M. Heald, T. K. Sham and M. J. Stillman, *J. Am. Chem. Soc.*, 1994, **116**, 11004.
101. Z. Gui, A. R. Green, M. Kasrai, G. M. Bancroft and M. J. Stillman, *Inorg. Chem.*, 1996, **35**, 6520.
102. M. J. Ryle, W. N. Lanzilotta, L. C. Seefeldt, R. C. Scarrow and G. M. Jensen, *J. Biol. Chem.*, 1996, **271**, 1551.
103. R. Riek, J. Fiaux, E. B. Bertelsen, A. L. Horwich and K. Wuthrich, *J. Am. Chem. Soc.*, 2002, **124**, 12144.
104. K. Wuthrich, *Nat. Struct. Mol. Biol.*, 2001, **8**, 923.
105. L. Banci, I. Bertini and S. Mangani, *J. Synchrotron Radiat.*, 2005, **12**, 94.
106. F. Arnesano, L. Banci, I. Bertini, S. Mangani and A. R. Thompsett, *Proc. Natl. Acad. Sci. U. S. A.*, 2003, **100**, 3814.
107. J. M. Guss and H. C. Freeman, *J. Mol. Biol.*, 1983, **169**, 521.
108. F. Arnesano, L. Banci, I. Bertini, I. C. Felli, C. Luchinat and A. R. Thompsett, *J. Am. Chem. Soc.*, 2003, **125**, 7200.
109. P. R. Carey, *J. Biol. Chem.*, 1999, **274**, 26625.
110. H. Peng, G. Zhou and C. Huang, *Chemistry*, 2005, **68**, 1.
111. V. Briois, D. Lutzenkirchen-Hecht, F. Villain, E. Fonda, S. Belin, B. Griesebock and R. Frahm, *J. Phys. Chem. A*, 2005, **109**, 320.
112. V. Briois, S. Belin, F. Villain, F. Bouamrane, H. Lucas, R. Lescouëzec, M. Julve, M. Verdaguer, M. S. Tokumoto and C. V. Santilli, *Phys. Scr.*, 2005, **115**, 38.
113. V. Briois, D. Vantelon, F. Villain, B. Couzinet, A. M. Flank and P. Lagarde, *J. Synchrotron Radiat.*, 2007, **14**, 403.
114. Q. Tang, P. E. Carrington, Y.-C. Horng, M. J. Maroney, S. W. Ragsdale and D. F. Bocian, *J. Am. Chem. Soc.*, 2002, **124**, 13242.
115. J. E. Wertz and J. R. Bolton, *Electron Spin Resonance: Elementary Theory and Practical Applications,* Chapman and Hall, New York, 1986.
116. S. R. Smith, K. Z. Bencze, K. Wasiukanis, T. L. Stemmler and M. Benore-Parsons, *Open Inorg. Chem. J.*, 2008, **2**, 22.

117. Y. Hu, M. C. Corbett, A. W. Fay, J. A. Webber, B. Hedman, K. O. Hodgson and M. W. Ribbe, *Proc. Natl. Acad. Sci. U. S. A.*, 2005, **102**, 13825.
118. J. C. Andrews, University of California, Berkeley, 1995, PhD thesis.
119. A. Volbeda, M. C. Feiters, M. G. Vincent, E. Bouwman, B. Dobson, K. H. Kalk, J. Reedijk and W. G. J. Hol, *Eur. J. Biochem.*, 1989, **181**, 669.
120. K. Flogeac, E. Guillon, E. Marceau and M. Aplincourt, *New J. Chem.*, 2003, **27**, 714.
121. J. P. Whitehead, R. J. Gurbiel, C. Bagyinka, B. M. Hoffman and M. J. Maroney, *J. Am. Chem. Soc.*, 1993, **115**, 5629.
122. J. Als-Nielsen, G. Grübel and B. S. Clausen, *Nucl. Instrum. Methods. Phys. Res. Sect. B*, 1995, **97**, 522.
123. T. Liu, Y. Xie and T. Hu, *Nucl. Technol.*, 2004, **27**, 401.
124. H. Rumpf, J. Hormes, A. Moller and G. Meyer, *J. Synchrotron Radiat.*, 1999, **6**, 468.
125. O. Kleifeld, A. Frenkel, J. M. L. Martin and I. Sagi, *Nat. Struct. Biol.*, 2003, **10**, 98.
126. R. Ortega, *Metallomics*, 2009, **1**, 137.
127. R. A. Scott, J. E. Shokes, N. J. Cosper, F. E. Jenney and M. W. W. Adams, *J. Synchrotron Radiat.*, 2005, **12**, 19.
128. H. Oyanagi, C. Fonne, D. Gutknecht, P. Dressler, R. Henck, M. O. Lampert, S. Ogawa, K. Kasai, A. Fukano and S. Mohamed, *Phys. Scr.*, 2005, **115**, 1004.
129. M. Benfatto, S. Della Longa and C. R. Natoli, *J. Synchrotron Radiat.*, 2003, **10**, 51.
130. X.-Y. Guo, W.-S. Chu, W.-M. Gong, Y.-H. Dong, Y.-N. Xie, F.-F. Yang, M. Benfatto and Z.-Y. Wu, *High Energy Phys. Nucl. Phys.*, 2007, **31**, 199.

CHAPTER 7
Protein Crystallography for Metalloproteins

ZENGQIANG GAO, HAIFENG HOU AND YUHUI DONG*

Beijing Synchrotron Radiation Facility, Institute of High Energy Physics, Chinese Academy of Sciences, Beijing 100049, China

7.1 Introduction

The structures of proteins are the key data for understanding the functions of proteins. There are three main techniques for measuring the structures of proteins: nuclear magnetic resonance (NMR), cryo-electron microscopy (cryo-EM) and protein crystallography (X-ray diffraction). The solved structures of macromolecules, including proteins, nucleic acids and their complexes, are deposited in the Protein Data Bank (PDB). Up to 23 June of 2009, a total of 58 414 structures of macromolecules had been released in the PDB, and these were mostly protein structures (Table 7.1). Among these structures, about 86% were determined by protein crystallography.

To solve the structure of proteins by using NMR, a series of nuclear magnetic resonant peak positions of proteins in solutions is measured. From the peak shifts due to chemical environments, the stereochemical configurations of the atoms which can be attributed to the peaks can be calculated, such as the bond lengths, bond angles, and dihedral angles. According to the stereochemistry information obtained, the structures of the macromolecules can be constructed.[1–3] However, due to the limitations in the resolutions of NMR

---

*Corresponding author: Email: dongyh@ihep.ac.cn; Tel: +86-10-88233090; Fax: +86-10-88233201.

Nuclear Analytical Techniques for Metallomics and Metalloproteomics
Edited by Chunying Chen, Zhifang Chai and Yuxi Gao
© Royal Society of Chemistry 2010
Published by the Royal Society of Chemistry, www.rsc.org

Table 7.1 The structures available in the Protein Data Bank (PDB) at 23 June 2009.

Technique used	Molecule type				
	Proteins	Nucleic acids	Protein/NA complexes	Other	Total
X-rays	46 778	1155	2168	17	50 118
NMR	6897	856	147	6	7906
Electron microscopy	168	16	59	0	243
Other	113	4	4	9	130
Total	53 970	2032	2379	33	58 414

Available at: http://www.pdb.org/pdb/statistics/holdings.do (Reprinted with permission from RCSB PDB.)

spectrometers, if the macromolecules are too big, or, in other words, contain too many atoms, the number of nuclear magnetic resonant peaks becomes too large and it becomes impossible to distinguish the individual peaks from the NMR spectrograms, therefore structure determinations become impossible. In general, the largest molecules that can be measured by NMR should be smaller than 40 kDa. Even though there is the limitation of molecular weight in NMR, this method is still widely applied in biological research due to the ability to study the dynamic processes of proteins in solutions.

In contrast, cryo-EM can handle much bigger molecules, even the huge complexes such as subcellular organelles.[4] It is worth noting that the cryo-EM technique referred to here is different from the EM technique used in materials research. By taking a photograph of a macromolecule at very low temperature (in order to avoid radiation damage) a projection of this molecule can be obtained. If enough projections are obtained, then the whole three-dimensional structure can be reconstructed by using computer tomography. However, radiation damage as a result of electron bombardment is very serious so the intensities of the electron beams used in cryo-EM studies of macromolecule can not be higher than 10 electrons per Å^2. Hence, the quality of the pictures is very poor. The result is that, if a high-quality high-resolution three-dimensional structure is required, a huge number of pictures has to be collected in order to increase the statistic. Usually thousands of particles can be distinguished from a series of pictures, and the resolutions of the reconstructed structures are typically 20–30 Å. For some high symmetrical particles, such as icosahedron or dodecahedron viruses, a sub-atomic resolution of about 3–4 Å becomes possible.

From the statistical data of structures released in the PDB, most of the high-quality and high-resolution structures have been obtained by protein crystallography. Although there are still some limitations in the structure determination of proteins by crystallography, the technique is the most widely used and precious method for studying the structures of macromolecules. The progress of protein crystallography is mainly a result of crystallography theory, cryo-protection techniques and the use of synchrotron radiation.

7.2 Structure Determination by Protein Crystallography

The most difficult and unsolved part of crystallography is the growth of well-diffracted crystals. Although, many years ago, it was shown that it was possible to crystallize proteins, there are still no systematic theories or instructions to guarantee the crystallization of one protein. We can only rely on the experience of crystallizing proteins of known structure. Every protein is unique, even if there is only a difference of one amino acid between one protein and another. Since the structures of proteins are unknown, the chemical properties are unclear; even the structures have been revealed, the chemical features of the proteins are still almost impossible to predict due to the complication of proteins. In such cases, the possibility of discovering suitable conditions for crystallization of a protein is really low. In general, the only way to obtain protein crystals is to test as many conditions as possible that have been successful for other known proteins. Even though the prediction of protein crystallization is still impossible, there have been many successful cases that can be referred to and some of these cases should be suitable for the proteins we are studying.

Assume that a protein has been crystallized and well-diffracted crystals are obtained. The next step is to place this crystal in an X-ray beam. Since the diffractions of protein crystals are very weak, a strong X-ray source, for example, synchrotron radiation, is necessary. Due to the periodicity of the protein molecular arrangement, a series of diffraction peaks can be detected. From the intensities of the diffraction peaks, $I(h,k,l)$, the structure of protein crystal can be solved.[5]

The theory of diffraction can be found in any crystallography book. The so-called structure factor $F(h,k,l)$, where h, k, and l are the Miller indexes for the diffraction spots, is the Fourier transformation of the electron densities inside the crystal cell, $\rho(x,y,z)$:

$$F(h,k,l) = \int_V \rho(x,y,z) \exp[2\pi i(hx + ky + lz)] \mathrm{d}x\mathrm{d}y\mathrm{d}z \qquad (7.1)$$

where V indicates the volume of crystal cell and the integration is applied to the whole cell. In other words, the electron densities inside the crystal cell can be reduced form the structure factors via inversed Fourier transformation:

$$\rho(x,y,z) = \frac{1}{V} \sum_{h,k,l} F(h,k,l) \exp[-2\pi i(hx + ky + lz)] \qquad (7.2)$$

Therefore, if many structure factors of a crystal can be obtained via any technique, the electron densities inside the crystal cell can be calculated. From the densities of electrons in the crystal cells, the positions of the atoms in the crystal cells can also be known, *i.e.* the structures of this crystal cell can be determine. Remember that the crystal cells are composites with several protein molecules, so the structures of the proteins inside the cells are known.

The intensities measured in detectors are proportional to the squares of the amplitudes of structure factors:

$$I(h,k,l) \propto |F(h,k,l)|^2 \qquad (7.3)$$

Unfortunately the structure factors are complex numbers, not only amplitudes $|F(h,k,l)|$, but also the phases of the structure factors, $\Psi(h,k,l)$, are necessary to determine this complex (Figure 7.1). Until now, the frequency of X-rays is so high that no instrument has been able to directly measure the phase of this type of electromagnetic wave.

It may be questioned whether such a parameter, the phase, is really important for structure determination in crystallography. The answer is not clear because the main information about structure is hidden in the phase not in the intensity of diffraction. In Fourier and inversed Fourier transformation phase, the phases are much more important than the amplitudes. This point is illustrated by the following examples.

Assume we have an object – a "duck"; that is, we construct an electron density distribution with a duck shape. The Fourier transformation of this duck, of course, gives the diffraction pattern. Only the amplitudes of the diffraction can be detected directly, so what we get is a diffraction pattern as in Figure 7.2. It must be kept in mind that the information of phases is lost during the detection of diffraction.

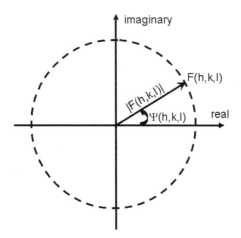

Figure 7.1 Sketch map of a structure factor $F(h,k,l)$. The structure factor is a complex number containing the real and imaginary parts. The amplitude $|F(h,k,l)|$ can be reduced from the measured intensity $I(h,k,l)$ but the phase $\Psi(h,k,l)$ is unknown.

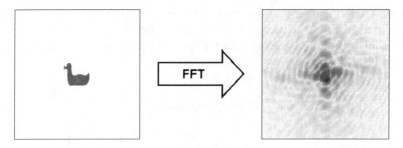

Figure 7.2 The structure of a duck and its diffraction pattern calculated by Fourier transformation. From http://www.ysbl.york.ac.uk/~cowtan/fourier/fourier.html. (Reprinted with permission from Kevin Cowtan.)

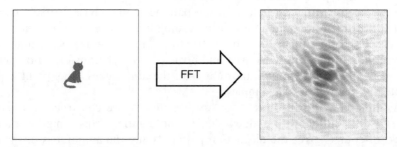

Figure 7.3 The structure of a cat and its diffraction pattern calculated by Fourier transformation. From http://www.ysbl.york.ac.uk/~cowtan/fourier/fourier.html. (Reprinted with permission from Kevin Cowtan.)

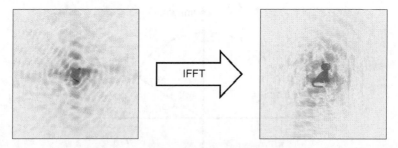

Figure 7.4 The combination of the amplitudes of the diffraction of duck with the phases of cat yields the main features of the cat. From http://www.ysbl.york.ac.uk/~cowtan/fourier/fourier.html. (Reprinted with permission from Kevin Cowtan.)

Another structure made of "cat" can be processed in the same way, and the diffraction pattern is also calculated by Fourier transformation (Figure 7.3).

So, if we combine the amplitudes of the diffraction of duck with the phases of cat, we can obtain a set of structure factors (Figure 7.4). The inverse Fourier

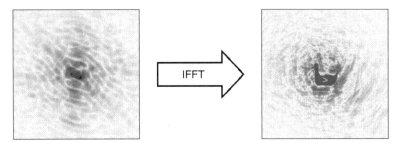

Figure 7.5 The combination of the amplitudes of the diffraction of cat with the phases of duck yields the main features of the duck. From http://www.ysbl.york.ac.uk/~cowtan/fourier/fourier.html. (Reprinted with permission from Kevin Cowtan.)

transformation of this set of structure factors should give some structure. Amazingly, the result of inverse Fourier transformation gives the main structure of cat, but not the duck.

In the same way, the structure factors combined with the amplitudes of cat and phases of duck, after inverse Fourier transformation, show the features of duck (Figure 7.5).

It is clear that the key elements for controlling the features we can get after inverse Fourier transformation are the phases of the diffraction, which can not be obtained directly in diffractions. The easily archived data, the amplitudes of structure factors, only play very weak roles in the structures. That is the so-called "phase problem". The main task of structure determinations by crystallography is to determine the lost phases.

After the phases of all observed diffraction spots have been solved, then the structure factors can be obtained. Therefore, the inverse Fourier transformation would give the electron densities inside the crystal cell. According to the electron densities, the positions of the atoms in the crystal cell can be decided then the initial models of the structures we need can be defined. One must be sure that these initial models are very rough since the errors in measuring the diffraction intensities, and also the calculations of phases would induce a great number of errors in the positions of atoms. The next step for structure determination is to refine the positions of atoms according to the experimental data of the amplitudes of structure factors via the stereochemistry constrains of proteins to obtained the final models of proteins.

For macromolecules such as proteins, the numbers of atoms that compose molecules are huge, therefore the crystal cells contain large numbers of atoms. It is not possible to apply the methods for small molecules, such as the direct method or Patterson map searching, in the structure determinations of proteins. The methods for retrieving the phases of protein crystal diffractions are molecular replacement, isomorphous replacement and anomalous scattering. In recent years, the direct method, which has been widely and successfully used in the determination of small-molecule structures, has also been applied in protein crystallography.

The concept of molecular replacement can be described as the following. Assuming that the structure of a protein (target structure) is required, we also can find another protein, whose structure is known and similar to the target structure (model structure). The molecular replacement method can find a set of rotation angles and translation positions to locate the model structure into the crystal of the target structure. Therefore the initial model for solving the target structure is obtained, and then the initial model can be refined via the experimental data to obtain the final target structure.

The arrangements of the protein molecules in the crystals, *i.e.* the symmetries of the crystals, determine the patterns of diffraction, and the structures of the protein molecules are reflected in the intensities of all diffraction intensities. Therefore similar arrangements or structures of proteins molecules should contribute similar diffraction patterns or intensities. In principle, from the known structures, the locations and translations of similar proteins inside different crystal cells can be deduced, and then the target proteins can be "replaced" by known model proteins to obtain initial models for the unknown crystal structures. This is the concept of molecular replacement. This method can be applied to proteins with similar structure; for example, high sequence homologies, mutant proteins, proteins after chemical modification or substrate-binding protein.

To determine the rotation angles of model proteins in an unknown crystal, Rossman and Blow[6] defined the rotation function:

$$R(\alpha, \beta, \gamma) = \int_u P_1(u) \times P_2(u') \, du \tag{7.4}$$

The rotations of model protein molecules are defined by three Eularian angles, α, β, and γ. $P_1(u)$ is the Patterson function of target crystal structure and $P_2(u')$ the Pattern function of the model crystal structure after the rotation defined by the three Eularian angles (Figure 7.6). The symbol u denotes the integration volume, usually a spherical space large enough to contain the whole model structure and target structure. By calculating the values of rotation function with different Eularian angles, the maximum of the rotation function can be found when the orientation of model molecules is the same as the unknown target protein molecules in the crystal. Therefore the orientation can be defined.

After the orientations have been solved, then we need to locate the model proteins in the unknown crystal cell, e.g. where the molecules are. Crowther[7] proposed the translation function:

$$T(t) = \int_v P(u) \times P_M(v, t) \, du \tag{7.5}$$

When the positions of model proteins are the same as unknown target proteins, the translation function also gets a maximum value (Figure 7.7).

The key point of such strategy for refining the rotation and translation values of model molecules in molecular replacement is to isolate the rotation and translation. Assuming that the structure of a crystal with cell dimensions of

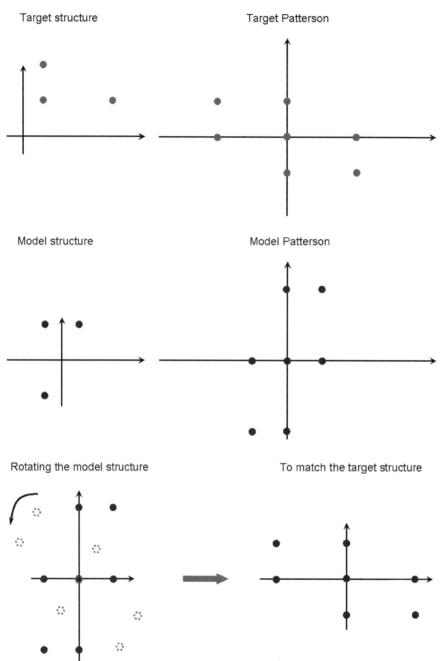

Figure 7.6 The sketch map of finding the rotation in molecular replacement. The target structure is unknown and a similar model structure is known. By rotating the Patterson map of model structure and calculating the overlap of the two Patterson maps, the rotation function is defined. The maximums of the rotation function yield the orientation of the target molecules in the crystals.

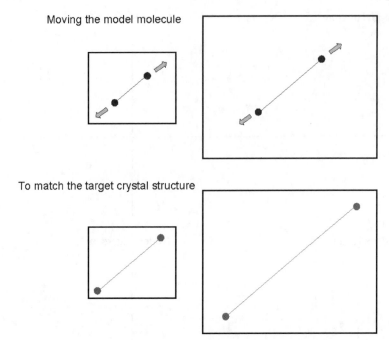

Figure 7.7 The sketch map of finding the position of target proteins in molecular replacement. By moving the model molecules and calculating the overlap of the two Patterson maps, the translation function is defined. The maximums of the translation function yield the positions of the target molecules in the crystals.

100 Å×100 Å×100 Å has to be solved, then we have to search the translation in 2 Å steps; therefore 50×50×50 possible combinations of translations should be tested. In every translation set, the Eularian angle can be any values in 360°×360°×180°; even if we select 5° as the steps, there are still 72×72×36 possible choices. Therefore a total of 72×72×36×50×50×50, about $2×10^{10}$ possible combinations of rotation and translation need to be tried, and the number of calculations is far too large to apply in any computer systems. By calculating the rotation first, only 72×72×36 possible combinations need to be calculated to find the corrected Eularian angles. By fixing these angles, and calculating 50×50×50 possible combinations of translations, the initial model is finally obtained. The order of calculation is decreased to the order of 10^5, which can be finished in hours even by a PC.

The application of molecular replacement relies on similar model proteins. If there are no known proteins that are similar to the proteins we want to study, then in such cases, what we face are the *de novo* structures, and molecular replacement is not valid any more. Isomorphous replacement or anomalous scattering methods have to apply in order to solve the phase problems.

Isomorphous replacement is the keystone of protein crystallography, by which the first protein structure was solved. This is also the first method to indirectly measure the diffraction phases, by experiments. The idea of isomorphous replacement is to induce some heavy atoms, such as Hg, Au, Pd, into the protein crystals without large distortions in the crystal structures of proteins. The crystals containing the heavy atoms are called "isomorphours replacement crystals". Since the heavy atoms contain much more electrons than the atoms of proteins, usually C, N, O, S, and P, the heavy atoms should present very strong peaks in the Patterson function and easily to be distinguished. Therefore, the complex structure determination of proteins is first simplified to the structure determinations of the heavy atoms, which is almost the same as small molecular structure determinations.

It is necessary that two data sets of diffractions are collected: one for the native protein crystals, without any heavy atoms inside; and another for the isomorphous replacement crystals. Since the phases of the diffractions are still unknown, only the amplitudes of all structure factors are known, for both native and isomorphous replacement crystals.

To find the positions of heavy atoms via these two sets of diffraction data, a different Patterson function is defined:

$$P_{\text{dif}}(u, v, w) = \frac{1}{V} \sum_{h,k,l} [I_{PH}(h,k,l) - I_P(h,k,l)] \exp[-2\pi i(hu + kv + lw)] \quad (7.6)$$

The intensities of the native protein crystal are donated as $I_P(h,k,l)$ and of isomorphous replacement crystals as $I_{PH}(h,k,l)$. The summation is applied to all diffractions. V is the volume of the crystal cell. Since the insertions of heavy atoms do not alter the structures of the proteins, so the cells of native and isomorphous replacement crystals are the same.

The Patterson function is the inverse Fourier transformation of the diffraction intensities without phases. This function provides information about the vectors between two atoms in the crystal cells, instead of the atomic positions. If there are N atoms inside a crystal cell, the Patterson function should contain $N(N-1)$ peaks, corresponding to $N(N-1)$ vectors between such atoms. The height of one Patterson peak is proportional to the product of the electron numbers of the two atoms connected by this vector. Due to the large numbers of atoms in the cells of protein crystals, it is not possible to distinguish the individual peaks that come from the atomic pairs related to two atoms. However, the different Patterson functions can provide very distinguishing peaks related to the heavy atoms, since the atomic numbers of the heavy atoms are much higher than the atoms forming the proteins. For example, the atomic number of Hg atom is 80; this number is much higher than the atomic number of C, $Z=6$. Therefore, in the different Patterson map, the information for the heavy atoms is distinguished from the relative weak peaks of other atoms of the native proteins. According to the Patterson peaks of heavy atoms, the positions of heavy atoms induced during isomorphous replacement can be defined. The structure factors of heavy atoms, F_H, can be calculated via the positions of the heavy atoms, including the amplitudes and also the phases:

$$F_H(h,k,l) = \sum_j f_j^H \exp\left[2\pi i\left(hx_j^H + ky_j^H + lz_j^H\right)\right] \qquad (7.7)$$

The atomic scattering factors of heavy atoms are donated as f_j^H, and the positions are x_j^H, y_j^H and z_j^H. The summation is applied to all the heavy atoms in an isomorphous replacement crystal cell.

In the case where the structures of heavy atoms are known, the relationship between the structure factors before and after isomorphous replacement can be written as:

$$F_{PH}(h,k,l) - F_P(h,k,l) = F_H(h,k,l) \qquad (7.8)$$

Note that this formula is correct only when the positions of the atoms forming the proteins do not change after the inclusion of heavy atoms. Of course, this is only an approximation. After all, the introduction of heavy atoms should alter the positions of the nearby atoms, but in some cases, this alteration could be so small that the space groups of the crystals do not change. That is the meaning of "isomorphous".

So, when the amplitudes of $F_P(h,k,l)$ and $F_{PH}(h,k,l)$, and the vectors $F_H(h,k,l)$, are known, a picture for every diffraction peak can be constructed, as shown in Figure 7.8.

Because the two circles give two possible solutions of phases, the phase problem is not completely solved. We can also prepare another isomorphous replacement, and then the third circle would give the unique solution. The direct method provides another way to distinguish which is the correct solution in Figure 7.8.

To solve protein structures by the isomorphous replacement method is quite difficult (Figure 7.9). The growth of the protein crystals is not easy, and it is necessary to search for the condition of isomorphous replacement or maybe more than two isomorphous replacements. It can be imagined how many trials need to be done during such process. Therefore, the anomalous scattering method is proposed to solve the phase problem of protein structure determination.

The method of anomalous scattering is based on the following phenomena. The scatterings of any atoms change dramatically under some energies of incident X-rays; that is, the absorption edge. For example, the scattering of X-rays by a Cu atom suffers a huge change around the energy of the incident X-ray of 8.9 keV, the K absorption edge of Cu. In such cases, the atomic scattering factors should be expressed as:

$$f = f_0 + f' + if'' \qquad (7.9)$$

There are two terms to modify the atomic factors, f' and f''. The values of f' and f'' change dramatically with the X-ray energy around the absorption edge.

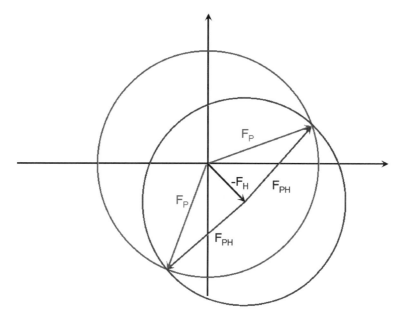

Figure 7.8 The relationship of structure factors in isomorphous replacement. The amplitudes of F_P and F_{PH} are known, but the phases are unknown. The vector F_P should be somewhere on the red circle. The vector F_H is fully known, therefore, another circle with radius of $|F_{PH}|$ and centering in the end point of vector $-F_H$ is drawn. The two intersection points give two possible solutions.

If we can alter the wavelengths or energies of the incident X-rays for diffraction around the absorption edge, the scattering of some atoms would make big changes. It is equivalent to replacing these atoms by another kind of atom. Therefore, we archive the total isomorphous replacements just by tuning the wavelengths of diffraction. This is the idea of anomalous scattering (Figure 7.10).

Certainly, to conduct anomalous scattering experiments, an X-ray source with tunable wavelengths is necessary. The in-house X-ray source, such as a rotating-anode X-ray generator, can only provide fixed-wavelength X-rays. Only synchrotron radiation can provide tunable wavelengths for diffraction. That is also one of reasons why synchrotron radiation has made such a large impact upon structural biology.

Anomalous-scattering experiments usually need to collect diffraction data under several wavelengths, so the time for data collection should be longer, but radiation damage may become severe. In such cases, cryo-protection of the sample becomes necessary, due to the very strong synchrotron radiation beam.

The wavelengths for diffraction usually are around 1 Å in order to archive the atomic resolution of protein structure. In another words, the energies of the X-rays used for structure determinations range from 6 to 16 keV, roughly. In such a range of energies the absorption edge of most transition metals is covered,

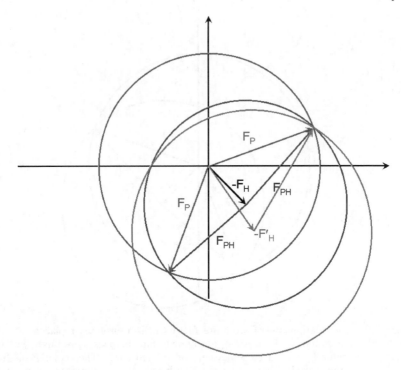

Figure 7.9 By another isomorphous replacement, the unique solution of phases can be defined.

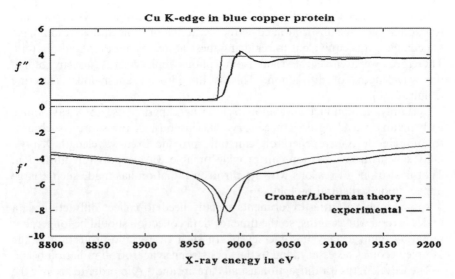

Figure 7.10 The atomic scattering factor of Cu around the K edge. The thin line is calculated by theory and the experimental result from a metalloprotein is also plotted for comparision. Cited from http://skuld.bmsc.washington.edu/scatter/AS_experiment.html. Reprinted with permission from the author Ethan Merritt.

such as Mn, Fe, Cu, Zn (the metals that occur most commonly in metalloproteins) and Se. In many cases, Se can be induced during protein expression, although there are some Se proteins in nature. Metalloproteins, due to the existence of metals like Mn, Fe, Cu, and Zn, provide a perfect change for structure determination by anomalous scattering.

7.3 Structure Determination Using the Multi-wavelength Anomalous Dispersion Method

So far, the most adopted methods for determining protein structures are molecular replacement (MR) and multi-wavelength anomalous dispersion (MAD). MR is used if the structures with high sequence homologous (greater than 30%) to the target protein are already known; otherwise, MAD is the routine technique. Moreover, with the development of technologies it becomes more and more convenient to obtain a heavy atom derivative (especially Se derivative) even a protein does not contain any metal ions. Therefore, MAD method can be applied. The phase from MAD is more accurate than that from MR, because the phase error from MR includes the one from the model structure except for that from the crystal itself, while the MAD phase error only contains this from the crystal. So, during the past decade, MAD has become the dominant method for determining the structures of novel proteins. In order to carry out MAD experiments proteins must contain some "anomalous" atoms (some heavy metal ions usually). As far as a metalloprotein is concerned, one character is that it contains at least one kind of metal ion; as a result it is much easier to carry out MAD experiments if the metal ion is suitable for structure determination, e.g. Fe, Zn, Cu.

7.3.1 Theoretical Background

The scattering of an X-ray beam by a crystal results from the interaction between the incident X-rays and the electrons in the crystal: this process is called Thomson scattering. In this process, the interaction of X-rays with electrons gives rise to elastic scattering of the incident radiation. Generally, in most protein crystallography experiments Thomson scattering dominates. However, when the energy of the incident X-rays is close to that of the electronic transition from a bound atom orbital, resonance scattering occurs and results in anomalous scattering. For X-rays in standard diffraction experiments with the wavelength around 1 Å, there are no electronic transitions for light atoms of biological macromolecules (H, C, N, O, S, and P) and the resonant component is negligible. The "anomalous" resonant scattering is, in practice, restricted to heavier atoms, which may be intrinsic (e.g. Zn in metalloproteins) or exogenous (e.g. Hg in a heavy-atom derivative); we will name them as anomalous scatters here.

If the protein has anomalous scatters in its molecule, the difference in intensity between the Bijvoet pairs, $|F_{hkl}(+)|^2$ and $|F_{hkl}(-)|^2$, can be used for the phase angle determination. In the MAD method the wavelength dependence of the anomalous scattering is used. The principle of this method is rather old,[8] but it was only after synchrotron radiation light was used in protein structure determination did it became a technically feasible method. Hendrichson and colleagues were the first to utilize this method to solve the structure of a protein.[9,10] In most cases, there is only one kind of anomalous scatter in the protein, and we will discuss this situation. This discussion is based on Karle's treatment of the problem,[11] which separates the non-anomalous scattering of all atoms in the structure from the wavelength dependent part. Therefore, the total atom scattering factor is written as equation 7.9:

$$f = f_0 + f' + if''$$

where f_0 is the normal scattering component, f' and f'' are the real and imaginary component of the anomalous scattering component. The magnitude of the anomalous scattering is independent of the scattering angle but strictly depends on the wavelength of the incident radiation while those of the normal scattering reduce quickly with increasing the scattering angle. The details of MAD method can be found in references 11, 12, and 13.

7.3.2 Experimental Strategies

7.3.2.1 Wavelength Selection

As its name indicates, MAD experiment should be carried out at different wavelengths; on the other hand, f' and f'' are both associated with λ, so it is essential to choose the proper λ in the experiment. In general, the selection of λ should be such that both the Bijvoet and dispersive difference absolute value are the maximum. Usually, the first one at which the f'' value is the largest, above the absorption edge generally, is named peak wavelength; the second one, at which the f' value is the smallest, at the inflection point of the edge or L_{III} in the case of L edge, is named inflection wavelength; the third one, which is remote from the second one, is called the remote wavelength.

From our experience, the peak or the inflection wavelength is essential for the determination of anomalous scattering atoms and the remote one is important to the model auto-building. The remote one should be such that it maximizes the quantity $f'' \times \Delta f'$, where $\Delta f'$ is the difference in f' between the remote and inflection wavelength. A high-energy remote wavelength is chosen mostly in practical experiments, because it provides a larger f'' value than the low-energy one and X-ray absorption is a little less at this wavelength. Generally, the farther away from the absorption edge, the more effective the remote wavelength will be. For example, for an Se edge, collecting the remote wavelength at 0.9 Å instead of 0.96 Å results in a 25% increase in $\Delta f'$ with only 13% decrease in f''.[14] For L edges, an energy about 200–300 eV above the L_I edge is the first choice; if this wavelength is not accessible, a long remote wavelength on the low energy side of the L_{III} edge is the second best choice.[14]

Although both f' and f'' can be calculated theoretically for isolated atoms, these magnitudes do not represent their true values in the proteins in which the anomalous scattering atoms are surrounded by different other atoms; that is to say, their chemical environments are not identical. Thus, it is important to measure the values of scattering factors experimentally.

Measurement of X-ray absorption can be used to obtain the anomalous scattering factors. The spectrum of atomic absorption coefficients $\mu(E)$, as a function of energy, E, is related directly to the imaginary component of anomalous scattering:

$$f''(E) = \frac{mc}{2he^2} E\mu(E) \tag{7.10}$$

where m, c, h, and e are the fundamental physical constants with the usual meaning. The real part is related to imaginary one by the Kramers–Kronig transformation:[15]

$$f'(E) = \frac{2}{\pi} \int_0^\infty \frac{E' f''(E')}{(E^2 - E'^2)} \, dE' \tag{7.11}$$

The program CHOOCH can be used to obtain the necessary values from the X-ray absorption spectrum, including the peak and inflection wavelengths, the f' and f'' values.[16]

7.3.2.2 Data Collection

As more reflections are successfully collected, more accurate phases will be obtained. In MAD experiments, the data collection strategy is the dominating ingredient that affects data quality after wavelength choice. Often, a unique completeness of at least 90% is needed and greater completeness is needed for poorly diffracting crystals or low anomalous signal.[17] Only high completeness for normal reflections is not enough for MAD, the data must contain a high percent of Friedel-related reflections. For most cases studied, 40–85% completeness for Friedel pairs has been found to be sufficient for MAD phasing, but it usually must be close to 100% for SAD phasing.[17]

Higher data redundancy, average 4 or higher, is useful for obtaining more accurate experimental phases by decreasing the error in the merged data and thus in the amplitude difference. It is important to ensure the exposure time and the distance from crystal to detector are adequate for collecting higher $I/\sigma(I)$ reflections, because a few good reflections are much better than many poor ones in phase measuring.

Diffraction resolution is also an important value that deserves attention. Higher resolution is not only useful for facilitating auto-model building[18] but also making it possible to observe fine structural details resulting from unbiased experimental phases.[19,20] However, higher resolution always means longer exposure time (of course, after reaching its maximum resolution, exposure time dose not affect the resolution) and unfairly diffusion of $I/\sigma(I)$, for example, very high $I/\sigma(I)$ value at lower resolution shells and very low at higher shells, thus the proper resolution should be decided on the basis of several factors. A more time-efficient strategy is to collect medium or low resolution data for phasing and use phase extension to extend the phases to the resolution limit of the crystal.

In MAD experiments, it is very important to collect each reflection under similar conditions at each wavelength.[12,21] If all the data are collected using the same crystal and oscillation range, it is easily achieved. At the earlier stage, three to five wavelengths were chosen to carry out the MAD experiment. However, with the development of software and the emergence of high-quality light sources, e.g. third-generation synchrotron radiation, two wavelengths anomalous dispersion even one wavelength (SAD) is suitable for the structure determination. Dauter provided an efficient data collection approach: the one-and-a-half wavelength approach.[22] After data collection at one wavelength, just continue to collect data at further wavelengths with MAD phasing in mind, but in parallel attempt to solve the structure against the first data set using the SAD approach. With currently available programs, this can be performed rapidly and, if successful, may the collection of further data to be abandoned. If the solution cannot be achieved by SAD, one can continue to perform the full MAD experiment. Dauter did not point out the sequence of wavelengths, *i.e.* which is the first and which is the second wavelength, however. According to some results, in two-wavelength MAD phasing, the best combination is the inflection and high remote wavelength.[20,23,24] But according to our practice, if the incident radiation is not so stable, the peak and the high remote wavelength combination is the best choice.

7.3.2.3 Data processing

A great deal of software, such as HKL2000,[25] Mosflm,[26] and XDS,[27] can be used to integrate the raw diffraction data. After integration, HKL2000 can also scale the data while Mosflm and XDS can not; these data must be scaled by other software, e.g. Scalait in the CCP4 program package.[28] The anomalous option must be included in the scaling process which separates the Friedel pairs. The quality of the scaled data is included in the scaling logfile; if it is reasonable, the scaled data are processed by software such as SOLVE[29] or SHELXD[30] to locate the anomalous scattering atoms. Then the software RESOLVE[31] or Arp/Warp[32] is performed for auto-model building. In general, there are some biases in this model, in order to obtain accurate phases. These biases must be removed furthest. This can be done manually by refinement program such as CNS[33] or refmac5 in CCP4.

The information mentioned above introduces the MAD method simply. For more details readers should refer to references 21, 34, and 35, and the references in them.

7.4 Crystal Structure Made Clear: Structure and Function of SmdCD

Approximately one-third of all proteins in the human body contain metals. These and other "metalloproteins" are essential for the basic processes of life, including DNA synthesis, metabolism photosynthesis detoxification, and the chemical transformations of nitrogen, oxygen, and carbon molecules required for life. Many diseases are due to metal imbalances or inactivity of critical metalloproteins. In fact, numerous essential biological functions require metal atoms. Thus, metalloproteins make life on Earth possible and the ability to understand and ultimately control the binding and activity of protein metal sites is of great biological and medical importance.

The cytidine deaminase (CDA) superfamily consists of a large number of metalloproteins, which contain the zinc atom, and function as the mononucleotide deaminases involved in the nucleotide metabolism, and RNA (DNA)-editing deaminases involved in gene diversity and in anti-virus defense. The mononucleotide deaminases include guanine deaminase (GD),[36] cytosine deaminase (CD),[37–39] cytidine deaminase (CDA),[40–43] dCMP deaminase (dCD),[44–47] dCMP-dCTP deaminase (bi-dCD),[48] riboflavin biosynthesis protein RIBG,[49–51] and the antibiotic-resistant blasticidin-S deaminase (BSD).[52–54] On the other hand, the RNA (DNA)-editing deaminase contains A-to-I tRNA-specific adenosine deaminases (TADs),[55–58] A-to-I adenosine deaminases acting on RNA,[59–61] and C-to-U cytidine deaminase acting on RNA (DNA).[62–64] The CDA family can be divided into two classes based on the active site residues. The first class utilizes one histidine and two cysteine residues for zinc co-ordination. The second class uses three cysteine residues. Their domain consists of a three-layered $\alpha/\beta/\alpha$ structure. Structural comparisons have demonstrated the central beta sheet and immediately surrounding helices are structurally conserved. A major challenge is to understand how nature has evolved the CDA fold of these deaminases to act on their substrates and how the regulators modulate the activity of the enzyme.

The synthesis of deoxythymidine-5'-triphosphate (dTTP) from deoxyuridine-5'-monophosphate (dUMP) is important in DNA synthesis and in cancer chemotherapy.[65,66] In most biological systems, deoxytrimethyl phosphate, the precursor of dTTP, is converted from dUMP; this conversion is catalyzed by thymidylate synthase. In Gram-positive bacteria and eukaryotic organisms, 2'-deoxycytidylate deaminases [or deoxycytidine-5'-monophosphate (dCMP) deaminases, dCDs; EC 3.5.4.12] catalyze the conversion of dCMP to dUMP in the pyrimidine salvage pathway. In contrast, in the Gram-negative bacteria *Escherichia coli* and *Salmonella typhimurium*, dUMP is synthesized via two steps; namely, the deamination of deoxycytidine-5'-triphosphate (dCTP) to deoxyuridine-5'-triphosphate (dUTP) by dCTP deaminase and the subsequent hydrolysis of dUTP to dUMP by dUTPase. Interestingly, dCD reduces the efficiency of anticancer and antiviral drugs via deamination at the nucleotide level;[67,68] this indicates that dCD inhibitors have potential applications in chemotherapy.

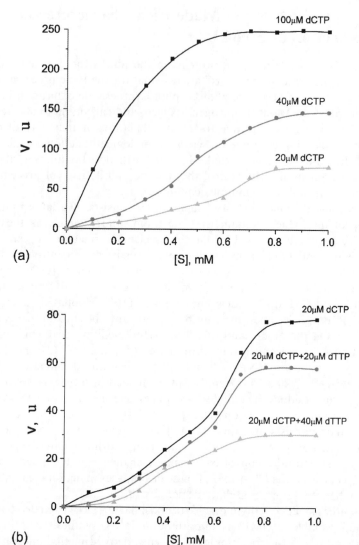

Figure 7.11 Kinetic assay for Sm-dCD. One unit of activity refers to the deamination of 1 μM dCMP per minute under the condition of the assay. (a) Activation of Sm-dCD by dCTP at increasing concentrations of dCMP. (b) Regulation of SmdCD by different ratios of dCTP to dTTP at increasing concentrations of dCMP. © 2007 Elsevier Ltd.

One of the interesting features of dCD is that its catalytic activity depends on the feedback regulation based on the ratio of dCTP to dTTP, which are both the end products of the pyrimidine salvage pathway, wherein dCTP functions as an activator and dTTP, as an inhibitor.[69] Both dCTP and dTTP appear to bind a single allosteric site located far from the active site,[70] and Mg^{2+} is involved in their binding. In some dCDs, Ca^{2+} plays the same role as Mg^{2+} in promoting activation by dCTP, although the enzyme activities of these dCDs are lower. Ni^{2+}, Zn^{2+}, Co^{2+}, and Cu^{2+} have no effect on dCTP and dTTP binding.[71]

Two classes of dCDs have been identified in mammals and Gram-positive bacteria, both of which exist as homohexamers wherein each subunit contains an identical active site. Mammalian dCD contains a single Zn^{2+} ion per monomer required for catalytic activity. In contrast, bacterial dCD contains two Zn^{2+} ions per monomer; one is involved in catalysis, and the other appears to be important for maintaining structural integrity and also for binding the phosphate group of the substrate.

In the past decade, dCDs from humans and T-even bacteriophage have been extensively studied for their enzyme activities and mutation effects.[44–46,72,73] All dCDs share two conserved signature sequences, namely, HXE and PCXXC. The crystal structures of T4-bacteriophage dCD (T4-dCD) containing an R115E mutation and of *E. coli* cytidine deaminase (CDA, EC3.5.4.5) revealed that the signature sequences contain a zinc-binding motif, wherein a histidine and two cysteine residues coordinate a Zn^{2+} ion, while the glutamate residue functions as a proton shuttle. The conservation of the active site suggests that these deaminases exhibit the similar catalytic mechanism.[42]

Mutations of R115E and F112A in T4-dCD lead to loss of activation by dCTP and convert the hexamer to dimer in solution, indicating that these two residues are very important for dCTP binding and consequent enzyme activation. Surprisingly, the dimer retains approximately 40–50% of the wild-type (WT) enzyme activity.[46] The structure of the R115E mutant of T4-dCD has been determined in complex with a substrate analog. The structure reveals several important features of the enzyme such as substrate binding and hexamer arrangement. Residues 46–79 are unique in T4-dCD; they mainly form a short helix and some loops near the active site. In addition, the helix α3 in T4-dCD is absent in *Streptococcus mutans* dCD (Sm-dCD) and is replaced by a loop. However, the F112 and R115 residues are not conserved in the CDA superfamily, and the T4-dCD structure lacks the dCTP-binding feature; therefore, the manner in which dCTP and dTTP regulate dCD activity remains unclear.

Hou and collaborators[47] chose *Streptococcus mutans* dCD as a model to investigate the mechanism of allosteric regulation. The recombined Sm-dCD has a similar feature to the T-even phage dCD, no activity was observed in the absence of dCTP or a metal ion at a concentration of 0.5 mM dCMP. As indicated in Figure 7.11a, dCTP significantly enhances the deamination of dCMP at low level of the substrate. The activation by dCTP is reversed on addition of dTTP (Figure 7.11b). In contrast to dCTP, dTTP enhances the sigmodal response as the enzyme is saturated by substrate. Although the predicted amino acid sequence of Sm-dCD exhibited only 34% identity to that reported for T4-dCD, the kinetic data indicate that recombined Sm-dCD has similar properties as wild-type T4-dCD with regard to the enzyme activity.

The crystal structure of Sm-dCD and its complex with substrate analog and allosteric regulator dCTP·Mg was solved by zinc multi-wavelength anomalous

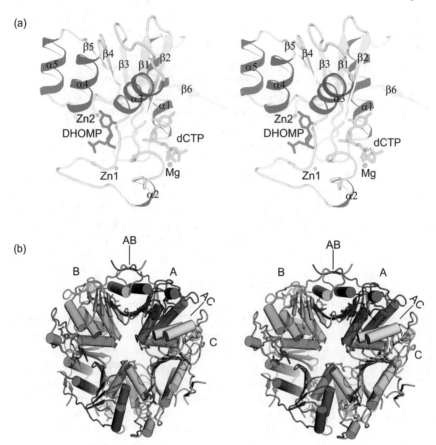

Figure 7.12 Stereoview of the complex structure of Sm-dCD with PdR and dCTP. (a) Overall structure of Sm-dCD monomer with the substrate analog and allosteric regulator bound. (b) Hexamer generated by crystallographic symmetry. © 2007 Elsevier Ltd.

dispersion (MAD) method and molecular replacement method. The fold of complex structure (Figure 7.12a) was quite similar to that of T4-dCD. Similar to T4-dCD, one substrate analog binds to one monomer, and the substrate-binding pocket contains two tightly bound Zn^{2+} ions located at opposite ends of the substrate analog. The similar configuration was also found in dCD from T4 phage[73] and CDA from *B. subtilis*.[43]

In the complex structure (Figure 7.12b), the allosteric regulator dCTP binds at the interface AB, while the hydrated substrate analog DHOMP binds close to the interface AC. The interaction between Sm-dCD and dCTP were analyzed using the LIGPLOT[74] program, as shown in Figure 7.13. The dCTP-binding pocket mainly comprises the polar residues of the three neighboring subunits. A magnesium ion is octahedrally coordinated to the α-, β-, γ-phosphate of the nucleotide and three water molecules. Mg^{2+} appears to compensate for

Protein Crystallography for Metalloproteins

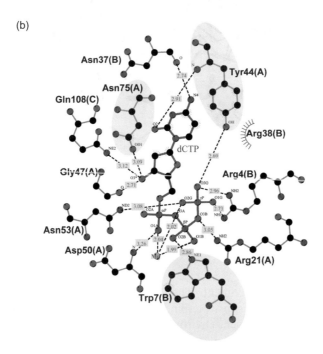

Figure 7.13 Schematic presentation of dCTP–protein interaction produced by the program LIGPLOT. Highly conserved residues for binding dCTP: Trp7, Tyr44, and Asn75 are emphasized. © 2007 Elsevier Ltd.

the negative charge due to the lack of positive amino acids in the binding pocket.

The overall tertiary structure of the complex is quite similar to that of the free-state enzyme. The most significant conformational changes occur at the substrate-binding site. The substrate-binding pocket become more packed in the complex structure, particularly at sites wherein the Arg26, His65, Tyr120 and Arg121 side chains interact with the phosphate group of DHOMP via salt bridges and hydrogen bonds (indicated in Figure 7.14). This is contrasting to the free-state enzyme structure, wherein these residues exhibit a more open conformation (indicated in Figure 7.14) and the substrate-binding pocket is

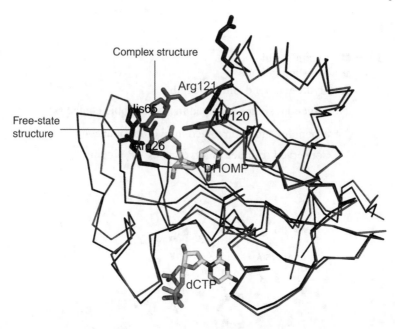

Figure 7.14 The superposition of the free-state enzyme structure and the complex structure on the residues coordinated with catalytic zinc ion. It is shown that the substrate-binding pocket of the free-state enzyme served as an "open state," while the complex served as a "closed state". The conformation of the labeled residues changed significantly to make the substrate analog binding more stable in the complex structure. © 2007 Elsevier Ltd.

exposed to the external environment. These residues envelope the complex, blocking the entrance to the substrate-binding pocket, and thus isolate the substrate from the external solvent environment during the catalytic reaction.

The binding of the allosteric regulator dCTP induces significant quaternary structural changes in the enzyme. dCTP molecules bind at interface AB, each in contact with the N-terminal α1 helix of the neighboring monomer (Figure 7.15). The phosphate groups of dCTP create a negatively charged environment that attracts the nearby positively charged residues, particularly Arg4 and Arg38. This attraction draws the two monomers closer to each other and results in a quaternary structural rearrangement. The association between the two monomers, A and B, becomes stronger, and the buried surface area increases significantly (from 1160 Å^2 to 1720 Å^2 per monomer).

Interface AC, which mainly comprises the substrate-binding pocket, undergoes a relatively small conformational change. The main changes involve shifts in the loops near the active site, which result in a closed conformation at the entrance to the pocket. Furthermore, the loops of the two monomers move closer together, and each may stabilize substrate binding in the other monomer. This conformational change may be partially caused by DHOMP binding and

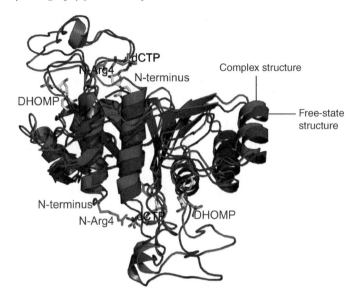

Figure 7.15 Quaternary structural changes caused by dCTP binding depicted in a dimer. The free-state enzyme structure and the complex structure were superposed on Ca atoms of one monomer; the g phosphate of the dCTP contacts tightly with N-terminal Arg4 of the neighboring monomer and thus pulls the a1 helix to form a tightly packing form in interface AB. © 2007 Elsevier Ltd.

even binding since the interactions between dCTP and the loop ahead of the $\alpha 2$ helix provide a torsion force to the $\alpha 2$ active-site loop motif. The quaternary structure of the complex is denser when compared with the free-state enzyme structure; *i.e.* the hexamer configuration is more stable when binding dCTP and PdR. This modification is favorable for the protein function.

References

1. N. J. Oppenheimer and T. L. James, eds. *Methods in Enzymology, Volume 177: Nuclear Magnetic Resonance, Part B, Structure and Mechanism*, Elsevier, 1989.
2. T. L. James, V. Dotsch and U. Schmitz, eds. *Methods in Enzymology, Volume 338, Nuclear Magnetic Resonance of Biological Macromolecules, Part A*, Elsevier, 2002.
3. T. L. James, V. Dotsch and U. Schmitz, eds. *Methods in Enzymology, Volume 339, Nuclear Magnetic Resonance of Biological Macromolecules, Part B*, Elsevier, 2001.
4. J. Frank, ed. *Three-Dimensional Electron Microscopy of Macromolecular Assemblies*, Oxford, Oxford University Press, 2006.

5. For protein crystallography, there are many reference books; for example, J. Drenth, *Principles of Protein X-ray Crystallography*, 3rd edn, Berlin, Springer, 2007.
6. M. G. Rossmann and D. M. Blow, *Acta Cryst.*, 1962, **15**, 24–31.
7. R. A. Crowther and D. D. Blow, *Acta Cryst.*, 1967, **23**, 544–548.
8. Y. Okaya and P. Pepinsky, *Phys. Rev.*, 1956, **103**, 1645–1647.
9. W. A. Hendrickson, J. L. Smith, R. P. Phizackerley and E. A. Merritt, *Proteins*, 1988, **4**, 77–88.
10. H. M. Krishna Murthy, W. A. Hendrickson, W. H. Orme-Johnson, E. A. Merritt and R. P. Phizacherley, *J. Biol. Chem.*, 1988, **263**, 18430–18436.
11. J. Karle and J. Int, *Quant. Chem. Symp.*, 1980, **7**, 357–367.
12. J. Drenth, *Principles of Protein X-ray Crystallography*, Berlin, Springer, 1994, pp. 207–211.
13. W. A. Hendrickson, *Trans. Am. Cryst. Assoc.*, 1985, **21**, 11–21.
14. A. Gonzalez, *Acta. Cryst. D*, 2003, **59**, 1935–1942.
15. J. Hoyt, D. de Fontaine and W. J. Warburton, *J. Appl. Cryst.*, 1984, **17**, 344–351.
16. G. Evans and K. S. Wilson, *Acta Cryst. D*, 1999, **55**, 67–76.
17. A. Gonzalez, *Acta Cryst. D*, 2003, **59**, 315–322.
18. R. J. Morris, A. Perrakis and V. S. Lamzin, *Acta Cryst. D*, 2002, **58**, 968–975.
19. F. T. Burling, W. I. Weis, K. M. Flaherty and A. T. Brunger, *Science*, 1996, **271**, 72–77.
20. A. Schmidt, A. Gonzalez, R. J. Morris, M. Costabel, P. M. Alzari and V. S. Lamzin, *Acta Cryst. D*, 2002, **58**, 1433–1441.
21. W. A. Hendrickson, *Science*, 1991, **254**, 51–58.
22. Z. Dauter, *Acta Cryst. D*, 2002, **58**, 1958–1967.
23. A. Gonzalez, J. D. Pedelacq, M. Sola, F. X. Gomis-Ruth, M. Coll, J. P. Samama and S. Benini, *Acta Cryst. D*, 1999, **55**, 1449–1458.
24. A. Gonzalez, F. V. Delft, R. C. Liddington and C. Bakolitsa, *J. Synchrotron Radiat.*, 2005, **12**, 285–291.
25. Z. Otwinowski and W. Minor, *Methods Enzymol.*, 1997, **276**, 307–326.
26. A. G. W. Leslie, *Joint CCP 4+ESF-EAMCB Newsletter Protein Crystallogr.*, 1992, **No. 26**.
27. W. J. Kabsch, *Appl. Cryst.*, 1993, **26**, 795–800.
28. Collaborative Computational Project, Number 4. *Acta Cryst. D*, 1994, **50**, 760–763.
29. T. C. Terwilliger and J. Berendzen, *Acta Cryst. D*, 1999, **55**, 849–861.
30. G. M. Sheldrick, *Acta Cryst. A*, 2008, **64**, 112–122.
31. T. C. Terwilliger, *Acta Cryst. D*, 2000, **56**, 965–972.
32. G. Langer, S. X. Cohen, V. S. Lamzin and A. Perrakis, *Nat. Protocols*, 2008, **3**, 1171–1179.
33. A. T. Brunger, P. D. Adams, G. M. Clore, W. L. Delano, P. Gros and R. W. Grosse-Kunstleve, *Acta Cryst. D*, 1998, **54**, 905–921.
34. M. A. Walsh, G. Evans, R. Sanishvili, I. Dementievaa and A. Joachimia, *Acta Cryst. D*, 1999, **55**, 1726–1732.

35. W. Hendrickson and C. Ogata, *Methods Enzymol.*, 1997, **276**, 494–523.
36. S. H. Liaw, Y. J. Chang, C. T. Lai, H. C. Chang and G. G. Chang, *J. Biol. Chem.*, 2004, **279**, 35479–35485.
37. D. J. Porter and E. A. Austin, *J. Biol. Chem.*, 1993, **268**, 24005–24011.
38. G. C. Ireton, G. McDermott, M. E. Black and B. L. Stoddard, *J. Mol. Biol.*, 2002, **315**, 687–697.
39. T. P. Ko, J. J. Lin, C. Y. Hu, Y. H. Hsu, A. H. J. Wang and S. H. Liaw, *J. Biol. Chem.*, 2003, **278**, 19111–19117.
40. Y. S. Teng, J. E. Anderson and E. R. Giblett, *Am. J. Hum. Genet.*, 1975, **27**, 492–497.
41. A. Vita, A. Amici, T. Cacciamani, M. Lanciotti and G. Magni, *Biochemistry*, 1985, **24**, 6020–6024.
42. L. S. Betts, S. Xiang, S. A. Short, R. Wolfenden and C. W. Carter Jr, *J. Mol. Biol.*, 1994, **235**, 635–656.
43. E. Johansson, N. Mejlhede, J. Neuhard and S. Larsen, *Biochemistry*, 2002, **41**, 2563–2570.
44. J. T. Moore, R. E. Silversmith, G. F. Maley and F. Maley, *J. Biol. Chem.*, 1993, **268**, 2288–2291.
45. J. T. Moore, J. M. Ciesla, L. M. Changchien, G. F. Maley and F. Maley, *Biochemistry*, 1994, **33**, 2104–2112.
46. R. G. Keefe, G. F. Maley, R. L. Saxl and F. Maley, *J. Biol. Chem.*, 2000, **275**, 12598–12602.
47. H. F. Hou, Y. H. Liang, L. F. Li, X. D. Su and Y. H. Dong, *J. Mol. Biol.*, 2008, **377**, 220–231.
48. Y. F. Zhang, F. Maley, G. F. Maley, G. Duncan, D. D. Dunigan and J. L. Van Etten, *J. Virol.*, 2007, **81**, 7662–7671.
49. G. Richter, M. Fischer, C. Krieger, S. Eberhardt, H. Luttgen, I. Gerstenschlager and A. Bacher, *J. Bacteriol.*, 1997, **179**, 2022–2028.
50. S. C. Chen, Y. C. Chang, C. H. Lin, C. H. Lin and S. H. Liaw, *J. Biol. Chem.*, 2006, **281**, 7605–7613.
51. P. Stenmark, M. Moche, D. Gurmu and P. Nordlund, *J. Mol. Biol.*, 2007, **373**, 48–64.
52. M. Kimura, T. Kamakura, Q. Z. Tao, I. Kaneko and I. Yamaguchi, *Mol. Gen. Genet.*, 1994, **242**, 121–129.
53. M. Kimura, S. Sekido, Y. Isogai and I. Yamaguchi, *J. Biochem.*, 2000, **127**, 955–963.
54. T. Kumasaka, M. Yamamoto, M. Furuichi, M. Nakasako, A. H. Teh, M. Kimura, I. Yamaguchi and T. Ueki, *J. Biol. Chem.*, 2007, **282**, 37103–37111.
55. A. Gerber, H. Grosjean, T. Melcher and W. Keller, *EMBO J.*, 1998, **17**, 4780–4789.
56. S. Maas, A. P. Gerber and A. Rich, *Proc. Natl. Acad. Sci. U. S. A.*, 1999, **96**, 8895–8900.
57. J. Wolf, A. P. Gerber and W. Keller, *EMBO J.*, 2002, **21**, 3841–3851.
58. M. Luo and V. L. Schramm, *J. Am. Chem. Soc.*, 2008, **130**, 2649–2655.
59. M. Schaub and W. Keller, *Biochimie*, 2002, **84**, 791–803.

60. K. M. Chen, E. Harjes, P. J. Gross, A. Fahmy, Y. J. Lu, K. Shindo, R. S. Harris and H. Matsuo, *Nature*, 2008, **452**, 116–119.
61. A. Phuphuakrat, R. Kraiwong, C. Boonarkart, D. Lauhakirti, T. H. Lee and P. Auewarakul, *J. Virol.*, 2008, **82**, 10864–10872.
62. K. M. Rose, M. Marin, S. M. Kozak and D. Kabat, *J. Biol. Chem.*, 2004, **279**, 41744–41749.
63. K. Krause, K. B. Marcu and J. Greeve, *Mol. Immunol.*, 2006, **43**, 295–307.
64. N. Navaratnam and R. Sarwar, *Int. J. Hematol.*, 2006, **83**, 195–200.
65. C. W. Carreras and D. V. Santi, *Annu. Rev. Biochem.*, 1995, **64**, 721–762.
66. T. L. Ferea, D. Botstein, P. O. Brown and R. F. Rosenzweig, *Proc. Natl. Acad. Sci. U. S. A.*, 1999, **96**, 9721–9726.
67. B. Hernandez-Santiago and L. Placidi *et al.*, *Antimicrob. Agents Chemother.*, 2002, **46**, 1728–1733.
68. J. Y. Liou, P. Krishnan, C. C. Hsieh, G. E. Dutschman and Y. C. Cheng, *Mol. Pharmacol.*, 2003, **63**, 105–110.
69. F. Maley and G. F. Maley, *Adv. Enzyme Regul.*, 1970, **8**, 55–71.
70. G. F. Maley and F. Maley, *Biochemistry*, 1982, **21**, 3780–3785.
71. G. F. Maley and F. Maley, *J. Biol. Chem.*, 1964, **239**, 168–176.
72. K. X. Weiner, R. S. Weiner, F. Maley and G. F. Maley, *J. Biol. Chem.*, 1993, **268**, 12983–12989.
73. R. Almog, F. Maley, G. F. Maley, R. MacColl and P. V. Roey, *Biochemistry*, 2004, **43**, 13715–13723.
74. A. C. Wallace, R. A. Laskowski and J. M. Thornton, *Protein Eng.*, 1995, **8**, 127–134.

CHAPTER 8
Applications of Nuclear Analytical Techniques for Iron-omics Studies

GUANGJUN NIE,* MOTAO ZHU AND BO NING

CAS Key Laboratory for Biological Effects of Nanomaterials & Nanosafety, National Center for Nanoscience and Technology of China & Institute of High Energy Physics, Chinese Academy of Sciences, Beijing 100190

8.1 Chemistry of Iron

Iron is the most versatile metal in all metallic elements utilized by organisms because of its active chemical properties. It is a component of ubiquitous heme-containing proteins and numerous non-heme iron-containing proteins and plays a fundamental role in vital processes such as oxygen transport, electron transfer, and DNA synthesis. However, because it easily catalyzes electron transfer between its two oxidation states, ferrous [Fe(II)] and ferric [Fe(III)], iron can participate in the Fenton reaction, which promotes the formation of toxic reactive oxygen species (ROS). The latter can attack all biological molecules.[1] At physiological pH and oxygen tension, ferrous iron is readily oxidized to ferric, which is prone to hydrolysis, forming insoluble ferric hydroxide and oxohydroxide minerals.[2]

---

*Corresponding author: Email: niegj@nanoctr.cn; Tel: +86-10-82545529; Fax: +86-10-62656765.

Nuclear Analytical Techniques for Metallomics and Metalloproteomics
Edited by Chunying Chen, Zhifang Chai and Yuxi Gao
© Royal Society of Chemistry 2010
Published by the Royal Society of Chemistry, www.rsc.org

8.2 Physiology of Iron

Iron represents 55 and 45 mg kg^{-1} of body weight in adult men and women, respectively. Within the total body iron, 60–70% of it is present in hemoglobin in red blood cells. Myoglobin, cytochromes, and other iron-containing enzymes comprise a 10% of body iron and the remaining 20–30% is stored in ferritin, the major cellular iron storage protein.

To fully utilize the metal as an electron donor and acceptor and diminish its potential toxicity and insolubility, organisms have evolved a special molecular mechanism to acquire, transport, store, and utilize iron in its biocompatible and safe forms to meet body iron requirements. Iron metabolism in organisms is highly regulated and coordinated processes which maintain proper levels of bioavailable iron for metabolism uses.[3–5] Transferrin (Tf) is a plasma glycoprotein and transports iron among the sites of absorption, storage, and utilization in the body. Tf binds iron very tightly but reversibly and is recognized by its receptor, transferrin receptor1 and 2 (TfR1 and 2). TfR1 is expressed on almost all types of cells, and is especially enriched on precursors of red blood cells because these cells have the highest demand for iron. Although iron bound to transferrin is less than 0.1% of total body iron, it is the most dynamic portion of body iron with a turnover rate of roughly 25 mg in 24 h.[2] Physiologically, 80% of transferrin iron is transported to the bone marrow for hemoglobin synthesis in developing erythroid cells.[2] Iron absorption occurs predominantly in the duodenum, with a daily amount of 1–2 mg. Iron from the senescent red blood cells will be recycled for metabolic uses, which is one of the most prominent features of iron metabolism in mammals.[6]

8.3 Cellular and Systemic Iron Metabolism Regulation

8.3.1 Cellular Iron Uptake

Cellular iron metabolism includes its absorption, transport, storage, utilization, and regulation for cellular metabolic processes. Individual cells must maintain intracellular iron homeostasis to ensure that iron is adequate for fundamental metabolism uses but no extra "free" iron that could promote oxidative damage occurred. Tf-bound ferric iron is the major physiological iron source for both erythroid and non-erythroid cells. The transferrin–TfR1 complex mediated iron uptake is the major cellular iron acquisition process. TfR1 is homodimeric glycoprotein of 180 kDa. Each subunit of TfR1 contains a short cytoplasmic tail of 67 amino acids, a single transmembrane domain and a large extracellular domain. After holo-transferrin (iron binding forms) recognized by TfR1, the complex is rapidly internalized by receptor-mediated endocytosis involving clathrin-coated pits (Figure 8.1). Within the cells, the internalized Tf-TfR1 complex localizes in endosomes. Acidification of the endosomes to approximately pH 5.5 via an ATP-dependent proton pump ensures conformation changes of Tf–TfR1 complex with the consequent alteration of binding affinity of iron to Tf and iron release from Tf. After releasing iron, the Tf–TfR1

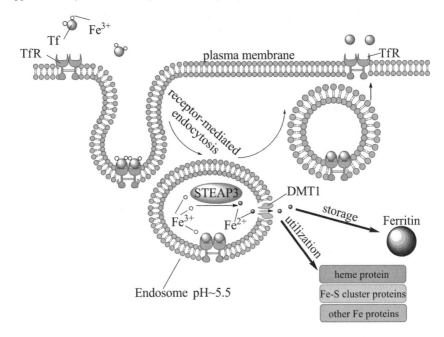

Figure 8.1 The Tf–TfR cycle. The plasma protein Tf binds to two ferric iron with high affinity. Once Tf-Fe is recognized by its receptor, TfR, it triggers endocytosis. The internalized vesicles become acidified by the proton pump. As the pH decreases in the endosomes, the structure of Tf–TfR complex changes and ferric iron releases. The newly identified endosomal reductase, STEAP3 converts ferric iron to the ferrous form, which is then transported out of endosomes. The fate of the iron after it leaves the endosomes is less well known. Eventually, iron will be used for metabolic processes, such as heme protein synthesis and iron–sulfur (Fe–S) assembly. The apo-Tf–TfR will be recycled to the cell membrane.

complex is recycled to the cell membrane. The endosomal ferric iron is reduced to ferrous iron by a ferri-reductase, identified as STEAP3,[7] followed by transport from the endosomes via an iron transporter called divalent metal transporter 1, DMT1.[8,9] It is not clear exactly what happens for iron transport after endosomes. It has been proposed that iron will be directly delivered to mitochondria by endosomes through organelle interaction and bypasses the cytosol.[10] This interaction has been observed at least in erythroid cells.[11]

8.3.2 Iron Storage

The tight control of iron metabolism is crucial for health, because not only iron deficiency but also iron overload is closely related to human diseases. Excess of cellular iron is stored and detoxified in ferritin. Ferritins are ubiquitous, highly conserved iron storage proteins which play a critical role in cellular and organismal iron homeostasis. In mammalian cells, cytosolic ferritin consists of two subunits, the H and L chains. Twenty-four subunits assemble to form a spherical shell which can contain up to 4500 atoms of iron in a mineral and compact form.[12,13] In ferritin shells, the H subunit has ferroxidase activity which converts soluble ferrous ions into inert ferric hydroxides. The L subunit lacks ferroxidase activity, but is more efficient than the H subunit in inducing iron nucleation and mineralization within the shells.[14,15]

Iron stored in ferritin can be mobilized during iron deficiency for metabolic needs, but the mechanism is incompletely characterized.[16,17] Two models regarding iron exit and re-utilization have been proposed: they are lysosome-mediated protein degradation and iron-releasing model and ferritin pores at the junctions of ferritin subunits as iron channel mediating iron release model.[17] *In vitro* studies support the concept that iron exits ferritin shell via the hydrophilic channel on the three-fold axes.[17] These studies have showed that the iron of the ferritin core can be mobilized by reductants in the presence of iron chelators.[18] Further investigations also demonstrate mutagenesis of conserved amino acids in gated pores[19,20] and treatment with chaotropic agents or specific binding peptides that control the pore opening[21,22] can accelerate iron release. However, other studies indicate that ferritin iron is mainly released after proteolytic degradation of the protein.[16] Evidence of lysosomal ferritin degradation and following iron release has been found in cells under different treatments. For example, lysosomal inhibitors, but not by proteasomal inhibitor block ferritin protein degradation induced by iron chelators in cell lines;[23,24] the lysosomal degradation of ferritin has also been seen in cells induced for autophagy by amino acid starvation,[18] in cells given a horse spleen ferritin derived cationic ferritin[25] and in cells exposed to the anti-cancer drug doxorubicin.[26] A recent study by Domenico *et al.* has shown that overexpression of the cellular iron exporter ferroportin causes ferritin degradation and iron release.[27] Interestingly, the mechanism of ferritin degradation has been found via a proteasomal pathway and iron can be mobilized without ferritin shell degradation.[27] Various other studies also showed that oxidation-induced ferritin turnover[28] and degradation of L-ferritin mutants linked to neuroferritinopathy[29] occur mainly via proteasomal ferritin degradation pathway. These studies indicate that multiple ferritin degradation pathways exist and the different degradation mechanisms may play different roles under various cellular conditions. However, it is not clear which ferritin degradation pathway is more physiological relevant iron release route and whether ferritin protein shell degradation is a necessary step for iron re-utilization (Figure 8.2).

8.3.3 Coordination of Iron Uptake and Storage

It is well characterized that iron acquisition and storage are co-ordinately regulated by iron at the post-transcriptional level.[30–33] mRNAs encoding TfR1 and ferritin (H and L chains) contain iron-responsive elements (IREs) in their untranslated regions (UTRs). The size of the labile iron pool (LIP) modulates

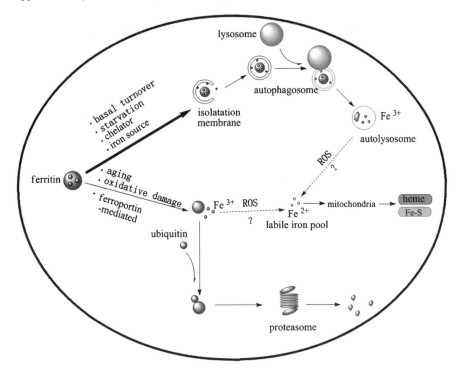

Figure 8.2 Mechanism of ferritin degradation and iron release. The degradation of ferritin nanocage probably occurs lysosomal and/or proteasomal. Ferritin basal turnover, iron chelation and nutrition starvation stimulates ferritin degradation through the lysosomal pathways. While ferroportin overexpression, aging and oxidative stress causes ferritin degradation via proteasome-mediated pathways. Once iron is released from the ferritin nanocage, it will used for various metabolic processes.

the binding of iron regulatory proteins (IRPs) to IREs present in the 5′-UTRs of ferritin mRNAs and 3′-UTRs of TfR1. When the level of iron in the LIP is low, translation of ferritin mRNA is suppressed by the binding of IRPs to IREs and consequently blocking ribosome assembly at the start codon, whereas the stability of TfR1 mRNA is increased by the binding of IRPs to IREs. The elevated TfR1 levels and decreased ferritin synthesis result in more iron acquired and less iron stored, consequently increase LIP iron levels. In the opposite scenario, iron supplementation upregulates ferritin synthesis and promotes TfR1 mRNA degradation through the dissociation of IRPs from IREs.[3–6] Therefore, the intracellular iron haemostasis maintains (Figure 8.3).

8.3.4 Iron Hormone Regulates Systemic Iron Homeostasis

Iron balance must be precisely regulated to provide sufficient iron for metabolic uses but avoid the toxicity associated with iron excess. And many elegant

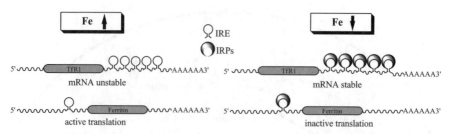

Figure 8.3 Regulation of cellular iron metabolism by the IRE/IRP system. Iron supplementation increases the iron supply and inactivates binding of IRPs to IREs, resulting in degradation of TfR mRNA and translation of the mRNAs of H- and L-ferritin. These responses lead to decreased iron uptake and elevated iron storage. Conversely, iron depletion activates binding of IRPs to IREs, resulting in stabilization of TfR mRNA and translational inhibition of the mRNAs of H- and L-ferritin. These responses lead to increased iron uptake and reduced iron storage.

regulatory processes evolved at different levels. Table 8.1 summarizes the major players at both cellular and systemic levels to facilitate iron acquisition, transport, regulation, storage, and utilization. These proteins mainly belong to the following functional categories: iron uptake, transporters, reductases and oxidases, storage and regulation. It can be found that proteins involving the conversion of iron between its oxidation and reduction status and iron metabolism regulation are the abundant groups. This feature may implicate a complicate regulatory network exiting for maintaining iron homeostasis and more players will be discovered.

It has long been proposed that the processes of the intestinal iron absorption pathway, macrophage iron recycling, and hepatocyte iron mobilization are regulated by an iron sensor for body iron stores and the requirement for erythropoiesis. There are many stimuli known to regulate the body iron balance; for example, erythropoiesis, hypoxia, iron deficiency, iron overload, and inflammation. A growing body of evidence suggests that an antimicrobial peptide, hepcidin, controls much of iron homeostasis.[34,35] Hepcidin is a circulating peptides synthesized predominantly in the liver and secreted into the plasma. The bioactive form of hepcidin is a 25-amino acid peptide containing eight cysteine residues that is derived from an 84-amino acid precursor.[36] Hepcidin negatively regulates iron transport by binding the iron exporter ferroportin. The binding of hepcidin causes ferroportin internalization and degradation in lysosomes[37] and consequently prevents iron export and increases cytosolic iron levels. Hepcidin levels are negatively regulated by hypoxia, anemia, and an iron-restricted diet; conversely, hepcidin synthesis is increased by iron loading and inflammation, causing iron to be sequestered in macrophages, hepatocytes, and enterocytes. The inflammation results in a decrease in plasma iron levels which eventually contributes to the anemia associated with infection and inflammation.[4]

Table 8.1 Some proteins involved in cellular and systemic iron metabolism, listed according to their functions.

Protein	Function
Iron uptake	
Transferrin (Tf)	Plasma Fe^{3+} carrier
Tf receptor (TfR)	Cognate membrane receptor for Tf
Tf receptor 2 (TfR2)	Unknown; similar to "classical" Tf receptor
Iron transporters	
DMT1	Membrane Fe^{2+} inward transporter
Ferroportin	Membrane Fe^{2+} outward transporter
Ferrireductases and ferroxidases	
Ceruloplasmin (Cp)	Plasma protein with ferroxidase activity involving in cellular iron export
Duodenal cytochrome b	Membrane ferric reductase involving in cellular iron uptake
Hephaestin	Membrane Cp homolog functioning in enterocyte iron export
Steap proteins	Ferrireductases required for iron uptake in Tf cycle
Storage and recycling	
Ferritin (H, L and M)	Cytosolic and mitochondrial Fe storage protein
Heme oxygenase 1 (HO-1)	Microsomal protein functioning in recycling Hb iron
Regulation of iron metabolism	
BMP	Involved in transcriptional activation of hepcidin
GDF15	May be erythroid regulator of iron acquisition by inhibition of hepcidin gene transcription
Hepcidin	Plasma peptide negatively regulates iron export and its deficiency leads to iron hyperabsorption
HFE	Binds TfR and its gene mutation is founded in >85% of hereditary hemochromatosis
Hemojuvelin	Membrane-GPI-linked protein and a BMP co-receptor
IRP (-1 and -2)	Cytosolic iron sensors; post-transcriptional regulation

DMT1, divalent metal transporter; Hb, hemoglobin; BMP, bone morphogenetic protein; GDF15, growth differentiation factor 15 ; IRP, iron regulatory protein.

8.4 Mitochondrial Iron Metabolism

Mitochondria play a key role in iron metabolism since heme and various iron–sulfur (Fe–S)-cluster containing proteins are synthesized in them.[38,39] The last step in heme biosynthesis, the insertion of Fe^{2+} into protoporphyrin IX by ferrochelatase, takes place in the mitochondrial matrix.[39] Fe–S clusters are synthesized mainly, if not entirely, in mitochondria and are combined with mitochondrial apo-proteins to form mature proteins or are exported from mitochondria for utilization by cytosolic and nuclear proteins.[38] Table 8.2 summarizes some known proteins involved in mitochondrial iron homeostasis and utilization and their deficiency-related human disorders.

Table 8.2 Some proteins involved in mitochondrial iron metabolism and the deficiency of these proteins are closely related with human (or putative) diseases.

Protein	Protein functions and deficiency related human diseases
ALAS2/eALAS	The first enzyme of heme synthesis in erythroid cells and its deficiency leads to x-linked sideroblastic anemia (XLSA)
ABCB7	Mitochodondrial membrane protein is thought to export components of ISC out of mitochondria to participate in maturation of cytosolic ISC proteins; its mutation leads to XLSA with ataxia (XLSA/A)
Ferrochelatase	The mitochondrial protein inserts Fe into the protoporphyrin IX ring to form heme
Frataxin	It is thought to be involved in mitochondrial iron export and formation of ISC and its mutation is found in patients with Friedrich ataxia
Glutaredoxin 5	A part of complex machinery for assembly and export of ISC and maturation of ISC proteins; its deficiency is associated with hypochromic sideroblastic anemia and iron overload
Mitochondrial ferritin	Mitochondrial Fe storage and highly expressed in "ring" sideroblasts in patients of sideroblastic anemias
Mitoferrin	The principal mitochondrial iron importer is essential for heme synthesis in vertebrate erythroblasts; its deficiency leads to profound hypochromic anemia and erythroid maturation arrest in zebrafish

ABCB7, ATP-binding cassette sub-family B, member 7; ALA-S2/eALAS, erythroid-specific 5-aminolevulinic acid synthase; ISC, iron–sulfur cluster.

Mitochondrial iron levels must be well regulated because an inadequate supply of iron would impair the metabolic and respiratory activities of the organelle, while excess "free" iron in mitochondria would promote the generation of harmful reactive oxygen species which are produced as a side reaction of mitochondrial electron transport.[40,41]

Defects in mitochondrial iron transport and utilization can result in mitochondrial iron overload. There is extensive iron accumulation in erythroblast mitochondria of both patients with X-linked sideroblastic anemia due to defective erythroid-specific 5-aminolevulinic acid synthase (eALAS)[39,42–44] and those with ring sideroblasts associated with myelodysplastic syndrome.[42,44] Mitochondrial iron overload has also been documented in patients with Friedreich's ataxia with defective frataxin[45–47] and in those with sideroblastic anemia with ataxia from defects in the Fe-S transporter ABC7.[48] In addition, studies with yeast, the best studied eukaryotic model of Fe–S cluster synthesis, showed that defects in any of the enzymes of the Fe–S cluster assembly pathway caused mitochondrial iron accumulation and lack of normal mitochondrial function.[49]

The recent discovery of mitochondrial ferritin (MtFt) has contributed to our understanding of mitochondrial iron metabolism[50–54] and sideroblastic anemia. Although most cells, including normal erythroid cells, express extremely low levels of MtFt, it is highly expressed in erythroblast mitochondria ("ringed sideroblasts") of patients with sideroblastic anemia[52,55] and it was shown that the iron in ringed sideroblasts is in MtFt.[55] In humans, MtFt is encoded by an intronless gene on chromosome 5q23.1. The protein is expressed as a 30 kDa precursor which is targeted to mitochondria. After intra-mitochondrial deletion of the N-terminal leader sequence, the 22 kDa MtFt subunits, which are highly homologous to H subunits, assemble into ferritin shells with ferroxidase activity.[50,52–54] Over-expression of both human and mouse MtFt, which incorporates iron, results in decreased cytosolic ferritin and increased transferrin receptor levels, suggesting cytosolic iron deficiency.[52,53] MtFt expression have been shown to prevent, at least in part, oxidative stress-induced damage of frataxin-defective yeast and HeLa cells,[56,57] it is tempting to speculate that the expression of MtFt may abrogate the toxic effects caused by oxidative stress *in vitro*.

8.5 Molecular Mechanism of Impaired Iron Homeostasis in Neurodegenerative Disorders

8.5.1 Iron Dysregulation and Neurodegenerative Diseases (NDs)

Iron is vitally important for metabolism because of its unsurpassed versatility as a biologic catalyst. Although living organisms are equipped with a multilevel regulatory mechanism to maintain iron homeostasis, dysregulation of iron can lead to iron deficiency or iron overload. Except for the common iron diseases, such as anemia and hemochromatosis, a growing body of evidence implicates that impaired iron homeostasis and pathological iron accumulation in the brain is associated with the progression of neurodegenerative diseases, such as Parkinson's disease (PD) and Alzheimer's disease (AD). Iron accumulation in the regions of brains has also been shown in many neurological diseases, including PD, AD, congenital aceruloplasminaemia, Friedreich's ataxia, and neuroferritinopathy. PD is a progressive disease with a mean age at onset of 55, and the incidence increases markedly with age. The pathological hallmarks of PD are the loss of the nigrostriatal dopaminergic neurons and the presence of intraneuronal cytoplasmic inclusions, Lewy bodies.[58] Despite recent genetic discoveries leading to a number of different genetic models of PD, there are still major gaps in our understanding of the molecular and cellular biology of PD. Since none of those models shows all typical degenerative features of dopaminergic neurons, and there is no apparent genetic linkage in 95% of PD cases, PD etiology remains mysterious. AD currently affects 12 million people worldwide; 4.5 million of these are in the USA alone. Pathologically, AD is characterized by synaptic loss, nerve cell loss, extracellular deposition of β-amyloid (Aβ) protein (forming senile plaques), and intracellular precipitation of hyperphosphorylated tau protein (forming neurofibrillary tangles).[59] The exact biochemical mechanism of the pathogenesis of AD is still unknown, but much attention is given to the possible implication of oxidative stress in its development. Although very little is known about why and how the neurodegenerative process begins and progresses, increasing amounts of research have demonstrated that there is pathological iron accumulation in affected region of patients of NDs.[60,61] It is speculated that impaired iron homeostasis either one of the initiators of pathogenesis of NDs or a major factor promoting the progression in those chronic disorders.

8.5.2 Iron Metabolism in the Central Nervous System

Neurons require iron for many aspects of their physiology; however, the mechanisms of iron uptake by neurons have not been completely understood.[60,61] Unlike Tf found in blood, Tf in the brain is fully saturated with iron. The excess iron will bind to other transporters. Hence, it is possible that there are two transport forms of iron in the brain: Tf-bound iron (Tf-Fe) and non-Tf-bound iron (NTBI).[60] The Tf-Fe will be taken up by neuronal cells via TfR1 and DMT1-mediated process. NTBI will be acquired by neurons probably via DMT1. Ferroportin is the only identified iron exporter and plays an important role in iron recycling by the reticulo-endothelium system.[4] The neuronal system needs significant amounts of heme for maintaining its high metabolic rate. Heme biosynthesis and degradation is finely tuned to these requirements. Heme catabolism is mediated by the heme oxygenases (HOs). In neuronal system, most cells express two isoforms of heme oxygenase, HO-1 and HO-2. Whereas HO-2 protein is widely distributed throughout the rodent CNS, basal HO-1 expression in the normal brain is confined to small groups of scattered neurons and neuroglia.[62]

8.5.3 Participation of Iron in Neurodegenerative Diseases

Although the brain has several characteristics that make it unique for utilizing iron (such as, it resides behind a vascular barrier – the blood–brain barrier, BBB – which limits its access to plasma iron), there are several clues indicating a deleterious role of iron in NDs:[47,57,63,64] there is higher accumulation of iron in affected regions as compared to those of age-matched controls in the brains of patients with neurodegenerative diseases; many neurons of the aging brain contain Fe^{3+} in contrast to neurons of younger brains; and mishandling of ferritin is also been found in the human brain. The translational activity of ferritin mRNA is regulated by IRPs by binding to IREs on ferritin mRNA in conditions with iron depletion to prevent its translation. IRPs are constitutively actively expressed in normal brains and those affected by Parkinson's disease, which is in accord with the absence of ferritin protein in dopaminergic neurons of the substantia nigra. Further evidence for the importance of normal ferritin function in neuronal iron metabolism comes from the newly described inherited disorder, neuroferritinopathy.[65]

Microglias, the resident immune cells in the brain, play an important role of immune surveillance under normal conditions and are critically involved in neurodegenerative diseases. These cells turn over rapidly as they scavenge dying cells and prevent the subsequent release of potentially hazardous molecules. In response to brain damage or injury, microglial cells become activated and undergo morphological as well as functional transformations. Activated microglia produce a wide array of cytotoxic factors, including ROS and nitric oxide (NO), which would induce neurodegeneration in some circumstances. Microglia contain abundant iron and activation of microglia is associated with an increase in iron accumulation and ferritin expression and consequently change of iron status profoundly affects other cellular pathways, such as ROS and NO production.[66] Many lines of evidence suggest that oxidative stress resulting in ROS generation play a pivotal role in the age-associated cognitive decline and neuronal loss in neurodegenerative diseases.[59,61] The majority of free radicals that damage biological systems are oxygen radicals, the main byproducts formed associated with mitochondria during electron transport in the oxidative phosphorylation. Mitochondria are thought to play a crucial role in the aging process not only due to their role as main intracellular generators of ROS, but also because they are targets of ROS attack. Although defenses against damage produced by ROS are extensive, including enzymatic and small molecule antioxidants as well as repair enzymes, an increased production of ROS or a poor antioxidant defense network can lead to a progressive damage in the cell with a decline in physiological function. The nervous system is particularly vulnerable to the deleterious effects of ROS: the brain contains relatively high concentrations of potential harmful iron in certain regions; it utilizes the highest amount of oxygen to produce energy; finally, the brain is relatively deficient in antioxidant systems with lower activity of glutathione peroxidase and catalase compared to other organs. Excess iron is extremely harmful by promoting highly reactive hydroxyl radical production and can result from iron transport across the BBB by DMT1, release from ferritin, other iron containing molecules and invaded red blood cells.[60,61] Serum iron cannot cross the BBB if it functions normally.

Dopaminergic neurotoxins 6-hydroxydopamine (6-OHDA) and N-methyl-4-phenyl-1,2,3,6-tetrahydropyridine (MPTP) induce accumulation of iron at sites where they initiate neurodegeneration.[60,61] This might result either from increased uptake of iron through DMT1, or iron release from ferritin. Iron is thought to enter a labile iron pool, where it is accessible to participate in the Fenton reaction with hydrogen peroxide, which is generated by dopamine metabolism. The resulting effect is oxidative stress as a consequence of depletion of cellular antioxidants, and increased membrane lipid peroxidation, DNA damage and protein oxidation and misfolding. Labile iron can also cause aggregation of α-synuclein and Aβ to form toxic aggregates, which, in turn, can initiate the generation of hydroxyl radicals, causing oxidative stress. α-Synuclein is a cytoplasmic protein with 140 amino acids, and its abnormal folding and aggregation is the main component of Lewy bodies, which is the important pathological characteristic of PD and AD patients. It is found that iron is involved in this process and promotes its progress.[67,68] β-Amyloid (Aβ) is the main component of senile plaque; moreover, it often deposits in the cerebral cortex, hippocampus, basal ganglia, and thalamus after combining iron and copper.[69,70]

8.6 Examples of Nuclear Analytical Techniques in Iron Metabolism Studies

Iron (Fe) is an essential trace element in the human body. As a transition metal, Fe has the capacity to accept and donate electrons readily, interconverting between ferric (Fe^{3+}) and ferrous (Fe^{2+}) forms. Most iron present in living organisms is tightly combined with proteins although some may be present as low molecular weight complexes in soluble pools.[71] Thus, the analysis of iron in biological environments should incorporate both bulk analysis and speciation studies.

Modern nuclear analytical techniques (NATs) have made significant achievements in trace elements analysis in biological and environmental sciences. The NATs, mainly including neutron activation analysis (NAA), Mössbauer spectroscopy, proton-induced X-ray emission (PIXE) spectroscopy, synchronous radiation-based analytical techniques, and isotope-based techniques, have attracted much interest as tools in metallomics and metalloproteomics to study trace element metabolism and biodistribution for understanding the mechanisms for toxicology, pathology, and pharmacology of metals of interest and their mechanisms. Furthermore, for speciation studies, a variety of physical, chemical, or biological separation procedures have been developed in combination with the NATs.[72] Thus, with the advantages of high sensitivity, high accuracy, sufficient detection limits, no or fewer interferences from other components in the sample, and having a reasonable cost-to-benefit ratio, NATs have been extensively applied in life sciences for trace element bulk analysis and chemical speciation studies.

The following examples of NATs in iron metabolism studies were described to illustrate their applications for different purposes. The features of different NATs for trace element quantitative analysis and chemical speciation studies were stated to elucidate the possibility, feasibility, and the potential of NATs in life sciences.

8.6.1 Synchronous Radiation-based Analytical Techniques

The most applied SR-based analytical technique for biological analysis is synchrotron radiation X-ray fluorescence (SRXRF), including edge X-ray absorption fine structure (EXAFS) and X-ray absorption near-edge structure (XANES). SRXRF is a widely applied technique for microscopic analysis of chemical elements, which could provide trace element imaging at tissue, cellular, and subcellular resolution. The high-resolution requirement can be achieved using microbeam synchrotron radiation X-ray fluorescence (μ-SRXRF). The detection limit of μ-SRXRF can reach 0.1–$1\,\mu g\,g^{-1}$. The spatial resolution is between 0.1 and 1 μm depending on the energy of the synchrotron radiation.[73,74] SRXRF microscopy can also provide information regarding the oxidation state and coordination environment of metals by XANES or micro-XANES spectroscopy. With these advanced techniques, we can identify and quantify metal directly in the context of their native environment, without destroying the samples.[75–77]

To address the potential mechanism of disturbed homeostasis of iron in Parkinson's disease (PD), Ortega et al.[78] developed an original experimental setup to perform chemical element imaging with a 90 nm spatial resolution using SRXRF to investigate the subcellular distribution of iron in dopamine-producing neurons to elucidate the role of dopamine on iron homeostasis. As dopamine can form stable complexes with iron *in vitro*, it has been suggested that dopamine may exert a protective effect by chelating iron in dopaminergic neurons. The loss of dopamine in the brain region of PD patients and consequent lack of protection might be faulty in PD.[79] The newly developed SRXRF nanoprobe was applied to determine whether the dopamine–iron complex occurs *in vivo*. PC12 rat pheochromocytoma cells, as the *in vitro* model of dopamine-producing cells, were treated with sub-cytotoxic concentrations of iron and/or α-methyltyrosine (AMT, an inhibitor of dopamine synthesis). The iron distribution in neurons in subcellular compartments (cell body, neurite outgrowths, and distal ends) was revealed. As exemplified in Figure 8.4, quantitative maps of potassium, iron, and zinc in cell bodies were obtained and expressed in $ng\,cm^{-2}$. Zinc, as a certified reference material, was chosen for calibrating the iron concentration in subcellular compartments using PyMCA software,[80] which enables data from selected zones of the scanned area to be extracted, and corresponding X-ray fluorescence spectra to be fitted. The study shows that iron accumulates in dopamine neurovesicles. In addition, when control cells are compared to cells exposed to an excess of iron, the same subcellular distribution is found but with a higher number of iron-rich structures in iron-exposed cells, the inhibition of dopamine synthesis results in a decreased vesicular storage of iron. These results indicate a new physiological role for dopamine in iron buffering within normal dopamine-producing cells.

Similar application of SRXRF was performed by Wang et al.[77] on iron micro-distribution in brain after intranasal instillation of ferric oxide (Fe_2O_3) particles. At day 14 after intranasal instillation of fine Fe_2O_3 particle (280 ± 80 nm) suspension, the iron micro-distribution and chemical state in the mice olfactory bulb and brain stem were analyzed by SRXRF. An obvious increase of Fe contents in the olfactory nerve and the trigeminus of brain stem could be observed on the micro-distribution map of iron in the olfactory bulb and brain stem (see Figures 8.5 and 8.6), suggesting that Fe_2O_3 particles were possibly transported via uptake by sensory nerve endings of the olfactory nerve and trigeminus. Iron K-edge XANES analyses were applied to study Fe chemical species in brain samples. Figure 8.7 indicates that the ratios of Fe(III)/Fe(II) were increased in the olfactory bulb and brain stem.

8.6.2 Particle-induced X-ray Emission

Particle-induced X-ray emission (PIXE) is another widely applied technique in which the surface of the specimen is scanned and thus provides information on the surface distribution of elemental species. PIXE could be applied for tissue and single-cell analysis by focusing ion beams down to a few μm^2 cross-section.

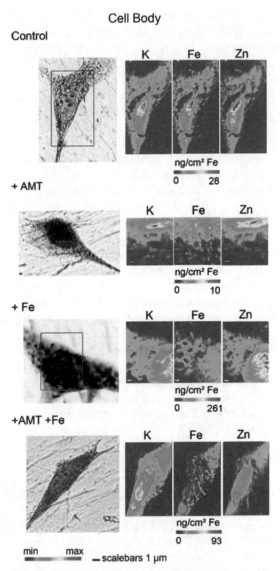

Figure 8.4 Nano-imaging of potassium, iron, and zinc in cell bodies. Each series of images are representative of the entire cell population for each condition (control, 1 mM AMT and/or 300 mM FeSO$_4$). The scanned area (the squares in the left images) is shown on a bright field microscopy view of the freeze dried cell. Iron is located within the cytosol in vesicles of 200 nm size or more (Control). In cells exposed to iron alone Fe, and to AMT + Fe, a larger number of iron-rich structures are observed in cell bodies. In the bodies of cells exposed to AMT alone, only a basal level of diffused iron is observed and almost no iron-rich structures. Min–max range bar units are arbitrary for potassium and zinc distributions. For iron distribution the maximum threshold values in micrograms per squared centimeter are shown in the shaded scale. Scale bars = 1 μm. (Reprinted from Solé et al.[80] © 2007 Ortega et al.)

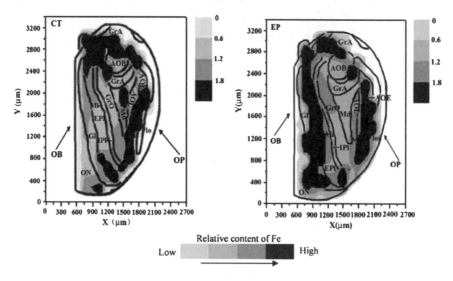

Figure 8.5 Fe distribution in the OB section. CT = Control group (n = 3); EP = exposed group (n = 3). OB = olfactory bulb; OP = olfactory peduncle; ON = olfactory nerve; Gl = glomerular layer; Epl = external plexiform layer; Mi = mitral cell layer; Ipl = internal plexiform layer; GrO = granule cell layer of olfactory bulb; Md = medullary layer; GrA = granule cell layer of accessory olfactory bulb; AOB = accessory olfactory bulb; AOE = anterior olfactory nucleus external part; AOL = anterior olfactory nucleus, lateral part; lo = lateral tract. (Reprinted from Wang et al.[77] © 2007 Humana Press Inc.)

The high-resolution requirement can be achieved using micro-PIXE (µ-PIXE) analysis. The detection sensitivity of µ-PIXE can reach 1–10 µg g^{-1}. The spatial resolution is typically in the micrometer range, but 0.2 µm is achievable for the latest high-resolution instruments.[81]

To study the cellular pharmacology of the anticancer agent 4-iodo-4-deoxy-doxorubicin, iodine (I) and Fe distributions were measured by PIXE, and carbon distribution was measured by Rutherford backscattering µ-spectrometry to normalize trace element concentrations in human carcinoma cells.[82] As shown in Figure 8.8, the iodine distribution map indicates the nucleus area, and 4-iodo-4-deoxy-doxorubicin is a DNA intercalating agent which accumulates in the nucleus. A modification of the Fe distribution was observed in cells exposed to the drug, demonstrating iron chelation by the anticancer agent and the redistribution of iron from the cytosol to the nucleus. From the PIXE study, the iron–doxorubicin complex triggered oxidative reactions in close vicinity to genomic DNA and were hypothesized as the anticancer pharmacology.

As a complement to SRXRF, micro-PIXE analysis could also be performed to obtain quantitative element concentration on groups of several hundred cells on the same samples to complete SRXRF imaging of single cells. In the above SRXRF study by Ortega et al.,[78] cellular iron and zinc concentrations (mg g^{-1} dry mass) could be obtained by PIXE quantitative micro-analysis by the mean

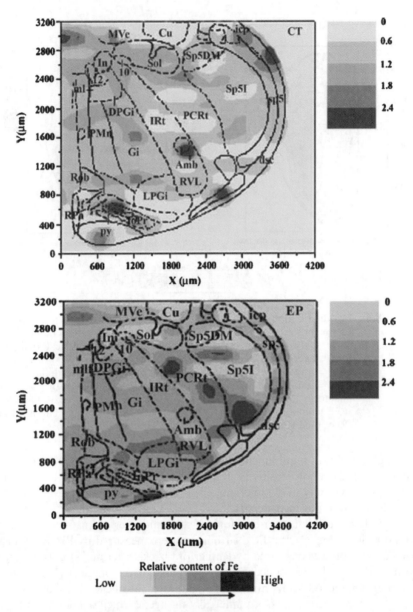

Figure 8.6 Fe distribution in the brain stem section. CT = Control group (n = 3); EP = exposed group (n = 3). Mve = medial vestibular nucleus; Cu = cuneate nucleus; icp = inferior cerebellar peduncle; 12 = hypoglossal nucleus; 10 = dorsal motor nucleus of the vagus; In = intercalated nucleus; Sol = solitary tract nucleus; Sp5 = spinal trigeminal tract; Sp5DM = dorsomedial part of Sp5; Sp5I = interpolar part of Sp5; PCRt = parvicellular reticular nucleus; Irt = intermediate reticular nucleus; Gi = gigantocellular reticular nucleus; DPGi = dorsal paragigantocellular nucleus; LPGi = lateral paragigantocellular nucleus; Amb = ambiguous nucleus; mlf = medial longitudinal fasciculus; Rob = raphe obscurus nucleus; PMn = paramedian reticular nucleus; IOPr = inferior olive, principal nucleus; py = pyramidal tract; dsc = dorsal spinocerebellar tract; RPa = raphe pallidus nucleus. (Reprinted from Wang et al.[77] © 2007 Humana Press Inc.)

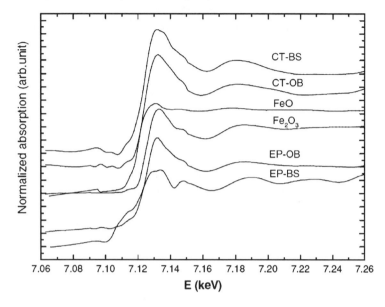

Figure 8.7 Fe K-edge XANES spectra of sample and standard powders. CT-OB = Olfactory bulb of control group; CT-BS = brain stem of control group; EP-OB = olfactory bulb of exposed group; EP-BS = brain stem of exposed group. FeO = FeO powder; Fe_2O_3 = Fe_2O_3 powder. (Reprinted from Wang et al.[77] © 2007 Humana Press Inc.)

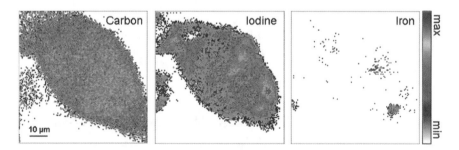

Figure 8.8 Carbon, iodine and iron chemical maps in a human ovarian carcinoma cell exposed *in vitro* to the anticancer agent 4-iodo-4-deoxy-doxorubicin. (Reprinted from Ortega.[82] © 2006 Elsevier Masson SAS.)

of six independent analyses performed on areas containing several hundreds of cells for each condition of culture.

8.6.3 Neutron Activation Analysis

Neutron activation analysis (NAA) is a powerful quantitative method for accurate determination of total amount of element with the detection limit from

1 ng g^{-1} to 1 µg g^{-1}. By using NAA, Zhu et al.[83] studied the particokinetics and extrapulmonary distribution of inhaled nano-^{59}Fe$_2$O$_3$ nanoparticles. Fe$_2$O$_3$ particles (22 nm) were irradiated in a heavy-water reactor at a neutron flux of 5×10^{13} n cm^{-2} s^{-1} to form ^{59}Fe$_2$O$_3$ by the (n,γ) reaction. The irradiated nano-^{59}Fe$_2$O$_3$ particle suspension was instilled intra-tracheally into rats, then the counts of ^{59}Fe in tissues and the metabolism of ^{59}Fe in blood, urine, and feces were determined. A well-type high-purity germanium detector (EG&G Ortec, Oak Ridge, TN) was used for the detection of ^{59}Fe radioactivities at 1099.25 and 1291.60 keV in samples. The intensity of radioactivity of each organ was presented as the percentage of instilled dose (ID) per gram of wet tissue (% ID g^{-1}). The initial ID was calculated by comparison with a diluted radioactive ^{59}Fe$_2$O$_3$ suspension solution and iron chemical standard. The quantitative analysis could be used to evaluate the potential risks and extrapulmonary translocation of respiratory exposed nanoparticles.

For understanding the mechanism of uptake of iron by lichens, isotopic distribution of iron in two kinds of lichens were studied.[84] Lichen samples were tentatively separated into two categorical sites, the active growing site and the inactive site. The isotopic ratio of ^{54}Fe to ^{58}Fe incorporated in the lichens was determined by neutron activation analysis, which was carried out at the Japan Atomic Energy Research Institute (JAERI) in Tokai-mura using a JRR-4 atomic reactor. The variance in the ratio of ^{54}Fe to ^{58}Fe of the sample was determined by measuring the gamma rays at 834.8 keV (^{54}Fe (n,p)→ ^{54}Mn, 100% emission, $t_{1/2}=312.5$ d) and 1099.3 k e V (^{58}Fe (n,γ)→ ^{59}Fe, 56.5% emission, $t_{1/2}=44.6$ days) for 3000 s after a cooling period of 45 days. Gamma-ray spectrometry was performed using a Ge(Li) detector with 4000 channels and a resolution of 0.5 keV per channel. The isotopic ratios (^{54}Fe /^{58}Fe) of the active growing site were significantly smaller than those of the inactive site or substratum for both lichens. An autonomous incorporation of iron with a kinetic isotope effect by these lichens can be considered.

8.6.4 Radioactive and Enriched Stable Isotope-based Techniques

Radioactive and enriched stable isotopes are used as classical methods for detection of the species of trace elements and their transformation in biological and environmental systems, as they can serve as specific and sensitive tracers. In comparison with radioactive isotopes, stable enriched isotopes have the evident advantage that they do not destroy the chemical bond during a labeling process. Iron has four stable isotopes, ^{54}Fe (5.8%), ^{56}Fe (91.7%), ^{57}Fe (2.2%), and ^{58}Fe (0.3%). The radioactive iron isotope is ^{59}Fe, which could easily be generated by nuclear reactors. The abundances and concentrations of stable iron isotopes could be determined by various isotope analytical methods. Several analytical approaches have been used in recent years for the determination of stable isotopes of minerals. These methods include NAA, inductively coupled plasma–mass spectrometry (ICP-MS), gas chromatography–mass spectrometry (GC-MS), electron impact mass spectrometry (EIMS), and magnetic sector, thermal ionization mass spectrometry (TIMS).[85–87] The ^{59}Fe content could be calculated using a gamma counter. Chemicals or medicines containing or labeled with isotopes could accordingly be detected and analyzed quantitatively.[83,88]

By using enriched stable isotopes, the absorption of zinc, copper, and iron in normal, healthy adults was studied.[89] Stable isotopes of zinc, copper, and iron were incorporated into diets. Complete fecal samples were collected and the unabsorbed isotopes remaining in the samples were measured by TIMS. Stable isotopes were also infused in five young men to evaluate mineral utilization and kinetics with stable isotopes. Isotopic enrichments can be measured in urine and blood, so kinetic studies of essential minerals utilization are now feasible with enriched stable isotopes.

8.6.5 Mössbauer Spectroscopy

Mössbauer spectroscopy is a spectroscopic technique based on the resonant emission and absorption of gamma rays in solids. Due to the high energy and extremely narrow line widths of gamma rays, it is one of the most sensitive techniques to provide precise information about the chemical, structural, magnetic, and time-dependent properties of a material. For the most common Mössbauer isotope, ^{57}Fe has a line width of 5×10^{-9} eV. Compared to the Mössbauer gamma-ray energy of 14.4 keV this gives a resolution of 1 in 10^{12}. The various biomedical applications of Mössbauer spectroscopy are grouped as quantitative analysis, qualitative analysis, the effects of environmental factors on biological molecules, Mössbauer elements metabolism, dynamic properties of biological subjects, and pharmaceuticals containing Mössbauer elements.[90]

To study the role of pigment neuromelanin (NM) on iron storage the dopaminergic neurons of the substantia nigra (SN) in Parkinson's disease, Double et al.[91] quantified and characterized the interaction between NM and iron using Mössbauer spectroscopy. NM was isolated from normal adult brain and synthetic melanin (DAM) was analyzed for comparison. NM and synthetic melanin were enriched with ^{57}Fe as this isotope has a typical Mössbauer effect. Spectra of NM were measured at 4.2, 120 and 300 K. Mössbauer spectra obtained from ^{57}Fe-loaded human NM exhibited a doublet at 300 K and a sextet (92%) at 4.2 K, characteristics of trivalent iron (Figure 8.9a and c). The spectrum at 120 K exhibited superparamagnetic relaxation, indicative of ferritin-like iron clusters (Figure 8.9b). The Mössbauer parameters describing the electronic structure of the iron complexes can be calculated from a least-squares fit using Lorentzian lines. The larger iron cluster size was demonstrated by Mössbauer spectroscopy in the native pigment compared with the synthetic melanin. This observation is consistent with the hypothesis that NM may act as an endogenous iron-binding molecule in dopaminergic neurons of the SN in the human brain.

Another example of iron uptake by *Escherichia coli* monitored by Mössbauer spectroscopy was performed by Matzanke et al.[92] As the iron deficiency under aerobic conditions is mediated by a highly stable ferric enterobactin [Fe(ent)$^{3-}$]

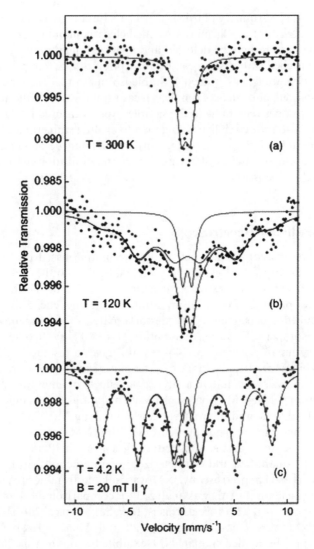

Figure 8.9 Mössbauer spectra obtained from ^{57}Fe-loaded human NM exhibits a doublet at 300 K (a) and a sextet (92%) at 4.2 K (c) characteristic of trivalent iron. The spectrum at 120 K exhibits superparamagnetic relaxation, indicative of ferritin-like iron clusters (b). The spectrum at 4.2 K was measured in an externally applied field of 20 mT, applied parallel to the gamma beam. (Reprinted from Double et al.[91] © 2003 Elsevier Science Inc.)

siderophore complex, the iron delivery by ^{57}Fe(ent) and a synthetic analog of enterobactin, 1,3,5-N,N',N'-tris(2,3-dihydroxybenzoyl)triaminomethylbenzene (MECAM) were studied. The Mössbauer samples were either frozen cells or frozen aqueous solutions. The Mössbauer spectrometer was of the constant acceleration type, operated in connection with a 512-channel analyzer in the time-scaling mode and in horizontal transmission geometry. A 50 mCi ^{57}Co source in a Rh matrix was kept at room temperature. A typical run lasted about 60 h and the spectrometer was calibrated against an iron foil. The results indicated that the transportation of Fe-MECAM and Fe(ent) across the outer membrane is approximately at the same rate; however, the behaviors of Fe(ent) and Fe-MECAM were very different. A major fraction of the iron originally absorbed as ferric enterobactin appeared as Fe(II), apparently in the cytoplasm of the cell after more than 30 min, but little iron was delivered to the cytoplasm by the MECAM complex.

8.6.6 Speciation Analysis by Pre-separation Procedures in Combination with Nuclear Analytical Techniques

Although NATs are thought to be unsuitable for speciation analysis, in recent years, more and more nuclear analysts have endeavored to use NATs for speciation studies by combining NATs with specific pre-separation procedures to provide chemical species information. A variety of physical, chemical, or biological separation procedures have been developed in combination with NATs. The general experimental methods of the hybrid NATs being used for chemical speciation studies are listed in Table 8.3.[75]

For instance, a combination of PIXE and biochemical separation has recently been established to perform on-line scanning species analysis of trace elements in proteins or enzymes. An example of the application of this technique is the determination of the distribution of Fe and Ni along a native PAGE electrophoresis belt containing hydrogenase from *Thiocapsa roseopersicina* and *Desulfovibrio gigas*, and showed that these metals were located at different polypeptides that formed the enzyme.[93,94] Similarly, by coupling electrophoresis and PIXE, Strivay *et al.*[95] investigated the nature and the quantity of metals (Cu, Fe, Zn) contained in proteins. The gel is dried after the electrophoresis and each track is scanned with a 2.5 MeV proton beam to induce X-ray emission. The metals contained in an electrophoretic band could

Table 8.3 General experimental procedures in the use of nuclear analytical techniques for chemical speciation studies.

	Pre-separation procedure	*Nuclear analytical technique*
Physical	Permeation	
	Ultracentrifugation	
	Phase separation	
Chemical	Solvent extraction	Neutron activation analysis
	Ion exchange	Proton-induced X-ray emission
	Precipitation + X-ray fluorescence	
	Step by step dissolution	Mass spectrometry
Biochemical	Gel chromatography	Synchronous radiation
	Gel filtration	Radioactive or stable isotopes
	Enzymatic	

be determined and quantified by comparing the characteristic X-ray peak area with those obtained with polyacrylamide gels doped with the same metal. Finally, the relative concentration of each protein is determined by densitometry in order to compute the protein/metal ratio to check if metals remain bound to proteins. As an example of pre-separation procedures combined with NAA, Jayawickreme and Chatt[86] used several bioanalytical techniques (including electrofocusing and isotachophoresis) in conjunction with NAA to characterize the protein-bound metal species. The dialysis of the homogenate bovine kidney showed that more than 90% of Ca, Cd, Cu, Fe, Mg, Mn, Se, V, and Zn, and about 20% of Br were bound to macromolecules, mainly proteins.

Taken together, nowadays we are able to study iron metabolism and distribution at the tissue, cellular, subcellular, or even molecular level by NATs independently or by combining NATs with a variety of pre-separation procedures. The improvement and upgrade of the qualitative and quantitative analytical techniques are the challenges for further progress on the species study at molecular level and the promotion for the development of metallogenomics, metalloproteomics, and metallomics of iron.

Acknowledgements

Guangjun Nie gratefully acknowledges the support of the Hundred Talents Program, Chinese Academy of Sciences. The work was supported partially by grants from the National Natural Sciences Foundation of China (10979011, 30900278). The authors also thank Dr Prem Ponka for his support.

References

1. B. Halliwell and J. M. Gutteridge, *Free Radicals in Biology and Medicine*, Oxford University Press, Midsomer Norton, Avon, 1999.
2. P. Ponka, *J. Trace Elem. Exp. Med.*, 2000, **13**, 73–83.
3. N. C. Andrews, *Nat. Rev. Genet.*, 2000, **1**, 208–217.
4. N. C. Andrews and P. J. Schmidt, *Annu. Rev. Physiol.*, 2007, **69**, 69–85.
5. T. Rouault and R. Klausner, *Curr. Top. Cell. Regul.*, 1997, **35**, 1–19.
6. M. J. Koury and P. Ponka, *Annu. Rev. Nutr.*, 2004, **24**, 105–131.
7. R. S. Ohgami, D. R. Campagna, E. L. Greer, B. Antiochos, A. McDonald, J. Chen, J. J. Sharp, Y. Fujiwara, J. E. Barker and M. D. Fleming, *Nat. Genet.*, 2005, **37**, 1264–1269.
8. H. Gunshin, B. Mackenzie, U. V. Berger, Y. Gunshin, M. F. Romero, W. F. Boron, S. Nussberger, J. L. Gollan and M. A. Hediger, *Nature*, 1997, **388**, 482–488.
9. F. Canonne-Hergaux, A. S. Zhang, P. Ponka and P. Gros, *Blood*, 2001, **98**, 3823–3830.
10. A. S. Zhang, A. D. Sheftel and P. Ponka, *Blood*, 2005, **105**, 368–375.
11. A. D. Sheftel, A. S. Zhang, C. Brown, O. S. Shirihai and P. Ponka, *Blood*, 2007, **110**, 125–132.

12. P. M. Harrison and P. Arosio, *Biochim. Biophys. Acta*, 1996, **1275**, 161–203.
13. F. M. Torti and S. V. Torti, *Blood*, 2002, **99**, 3505–3516.
14. S. Levi, P. Santambrogio, A. Cozzi, E. Rovida, B. Corsi, E. Tamborini, S. Spada, A. Albertini and P. Arosio, *J. Mol. Biol.*, 1994, **238**, 649–654.
15. P. Santambrogio, S. Levi, A. Cozzi, E. Rovida, A. Albertini and P. Arosio, *J. Biol. Chem.*, 1993, **268**, 12744–12748.
16. P. Arosio, R. Ingrassia and P. Cavadini, *Biochim. Biophys. Acta*, 2008.
17. X. Liu and E. C. Theil, *Acc. Chem. Res.*, 2005, **38**, 167–175.
18. P. M. Harrison, T. G. Hoy, I. G. Macara and R. J. Hoare, *Biochem. J.*, 1974, **143**, 445–451.
19. W. Jin, H. Takagi, B. Pancorbo and E. C. Theil, *Biochemistry*, 2001, **40**, 7525–7532.
20. H. Takagi, D. Shi, Y. Ha, N. M. Allewell and E. C. Theil, *J. Biol. Chem.*, 1998, **273**, 18685–18688.
21. X. Liu, W. Jin and E. C. Theil, *Proc. Natl. Acad. Sci. U. S. A.*, 2003, **100**, 3653–3658.
22. X. S. Liu, L. D. Patterson, M. J. Miller and E. C. Theil, *J. Biol. Chem.*, 2007, **282**, 31821–31825.
23. T. Z. Kidane, E. Sauble and M. C. Linder, *Am. J. Physiol Cell Physiol.*, 2006, **291**, C445–C455.
24. J. Truty, R. Malpe and M. C. Linder, *J. Biol. Chem.*, 2001, **276**, 48775–48780.
25. D. C. Radisky and J. Kaplan, *Biochem. J.*, 1998, **336**(Pt 1), 201–205.
26. J. C. Kwok and D. R. Richardson, *Mol. Pharmacol.*, 2004, **65**, 181–195.
27. I. D. Domenico, M. B. Vaughn, L. Li, D. Bagley, G. Musci, D. M. Ward and J. Kaplan, *EMBO J.*, 2006, **25**, 5396–5404.
28. J. Mehlhase, G. Sandig, K. Pantopoulos and T. Grune, *Free Radic. Biol. Med.*, 2005, **38**, 276–285.
29. A. Cozzi, P. Santambrogio, B. Corsi, A. Campanella, P. Arosio and S. Levi, *Neurobiol. Dis.*, 2006, **23**, 644–652.
30. P. Aisen, M. Wessling-Resnick and E. A. Leibold, *Curr. Opin. Chem. Biol.*, 1999, **3**, 200–206.
31. R. S. Eisenstein, *Annu. Rev. Nutr.*, 2000, **20**, 627–662.
32. E. A. Leibold and H. N. Munro, *Proc. Natl. Acad. Sci. U. S. A.*, 1988, **85**, 2171–2175.
33. W. Mikulits, M. Schranzhofer, H. Beug and E. W. Mullner, *Mutat. Res.*, 1999, **437**, 219–230.
34. G. Nicolas, M. Bennoun, I. Devaux, C. Beaumont, B. Grandchamp, A. Kahn and S. Vaulont, *Proc. Natl. Acad. Sci. U. S. A.*, 2001, **98**, 8780–8785.
35. C. Pigeon, G. Ilyin, B. Courselaud, P. Leroyer, B. Turlin, P. Brissot and O. Loreal, *J. Biol. Chem.*, 2001, **276**, 7811–7819.
36. E. Nemeth, G. C. Preza, C. L. Jung, J. Kaplan, A. J. Waring and T. Ganz, *Blood*, 2006, **107**, 328–333.

37. E. Nemeth, M. S. Tuttle, J. Powelson, M. B. Vaughn, A. Donovan, D. M. Ward, T. Ganz and J. Kaplan, *Science*, 2004, **306**, 2090–2093.
38. R. Lill and G. Kispal, *Trends Biochem. Sci.*, 2000, **25**, 352–356.
39. P. Ponka, *Blood*, 1997, **89**, 1–25.
40. E. Cadenas and K. J. Davies, *Free Radic. Biol. Med.*, 2000, **29**, 222–230.
41. G. Lenaz, *Biochim. Biophys. Acta*, 1998, **1366**, 53–67.
42. T. Alcindor and K. R. Bridges, *Br. J. Haematol.*, 2002, **116**, 733–743.
43. M. D. Fleming, *Semin. Hematol.*, 2002, **39**, 270–281.
44. A. May and E. Fitzsimons, *Baillieres Clin. Haematol.*, 1994, **7**, 851–879.
45. M. Babcock, D. de Silva, R. Oaks, S. Davis-Kaplan, S. Jiralerspong, L. Montermini, M. Pandolfo and J. Kaplan, *Science*, 1997, **276**, 1709–1712.
46. M. Pandolfo, *Blood Cells Mol. Dis.*, 2002, **29**, 536–547discussion 548–552.
47. P. Ponka, *Semin. Hematol.*, 2002, **39**, 249–262.
48. R. Allikmets, W. H. Raskind, A. Hutchinson, N. D. Schueck, M. Dean and D. M. Koeller, *Hum. Mol. Genet.*, 1999, **8**, 743–749.
49. R. Lill, K. Diekert, A. Kaut, H. Lange, W. Pelzer, C. Prohl and G. Kispal, *Biol. Chem.*, 1999, **380**, 1157–1166.
50. B. Corsi, A. Cozzi, P. Arosio, J. Drysdale, P. Santambrogio, A. Campanella, G. Biasiotto, A. Albertini and S. Levi, *J. Biol. Chem.*, 2002, **277**, 22430–22437.
51. J. Drysdale, P. Arosio, R. Invernizzi, M. Cazzola, A. Volz, B. Corsi, G. Biasiotto and S. Levi, *Blood Cells Mol. Dis.*, 2002, **29**, 376–383.
52. S. Levi, B. Corsi, M. Bosisio, R. Invernizzi, A. Volz, D. Sanford, P. Arosio and J. Drysdale, *J. Biol. Chem.*, 2001, **276**, 24437–24440.
53. G. Nie, A. D. Sheftel, S. F. Kim and P. Ponka, *Blood*, 2005, **105**, 2161–2167.
54. G. Nie, G. Chen, A. D. Sheftel, K. Pantopoulos and P. Ponka, *Blood*, 2006, **108**, 2428–2434.
55. M. Cazzola, R. Invernizzi, G. Bergamaschi, S. Levi, B. Corsi, E. Travaglino, V. Rolandi, G. Biasiotto, J. Drysdale and P. Arosio, *Blood*, 2003, **101**, 1996–2000.
56. A. Campanella, G. Isaya, H. A. O'Neill, P. Santambrogio, A. Cozzi, P. Arosio and S. Levi, *Hum. Mol. Genet.*, 2004, **13**, 2279–2288.
57. I. Zanella, M. Derosas, M. Corrado, E. Cocco, P. Cavadini, G. Biasiotto, M. Poli, R. Verardi and P. Arosio, *Biochim. Biophys. Acta*, 2008, **1782**, 90–98.
58. W. Dauer and S. Przedborski, *Neuron*, 2003, **39**, 889–909.
59. D. A. Butterfield, B. J. Howard and M. A. LaFontaine, *Curr. Med. Chem.*, 2001, **8**, 815–828.
60. Y. Ke and Q. Z. Ming, *Lancet Neurol.*, 2003, **2**, 246–253.
61. L. Zecca, M. B. Youdim, P. Riederer, J. R. Connor and R. R. Crichton, *Nat. Rev. Neurosci.*, 2004, **5**, 863–873.
62. H. M. Schipper, *Free Radic. Biol. Med.*, 2004, **37**, 1995–2011.
63. M. E. Gotz, K. Double, M. Gerlach, M. B. Youdim and P. Riederer, *Ann. N.Y. Acad. Sci.*, 2004, **1012**, 193–208.
64. P. Ponka, *Ann. N.Y. Acad. Sci.*, 2004, **1012**, 267–281.

65. A. R. Curtis, C. Fey, C. M. Morris, L. A. Bindoff, P. G. Ince, P. F. Chinnery, A. Coulthard, M. J. Jackson, A. P. Jackson, D. P. McHale, D. Hay, W. A. Barker, A. F. Markham, D. Bates, A. Curtis and J. Burn, *Nat. Genet.*, 2001, **28**, 350–354.
66. T. Moos and E. H. Morgan, *Ann. N.Y. Acad. Sci.*, 2004, **1012**, 14–26.
67. N. B. Cole, D. D. Murphy, J. Lebowitz, L. Di Noto, R. L. Levine and R. L. Nussbaum, *J. Biol. Chem.*, 2005, **280**, 9678–9690.
68. S. Mandel, G. Maor and M. B. Youdim, *J. Mol. Neurosci.*, 2004, **24**, 401–416.
69. M. B. Youdim, M. Gassen, A. Gross, S. Mandel and E. Grunblatt, *J. Neural Transm. Suppl.*, 2000, **58**, 83–96.
70. B. Zhao, *Mol. Neurobiol.*, 2005, **31**, 283–293.
71. N. C. Andrews, *N. Engl. J. Med.*, 1999, **341**, 1986–1995.
72. J. S. Garcia, C. S. Magalhaes and M. A. Arruda, *Talanta*, 2006, **69**, 1–15.
73. R. Ortega, S. Bohic, R. Tucoulou, A. Somogyi and G. Deves, *Anal. Chem*, 2004, **76**, 309–314.
74. B. S. Twining, S. B. Baines, N. S. Fisher, J. Maser, S. Vogt, C. Jacobsen, A. Tovar-Sanchez and S. A. Sanudo-Wilhelmy, *Anal. Chem.*, 2003, **75**, 3806–3816.
75. Z. Chai, X. Mao, Z. Hu, Z. Zhang, C. Chen, W. Feng, S. Hu and H. Ouyang, *Anal. Bioanal. Chem.*, 2002, **372**, 407–411.
76. Y. Gao, N. Q. Liu, C. Y. Chen, Y. F. Luo, Y. F. Li, Z. Y. Zhang, Y. L. Zhao, B. L. Zhao, A. Iida and Z. F. Chai, *J. Anal. Atom Spectrom.*, 2008, **23**, 1121–1124.
77. B. Wang, W. Feng, M. Wang, J. Shi, F. Zhang, H. Ouyang, Y. Zhao, Z. Chai, Y. Huang, Y. Xie, H. Wang and J. Wang, *Biol. Trace Elem. Res.*, 2007, **118**, 233–243.
78. R. Ortega, P. Cloetens, G. Deves, A. Carmona and S. Bohic, *PLoS ONE*, 2007, **2**, e925.
79. I. Paris, P. Martinez-Alvarado, S. Cardenas, C. Perez-Pastene, R. Graumann, P. Fuentes, C. Olea-Azar, P. Caviedes and J. Segura-Aguilar, *Chem. Res. Toxicol.*, 2005, **18**, 415–419.
80. V. A. Sole, E. Papillon, M. Cotte, P. Walter and J. Susini, *Spectrochim. Acta B*, 2007, **62**, 63–68.
81. R. Lobinski, C. Moulin and R. Ortega, *Biochimie*, 2006, **88**, 1591–1604.
82. R. Ortega, *Polycyclic Aromat. Compd.*, 2000, **21**, 99–108.
83. M. T. Zhu, W. Y. Feng, Y. Wang, B. Wang, M. Wang, H. Ouyang, Y. L. Zhao and Z. F. Chai, *Toxicol. Sci.*, 2009, **107**, 342–351.
84. A. Nakamura and M. Inoue, *Proc. NIPR Symp. Polar Biol.*, 1991, **4**, 82–90.
85. D. L. Hachey, J. C. Blais and P. D. Klein, *Anal. Chem.*, 1980, **52**, 1131–1135.
86. C. K. Jayawickreme and A. Chatt, *J. Radioanal. Nucl. Ch Ar*, 1988, **124**, 257–279.
87. P. E. Johnson, *J. Nutr.*, 1982, **112**, 1414–1424.
88. T. Walczyk, L. Davidsson, N. Zavaleta and R. F. Hurrell, *Fresenius' J. Anal. Chem.*, 1997, **359**, 445–449.
89. J. R. Turnlund, *Biol. Trace Elem. Res.*, 1987, **12**, 247–257.

90. M. Oshtrakh, *Hyperfine Interact.*, 2005, **165**, 313–320.
91. K. L. Double, M. Gerlach, V. Schunemann, A. X. Trautwein, L. Zecca, M. Gallorini, M. B. Youdim, P. Riederer and D. Ben-Shachar, *Biochem. Pharmacol.*, 2003, **66**, 489–494.
92. B. F. Matzanke, D. J. Ecker, T. S. Yang, B. H. Huynh, G. Muller and K. N. Raymond, *J. Bacteriol.*, 1986, **167**, 674–680.
93. Z. Szökefalvi-Nagy, C. Bagyinka, I. Demeter, K. Hollós-Nagy and I. Kovács, *Fresenius' J. Anal. Chem.*, 1999, **363**, 469–473.
94. Z. Szokefalvi-Nagy, C. Bagyinka, I. Demeter, K. L. Kovacs and L. H. Quynh, *Biol. Trace Elem. Res*, 1990, **26–27**, 93–101.
95. D. Strivay, B. Schoefs and G. Weber, *Nucl. Instrum. Methods Phys. Res. Sect. B*, 1998, **137**, 932–935.

CHAPTER 9
Nuclear-based Metallomics in Metal-based Drugs

RUIGUANG GE,[a,b] IVAN K. CHU[a] AND HONGZHE SUN[a,*]

[a] Department of Chemistry and Open Laboratory of Chemical Biology, The University of Hong Kong, Hong Kong, P. R., China; [b] The Laboratory of Integrative Biology, College of Life Sciences, Sun Yat-Sen University, Guangzhou 510006, P. R., China

9.1 Introduction

Metallomics is an emerging scientific area integrating the research fields related to the understanding of the molecular mechanisms of metal-dependent life processes and the entirety of metal and metalloid species within a cell or tissue type.[1] This area of science requires dedicated analytical approaches for the *in vivo* detection, localization, identification and quantification, the *in vitro* functional analysis, and *in silico* prediction using bioinformatics. Metallomics is thus an interdisciplinary research area with an ultimate goal to provide a global and systematic understanding of the metal uptake, trafficking, and function in biological systems. It is expected to develop as an interdisciplinary science complementary to genomics, proteomics, and metabolomics, with some specific areas of interest including (1) structure–function analysis of metalloproteins and their models; (2) survey and identification of metalloproteins and chemical speciation of bio-trace elements in the environment and biological systems; (3) biological regulation of metals and their metabolisms; and (4) medical diagnosis of health and disease relevant to trace elements as well as the (pre)clinical use of metallodrugs.

---

*Corresponding author: Email: hsun@hkucc.hku.hk; Tel: +852-28598974; Fax: +852-28571586.

Nuclear Analytical Techniques for Metallomics and Metalloproteomics
Edited by Chunying Chen, Zhifang Chai and Yuxi Gao
© Royal Society of Chemistry 2010
Published by the Royal Society of Chemistry, www.rsc.org

Advanced nuclear analytical techniques have been playing increasingly important roles in the studies of metallomics and metalloproteomics primarily due to their relatively high sensitivity, excellent accuracy, low matrix effects, and non-destructiveness,[2] allowing for large-scale and system-based research to comprehensively analyze biological processes and disease states. Various nuclear analytical techniques, such as neutron activation analysis (NAA), X-ray emission spectroscopy (XE), X-ray fluorescence (XRF), isotope dilution, and tracing techniques have been widely used for chemical speciation analysis of metalloproteins and metallodrugs, either alone or linked with one or two of the traditional techniques. For example, separation and enrichment methods such as liquid chromatography (LC), two-dimensional gel electrophoresis (2-DE), immobilized metal affinity chromatography (IMAC), and detection methods such as inductively coupled plasma–mass spectrometry (ICP-MS), matrix-assisted laser desorption/ionization mass spectrometry (MALDI-MS), electrospray ionization mass spectrometry (ESI-MS/MS), whereas Mössbauer spectroscopy, X-ray absorption (XAS), neutron scattering (NS), electron paramagnetic resonance (EPR), X-ray diffraction (XRD), and nuclear magnetic resonance (NMR) have found their extensive usage in probing structure–function relationships of metalloproteins and metallodrugs. The principles of advanced nuclear analytical techniques can be referred to in previous chapters.

The use of metal-based drugs in medicine can be traced back to 5000 years ago, when Egyptians used copper for sterilizing water.[3] With metallodrugs playing an increasingly significant role in therapeutic and diagnostic medicine, especially after the discovery of the anticancer activity of cisplatin, cis-$(NH_3)_2PtCl_2$,[4] metal-based drugs have been extensively investigated and used therapeutically, such as platinum(II)/(IV), gallium(III) and arsenic(III) complexes in cancer therapy, gold(I) complexes as anti-arthritis and asthma agents, bismuth(III) complexes in anti-ulcer and antimony(IV) complexes in anti-parasite treatment (Figure 9.1).[3,5–8] In addition to cisplatin, two other platinum drugs have been approved for clinical use: cis-diammine-1,1-cyclobutanedicarboxylatoplatinum(II) (carboplatin) which has received worldwide approval and $trans$-(R,R)-1,2-diaminocyclohexaneoxalatoplatinum(II) (oxaliplatin) approved in Europe and several other countries, including the USA. However, in view of the narrow spectrum of treatable tumors and microbial infections, increasing occurrence of drug resistance and undesirable side effects of most of the metal-based drugs, detailed information on the structure–function relationship of metallodrug–biological macromolecule adducts and metabolites of these metallodrugs should be obtained in order to design new effective metal-based drug leads, which become achievable as a result of development in traditional proteomic methods as well as in advanced nuclear analytical techniques. This chapter will deal with the application of advanced nuclear analytical techniques in metallodrug research, with a focus on metabolism and metallodrug–biomolecule interactions for the most commonly studied platinum(II)- and ruthenium(III)-containing anticancer drugs.

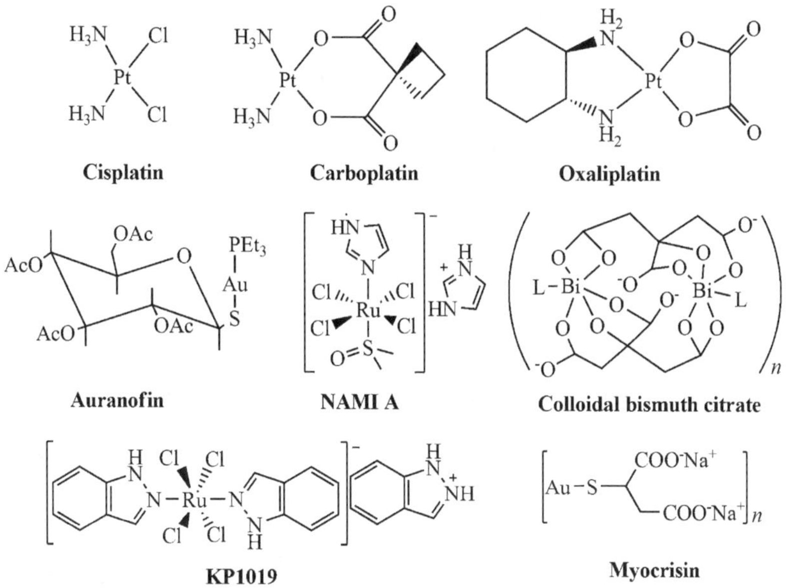

Figure 9.1 Schematic drawings of classical platinum-, ruthenium-, gold-, and bismuth-based pharmacological agents.

9.2 Cellular Distribution and Metabolism of Metallodrugs

Analysis of a metallodrug in a biological tissue is a challenging task in analytical chemistry, primarily because the traditional methods used are usually indirect and semi-quantitative to a large extent, and are unable to visualize the metal ions *in vivo*. Advanced nuclear analytical techniques, such as X-ray fluorescence, neutron activation analysis, X-ray emission, X-ray absorption near-edge structure spectroscopy, nuclear magnetic resonance, and isotope tracing/dilution techniques offer some means by which elemental distribution, oxidation states, and species structural information can be studied.[9]

9.2.1 Hydrolysis of Platinum Compounds

Although the formal edge energies between the XANES spectra of platinum(II) and platinum(IV) complexes vary only by 2.2 eV, the relative heights of the absorption edges differ substantially regardless of the coordination environment and can be used as a diagnostic indicator of the platinum oxidations states and to determine their proportions by the peak–height ratio of micro-XANES spectra.[10] Micro-XANES spectra of platinum(IV)-treated cells confirmed the reduction to the corresponding platinum(II) complexes, thus strongly supporting the proposed mechanism of action for platinum(IV) complexes.[11]

In cells treated with a most difficult to reduce platinum(IV) complex and also the most important, cis,trans,cis-(PtCl$_2$(OH)$_2$(NH$_3$)$_2$), platinum(IV) was detected along with platinum(II), an observation attributed to a positive relation between the relative ease of reduction of the platinum(IV) complexes used and the potentness against cancer cells.

Metallodrugs are known to behave, at least in some cases, as a "prodrug" and an activation step, either via a ligand exchange or a redox process, is required before they can exert their pharmacological effects.[5,12–14] The resulting species are the "chemical entities" responsible for the observed biological actions, and manifest a high propensity to react with biomolecules and to transfer the "metal-containing molecular fragments" commonly through simple ligand substitution reactions. For example, when entering a cellular environment, the anticancer drug cisplatin is activated by the substitution of one or both chloride (Cl$^-$) with water molecules (water exchange) to give cis-(Pt(NH$_3$)$_2$(H$_2$O)$_2$)$^{2+}$ (diaqua) or cis-(Pt(NH$_3$)$_2$(H$_2$O)Cl)$^+$ (chloro-aqua) cation,[13,15,16] which is promoted by a lower chloride concentration (3–20 mM) in the cellular environment than previously in the blood stream with a high concentration of chloride ($\sim$100 mM), suppressing cisplatin hydrolysis and maintaining the compound intact. Notably, the introduction of kinetic restrictions to the production of these metallic fragments and to their transfer to target biomolecules may lead to a substantial loss of biological activities for these metallodrugs, as demonstrated in the case of some representative Au(III)[17] and Ir(III)[18] anticancer agents.

The structural information for the hydration and hydrolysis of platinum(II) in aqueous solution have been studied by extended X-ray absorption fine structure (EXAFS) and large-angle X-ray scattering (LAXS),[19,20] which concluded that the hydrated platinum(II) in acidic aqueous solution coordinates four water molecules in a square-planar geometry with the Pt distances of 2.01–2.02 Å whereas another study with Pt L$_3$-edge EXAFS spectra confirmed the presence of one or two weakly bound water molecules (Pt–O distance of 2.39 Å) in axial positions of the hydrated platinum(II) in acidic solution (HClO$_4$).[21] The presence of axial water molecule(s) would explain the previously reported small activation volume for the slow water exchange in the (Pt(H$_2$O)$_4$)$^{2+}$ entity,[22] consistent with an interchange mechanism of relocating an axial water molecule in a five-coordinate tetragonal pyramidal reactant through a trigonal bipyramid-like transition state to an equatorial position. The hydrated cis-diammineplatinum(II) complex was found to have a similar coordination sphere in the latter report, with two ammine and two aqua ligands strongly bound (Pt–O/N bond distances of $\sim$2.01 Å) and one or two axial water molecules at $\sim$2.37 Å.[21] Figure 9.2A shows the EXAFS curve-fitting for the hydrated platinum(II) ion, with and without the Pt–O* scattering path (Figure 9.2B) included. As for the hydrated cis-diammineplatinum(II) ion, the corresponding Fourier transform for the model without axial Pt–O* shows a small peak in the residue, which disappears when the axial Pt–O* path is included. Similar improvement was obtained by including axial water in the fitted model when analyzing the EXAFS spectrum of the hydrated platinum(II)

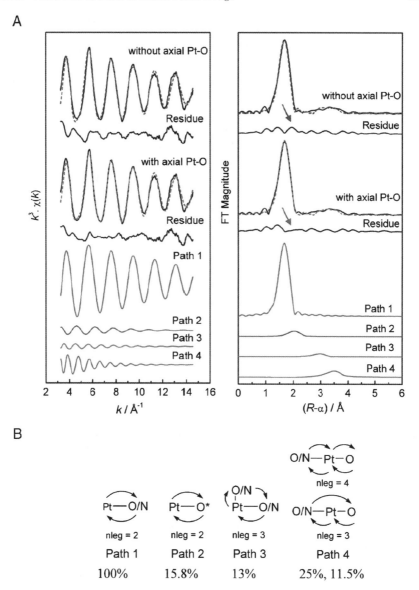

Figure 9.2 (A) k^3-Weighted EXAFS functions (solid lines) for 0.16 M Pt^{2+} (aq) in 7.5 M $HClO_4$: (left, top) Comparison between two models (dashes) without and with axial Pt–O* scattering path included, respectively, and corresponding residues (= exp – fit); (left, below) Individual contributions for model with axial Pt–O, based on the scattering paths shown in (B); (right) Corresponding Fourier transforms. Noted for a small peak in the residue in the Fourier transform for the corresponding hydrated cis-diammineplainium(II) ion, whereas the peak disappears when the axial Pt–O* path is included. (B) Single- and multiple-scattering pathways considered in EXAFS model fittings of $(Pt(H_2O)_n)^{2+}$ and $(Pt(NH_3)_2(H_2O)_m)^{2+}$ complexes, with their relative amplitude ratios calculated in FEFF for stationary atoms (disorder parameter $\sigma^2 = 0.0 \text{ Å}^2$). Adapted with permission from Jalilehvand and Laffin.[21] © 2008 American Chemical Society.

ion by means of EXAFSPAK software. The resulting parameters were not significantly different: 4 Pt–O 2.01(2) Å ($\sigma^2 = 0.0026(5)$ Å^2) and 1 Pt–O* 2.39(2) Å ($\sigma^2 = 0.009(1)$ Å^2), or 2 Pt–O* 2.41(2) Å ($\sigma^2 = 0.0015(2)$ Å^2).[21]

Nuclear magnetic resonance is an excellent technique for monitoring the hydrolysis of platinum compounds with DNA and/or protein.[23,24] Pt drugs generally consist of some of the following atoms: platinum, nitrogen, hydrogen, chloride, oxygen, and carbon, among which the most useful isotopes for NMR are ^{1}H, ^{15}N, and ^{195}Pt.[25] The combined use of ^{15}N and ^{1}H in a two-dimensional inverse NMR experiment, such as HSQC or HMQC, is especially useful, since both the ^{15}N chemical shift and the one-bond coupling constant $^1J(^1\text{H}-^{15}\text{N})$ are diagnostic of the *trans*-ligand in the platinum complex[25–27] and the sensitivity is improved by a theoretical maximum of 306 with respect to direct ^{15}N detection.[28] The rate determination step *in vitro* is the aquation of cisplatin dichloro to choloro-aqua complex following the first-order kinetics, which was demonstrated first by Horacek and Drobnik by simple comparison of the rate constants[29] and later confirmed by Sadler and co-workers with 2D [^{1}H,^{15}N]-HSQC NMR spectroscopy.[30] *cis*-(Pt(NH$_3$)$_2$(H$_2$O)$_2$)$^{2+}$, formed in the second hydrolysis step, can be a competing species in the *in vitro* platination of DNA, which is two orders magnitude faster than the cholor-aqua complex.[31] In a 2D [^{1}H,^{15}N]-HSQC NMR study, the hydrolysis of cisplatin at 298 K, pH 5.9 was determined in 9 mM NaClO$_4$ and 9 mM phosphate buffer, respectively.[32] In the former case, the chloro-aqua is the major species throughout the reaction of 40 h and the diaqua species appears after 3.5 h in a fraction no more than 7% of total Pt. Besides chloro-aqua and diaqua species, two hydroxo bridged dimers were observed which accounted for ~5% of the total Pt after 40 h. In phosphate buffer, the presence of phosphate-bound aquated Pt species was confirmed. The rate constants could not be accurately determined due to the complex reaction pattern and the lack of data in the time period of 10–24 h. Centerwall and co-workers observed that cisplatin chloro-aqua readily coordinated to a carbonate in the buffer and further speculated that the carbonato form of cisplatin may be the species transported through the cell membrane and subsequently reacting with DNA.[33]

Carboplatin has fewer side effects than cisplatin due to the lower reactivity of the chelating cyclobutanedicarboxylate (CBDCA) ligand. Carboplatin hydrolysis gives the same DNA adducts as cisplatin as expected for the leave of the dicarboxylate ligand,[34] with an aquation rate 2–4 orders of magnitude lower than cisplatin.[35,36] Ring-opening may be facilitated by enzymes, reaction with sulfur nucleophiles or direct reaction with DNA. The NMR experiment showed that the reaction of carboplatin with 5'-GMP (4.1×10^{-6} s^{-1}) was faster than with phosphate (4.3×10^{-7} s^{-1}), phosphate and chloride (1.2×10^{-6} s^{-1}) or water ($<5 \times 10^{-9}$ s^{-1}), suggesting the direct attack of nucleotides on carboplatin.[35] The stable ring-opened adduct (Pt(CBDCA-*O*)(NH$_3$)$_2$(L-Met-*S*))$^+$ may be formed between carboplatin and a variety of sulfur-containing amino acids, as demonstrated by ^{1}H and 2D [^{1}H,^{15}N]-HMQC NMR and HPLC.[37,38] This complex is slowly converted (half-life of 28 h at 310 K) into an S,N-chelated adduct *cis*-(Pt(NH$_3$)$_2$(L-Met-*S,N*))$^+$.

Platinum(II) iminoethers consist of a group of interesting metallodrugs with innovative and well-documented antitumor properties.[39,40] Among them, *trans*-PtCl$_2$(*E*-HN=C(OMe)Me)$_2$ (*trans*-EE) was found to be as active as cisplatin toward P388 leukemia and Lewis lung carcinoma in mice through formation of monofunctional DNA adducts.[41] ^{1}H NMR studies were performed on *trans*-EE to elucidate the hydrolysis reactions in ammonium carbonate buffer, pH 7.4 at 310 K.[42] One hour after dissolution of *trans*-EE into ammonium carbonate buffer, one of the two chlorides from the starting complex (**A** in Figure 9.3) was replaced by a water molecule to afford a new species **B**. In 9–24 h, two new species (**C1** and **C2**) formed and showed proton signals at 2.55 and 2.51 ppm (assigned to CH$_3$) and 3.77 and 3.74 ppm (OCH$_3$), respectively, corresponding to the two iminoether ligands in the original *E* configuration. Species **D** showed peaks at 2.46 (CH$_3$), 2.30 (CH$_3$) and 4.00 ppm (OCH$_3$), assigned to a new complex containing two iminoether ligands, one in *E* and the other in *Z* configurations. After 60 h incubation, free methanol at 3.26 ppm formed, and new methyl signals (**E**) and the characteristic platinum-coordinated amide signal appeared at around 1.85 and 5.50 ppm, respectively. Two weeks later, the iminoether signals almost disappeared, while **E** signals further increased in intensity and new proton signals (**F**) at 2.09 (CH$_3$) and 6.00 ppm (NH) appeared. Species **F** was assigned to be the amidine complexes when compared to the ^{1}H-NMR characteristics of the established platinum-amidine compounds.[42] Therefore, NMR spectroscopy is proven to be successfully applied in tracking the hydrolysis/aminolysis reactions of *trans*-EE, and may be used in similar studies.

9.2.2 Cellular Localization of Metallodrugs

Neutron activation analysis is the process whereby free neutrons are captured by atomic nuclei, resulting in the formation of new and most frequently

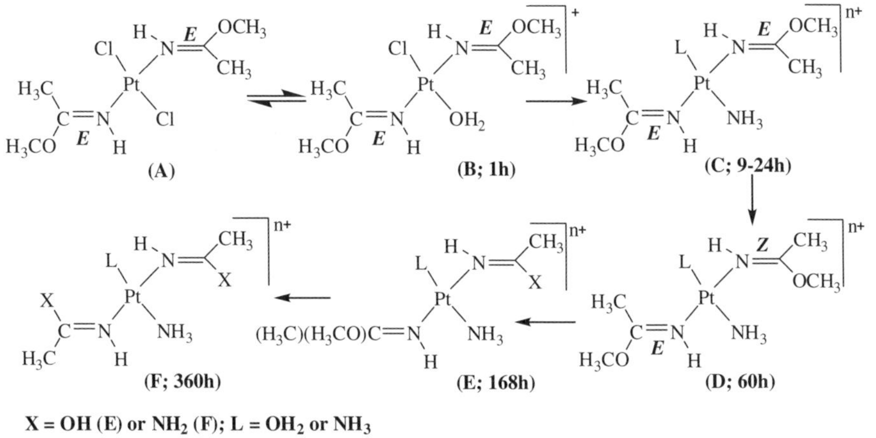

Figure 9.3 Proposed reaction pathways for *trans*-EE in ammonium carbonate buffer, pH 7.4 and 310 K. Adapted with permission from Casini *et al.*[42] © 2007 American Chemical Society.

radioactive nuclei, which will decay with time and a proportion of the energy is released as gamma radiation. The emitted gamma radiation in a particular energy are detected and indicated as the presence of a specific radionuclide, which allows for the determination of the concentration of various elements in the sample. This technique allows the non-destructive quantification of multiple major and trace elements in a sample, and can be used to penetrate and pass through most samples due to uncharged neutrons. Gamma rays released from the activated sample are also penetrating and can escape the sample for efficient detection, which makes NAA an error-proof technique suitable for a sample directly requiring only minimal sample preparations. NAA has been successfully applied in the ionomic studies of breast cancer, colorectal cancer, and brain cancer, which showed that ionome was perturbed in the diseased tissues or organisms.[43–45] Lux and co-workers compared the cytotoxicity, platinum accumulation and DNA platination in MCF-7 breast cancer cells treated with cisplatin and diaqua(1,2-diphenylethylenediamine)platinum(II) sulfate (4F-PtSO$_4$) drug leads, which showed that cytotoxicity was correlated neither with the extent of cellular platinum enrichment nor with the degree of genomic DNA platination.[46] Although the cytotoxicity and DNA-associated platinum were similar for cisplatin and 4F-PtSO$_4$, NAA showed that a 24 h treatment of the MCF-7 cells with raceme-4F-PtSO$_4$ and meso-4F-PtSO$_4$ caused a 22.3- and 10.3-fold accumulation, respectively, whereas the accumulation factor for cisplatin was only 2.55.[46] NAA was also used in the determination of traces of platinum and gold in different tissues of Wistar rats from the following neural tissues: the dorsal root ganglions as well as the dorsal and ventral part of the spinal cord. The highest level of platinum was found in dorsal root ganglions.[47] The drawbacks with the NAA technique are the low sensitivity and neutron radiation exposure (0.25–10 mSv for a normal delayed neutron activation),[48] which prevent its extensive uses in the human subjects of cellular metallodrug location and metallodrug metabolism, especially for children and pregnant women, whereas NAA has still been used frequently in the studies of environmental toxicology and plant ionomics.[49]

Most of the platinum drugs in clinical use are in the more reactive divalent oxidation state and many severe side effects result from the incidental reactions of these complexes with proteins. Platinum(IV) complexes are far more inert and do not react with proteins, at least in the blood stream,[50] but are still highly effective anticancer agents. The anticancer activity of the kinetically inert platinum(IV) complexes is thought to involve intracellular reduction to platinum(II) upon cellular uptake.[51] The platinum oxidation states (II) and (IV) (d^8 and d^6, respectively) do not have "electronic handles" which make direct monitoring of the oxidation state *in vivo* with spectroscopic methods difficult. Traditional techniques, e.g. IR, HPLC, or MS, do not allow for facile *in situ* determination of the average or component oxidation states of a complex system.[52,53] XRF and synchrotron radiation-induced X-ray emission (SRIXE) are uniquely suited for the study of biological speciation of metal-based drugs, especially in the determination of cellular location and oxidation states of metal ions, because they can be directly detected without labeling. During X-ray fluorescence experiment, when a primary X-ray with sufficient energy attacks an atom, the inner electrons may be ejected and thus the atoms are excited. The excited atoms can emit characteristic radiation during the subsequent process of de-excitation, which carry information about the elemental composition of the specimen in the irradiated region.[54] In XE, the X-ray microbeams with high spatial resolutions ($\sim 0.25\,\mu m$ for a microparticle-induced X-ray emission (micro-PIXE))[55] can penetrate the sample in depth of $\sim 1\,mm$, and can be used to determine the intracellular trace elements distribution. The recently developed scanning X-ray fluorescence microtomography (XRFM) has made it possible to detect elements of interest by a single measurement and give a profile of these elements at the single cell level.

SRIXE mapping has been used on thin sections of human ovarian (A2780) cancer cells to gain insight into the cellular location and metabolism of platinum(II) and platinum(IV) complexes in cells and tumors treated with bromine-containing anticancer platinum compounds, cis-PtCl$_2$(NH$_3$)(3-Brpyr) (3-Brpyr = 3-bromopyridine), and $cis,trans,cis$-PtCl$_2$(NH$_3$)$_2$(OAcBr)$_2$ (OAcBr = bromoacetate) or a platinum complex attached with an intercalator cis-PtCl$_2$(2-(3-aminopropyl)amino-9,10-anthracenedione)(NH$_3$).[56] After 24 h treatment, the complexes were found to be mainly localized in the cell nucleus with a lower fraction in the surrounding cytoplasm. The localization of cisplatin to the cell nucleus was supported by a study with the combined use of an ion microbeam and micro-PIXE on human lung cancer cells,[55] which demonstrated increased cisplatin uptake in cell nucleus after longer exposure periods in a time-course study. In cells treated with cis-PtCl$_2$(NH$_3$)(3-Brpyr), the concentration of bromine was substantially higher than in control cells and the bromine was co-localized with the platinum, suggesting the integrity of the platinum complexes.[56] The cells treated with $cis,trans,cis$-PtCl$_2$(OAcBr)$_2$(NH$_3$)$_2$ also showed an increased level of bromine, although to a much lesser extent than for those treated with cis-PtCl$_2$(3-Brpyr)(NH$_3$), indicating substantial reduction of the platinum(IV) complex, which was supported by a similar study that the cellular distribution of platinum(IV) (such as cis-PtCl$_4$(NH$_3$)$_2$ and $cis,trans,cis$-PtCl$_2$(OAc)$_2$(NH$_3$)$_2$) after 24 h treatment was similar to that of cisplatin.[11]

Limited penetration of cytotoxic drugs into tumors is a significant contributing factor to multicellular resistance and the limited effectiveness of cancer chemotherapy.[57–59] Clinically important drugs such as doxorubicin diffuses only 40–100 μm from blood vessels and reach a fraction of the viable cells that make up a solid tumor.[60,61] The techniques used to study platinum penetration, such as the extent of cell killing in spheroids, cell sorting techniques utilizing radiolabeled complexes, are usually indirect and semi-quantitative, and cannot visualize these compounds in tissue. To understand the underlying mechanism of multidrug resistance of malignant melanomas, Chen et al. used synchrotron radiation XRFM to map the intracellular distribution of cisplatin, which revealed that cisplatin was sequestered in melanosomes, a process that significantly reduced nuclear localization of the drug when compared with non-melanoma epidermoid carcinoma cells.[62] Shimura et al. applied the same technique for the analysis of intracellular platinum and zinc

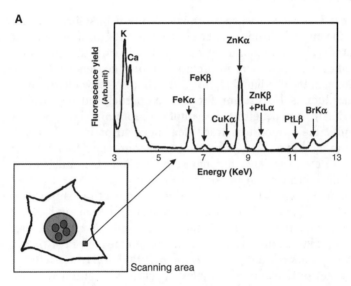
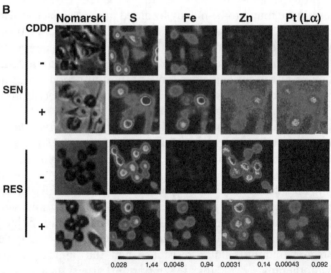
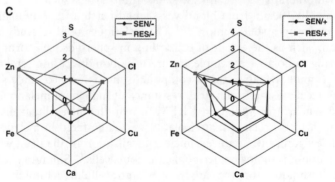

concentration in cisplatin-sensitive and cisplatin-resistant cancer cells (Figure 9.4).[63] At 12 h after cisplatin treatment (1 μM), the Pt level was found to increase in PC/SEN (PC/Sensitive) cells, whereas little changes in the PC/RES (PC/resistance) cells (Figure 9.4A and B). Based on the mean signal intensity obtained by SXFM, element array analysis was carried out (Figure 9.4C), which facilitates the identification of the elements related to the mechanism of drug resistance to cisplatin. The average platinum content of cisplatin-resistant cells was 2.6 times less than that of platinum-sensitive cells as confirmed by ICP-MS, whereas the zinc content was inversely correlated with the platinum content, indicating that zinc-related detoxification is responsible for the resistance to cisplatin. A combined treatment of cisplatin and zinc chelator N,N,N',N'-tetrakis-(2-pyridylmethyl)-ethylenediamine (TPEN) resulted in increased platinum uptake and significantly impaired the growth of PC/RES cells, indicative of the potential of the combined use of these two drugs in eliminating tumors even if they include a cisplatin resistant population of cells with high zinc content. Therefore SXFM has been demonstrated to be particularly useful in examining a mechanism of cisplatin resistance. Although Cu has been suggested to be a necessary factor for cisplatin incorporation,[64] recent work by Shimura et al.[63] does not reveal any evidence of Cu being involved in PC/RES cells.

Besides platinum-based chemotherapeutics, several other transition metal compounds have been found to yield anticancer activities.[12] Waern and co-workers used synchrotron radiation XRF to explore the subcellular distribution of metallocene dihalide anticancer complexes. They found that incubation of Chinese hamster lung cells with subtoxic doses of molybdocene dichloride (Cp_2MoCl_2) led to efficient cellular uptake of Mo that diffused throughout the cell.[65] In contast, the Nb form, Cp_2NbCl_2, revealed localization of Nb in small concentrated spots, in agreement with a chemical speciation study suggesting different mechanisms of actions of these two drugs.[66]

Metal-containing nanoparticles and clusters are promising in therapeutic and diagnostic applications in the emerging area of nanomedicine.[67] XRFM is quite useful in the exploration of the inorganic physiology of such nanomaterials in

---

Figure 9.4 Element array by SXFM. (A) Scheme of imaging cellular elements by SXFM. Coherent X-rays are focused on each area and the X-ray fluorescence from each element is detected. (B) SXFM analysis after cisplatin treatment. Cell morphologies are shown at ×100 magnifications (left). Each field of view is equivalent to an area of 70×70 μm. Results are shown for PC/SEN (top) and PC/RES cells (bottom). (C) Element array based on SXFM analysis. The mean signal intensity of each element obtained by SXFM analysis was calculated, and the fold increase of elements in PC/RES cells was depicted by using the intensity in PC/SEN cells as a standard (left). A part of analyzed elements is shown. The fold increase of elements in PC/SEN and PC/RES cells after cisplatin treatment was also shown by using the intensity in PC/SEN before cisplatin treatment as a standard (right). Adapted with permission from Shimura et al.[63] © 2005 American Association For Cancer Research.

tissues and cells. For example, TiO_2–oligonucleotide nanocomposites with light-inducible nucleic acid endonuclease activity were found to be present as nanoparticles (4.5 nm) inside cell nuclei by means of synchrotron radiation XRFM on the basis of the Ti-specific Kα X-ray emission.[68] To co-localize the elemental distribution in synchrotron radiation XRFM with the location of specific cellular structures and organelles, McRae et al. developed a label technique based on the ultra-small gold particles conjugated with a secondary antibody and a small organic fluorophore to correlate synchrotron radiation XRFM with immunofluorescence microscopy, which provided two-dimensional maps of the gold-labeled organelles as well as the subcellular distribution of the biologically important trace elements upon the excitation of the Au Lα line.[69] Q-dot approaches are another class of important and widely used molecular imaging tool readily detectable by XRF, providing information complementary to optical fluorescence approaches.

9.2.3 Pharmacokinetics of Metallodrugs

The drug development process is scientifically complex and financially risky,[70] as it has been estimated that for every 5000 new chemical entities (NCEs) evaluated in a discovery program, only one is approved for market.[71] The major reasons for NCE failure, other than poor clinical efficacy, are serious undesirable side effects, adverse drug reactions, and unfavorable drug metabolism and pharmacokinetics.[72,73] The importance of in-depth knowledge about the pharmacokinetics of a drug is evident due to the close correlation of pharmacokinetic behavior with activity and toxicity. The most elaborate pharmacokinetic investigation is a so-called mass balance study using a radioactive tracer, which allows investigation of the plasma pharmacokinetics and excretion of both the intact drug and the total radioactivity (drug and metabolites), thus elucidating the metabolic fate of a drug. The major advantage of a radioactive tracer is that the total concentration of drug and metabolites can be quantified relatively easily in a variety of matrices by determining the total radioactivity. Radioactive isotopes used in studying the metabolic disposition of compounds include ^{3}H, ^{14}C, ^{32}P, ^{35}S, ^{131}I, and $^{191/193}$Pt. However, the decay half-life of the radioactive isotopes affects the extensive uses, as isotopes with short half-lives (e.g. ^{191}Pt, $\sim$2.9 days; ^{131}I, $\sim$8.0 days; ^{32}P, $\sim$14.3 days) require delicate correction for radioactive decay between preparation and analysis of the radioisotope which normally decrease the accuracy of the analysis and isotopes with long decay half-lives (e.g. ^{14}C, $\sim$5730 years) often cause some environmental concerns.[74] Although radiotracer technology (^{14}C or ^{3}H) is still the most commonly used method in studying the in vivo disposition of a new drug,[75,76] due to ethical reasons or cost concerns, stable isotopes coupled with mass spectrometry have been used increasingly.

Since the introduction of cisplatin in human cancer treatment, numerous studies have dealt with the exploration of its binding to DNA molecules.[16,77] It is well known that cisplatin forms covalent bonds to the bases of guanine and adenine, with a preference for stretches of two adjacent guanines (N7) from the same DNA strand, which results in the so-called cross-links between adjacent nucleobases that block DNA replication and transcription and, ultimately cell division.[78] Up to now, only some of the mono- and bifunctional covalent adducts formed by interaction of cisplatin with the DNA nucleobases[79] have been studied in detail through the use of ^{32}P-postlabeling after enzymatic hydrolysis of DNA and deplatination of the adducts,[80] which otherwise cannot monitor continuous changes in the binding reactions and requires numerous and cumbersome steps although with high sensitivity.

Therefore, analytical tools are highly demanded to investigate the complexes of cisplatin with intact DNA and relevant biomolecules that permit a continuous monitoring of the events occurring during the whole reaction.[81] In this regard, ICP-MS has proved to be a versatile tool for the qualitative and quantitative detection of biomolecules and, as a consequence, an increase in the number of applications in this emerging field can be found.[82-84] With the development of atmospheric pressure ionization (API) sources, especially the most "soft" electrospray ionization technique, mass spectrometry has become one of the preferred analytical tools for the detection and identification of metabolites.[85-87] ESI is the method of choice for polar to ionic compounds, and this technique enables the soft ionization of phase II metabolites, providing reliable information on the size of these conjugates. For parent drug and phase I metabolites with a lower polarity, API may provide better ionization efficacy and sensitivity.[88] Since ICP-MS detection does not provide structural information, the combination of ICP-MS with ESI-MS detection can be deployed to one single HPLC system for simulataneous quantification and structure identification of the formed complex. The complexity of the biological sample and the multistep character of many bioinorganic speciation analytical procedures stimulate interest in isotope dilution analysis (IDA) to increase the precision and accuracy of the quantification of metallodrug metabolites. For example, the precision of butyltin measurement results is increased roughly by one order of magnitude when isotope dilution is used as opposed to standard additions.[89,90] IDA is a method that internalizes the standard to the sample, *i.e.* adding a known quantity of a rare isotope (a spike) into each sample. The concentration of the element of interest is calculated using the known natural abundance and the sample isotope ratio which can readily be measured with MS. One of the main advantage of IDA is that once complete isotope equilibration between the sample and the spike is achieved, subsequent processing, such as concentration, extraction, dilution, separation, *etc*, will not affect the final results.[2]

The interactions of biomolecules and cytostatic Pt-containing compounds have been explored by means of the coupling of LC to ICP-MS for monitoring DNA adducts *via* ^{31}P detection.[91-93] Hann *et al.* applied simultaneous ^{31}P and ^{195}Pt detection for the studies of cisplatin and guanosine monophosphate and proposed an intermediate [Pt(NH$_3$)$_2$Cl(GMP)]$^-$ without any further elucidation.[94] Garcia Sar and co-workers applied the coupling of LC and ICP-MS equipped with a collision cell to simultaneous ^{31}P and ^{195}Pt detection for monitoring the formation of reaction products of cisplatin with an oligonucleotide (5'-TCCGGTCC-3) and with calf thymus DNA.[95] Oe et al. developed a stable isotope (^{2}H) dilution LC-MS/MS assay in the quantitative analysis of platinum compound cis-amminedichloro(2-methylpyridine)platinum(II) (ZD0473) in human plasma ultrafiltrate,[96] and subsequently used in the study of metabolites of ZD0473 in the human urine.[97] A novel platinum adduct formed during the storage of ZD0473 in human urine, which did not correspond to any of the typical sulfhydryl adducts identified previously and could be counteracted by the addition of 50% (w/v) NaCl to the urine.[97] This method has been used in a systematic analysis of fifteen pharmaceuticals, four metabolites of pharmaceuticals, three potential endocrine disruptors, and one healthcare product in water.[98] The combination of ICP-MS and ESI-MS has been shown to be a very powerful tool in the metabolite profiling and metabolite identification of bromine-containing compounds when coupled with ^{81}Br isotope dilution.[99] In a similar study, Brüchert and co-workers coupled continuous elution gel electrophoresis to ID-ICP-MS and MALDI-MS to monitor the interaction process between cisplatin and oligonucleotides (5'-TCCGGTCC-3' and 5'-TCCTGTCC-3') through ^{194}Pt labeling,[100] which allows for the determination of the binding kinetics of cisplatin to these model nucleotides as well as the observation of dominant intermediates.

Chaney and co-workers investigated the in vitro biotransformations of oxaliplatin and other Pt-DNA adducts containing trans-(R,R)-1,2-diaminocyclohexane carrier ligand in rat blood by ^{3}H-isotope tracing and found that their decay in the plasma ultrafiltrate occurred rapidly ($t_{\frac{1}{2}} < 1$ h).[101] Size-exclusion HPLC coupled on-line to ICP-MS was utilized to directly monitor early protein biotransformations of ^{195}Pt-labeled oxaliplatin following its intravenous administration to cancer patients.[102] It was found that oxaliplatin is almost equally bound to γ-globulins and albumin (40% of the Pt bound respectively), in consistence with a previous in vitro study.[103]

9.3 Metallodrug–Biomolecule Interactions

The major cellular processes by which cisplatin suppress cancer cells include uptake and transport, DNA adducts formation and recognition by damage-response proteins and signal transduction pathways.[13] Any factors interfering with these pathways may lead to drug resistance. Therefore it is evident that probing the metallodrug–biomolecule interactions will help to understand the side effects of the metallodrugs and the inherent or acquired resistance of cancer cells. Nowadays, the study of the interactions occurring between metallodrugs and proteins may take considerable advantage of the availability of very sophisticated and advanced analytical tools. For instance, a number of studies have highlighted the great potential of modern mass spectrometry ionization methods, in particular ESI and MALDI MS, to characterize metal–protein adducts at the molecular level, especially in obtaining information from the transient state of metallodrug–protein binding process and the residual reactivity of the bound metal fragments. On the other hand, X-ray diffraction[104] and nuclear magnetic resonance (NMR)[105] may provide invaluable information of the metallodrug–biomolecule species precisely, at least accurately in a static vision of the final binding species. NMR spectroscopy has enjoyed exceptional development (e.g. magnets with increased field strength, shielded magnets, cryogenic probes) during the last decade, which makes this technique useful in identification of drug lead structures, optimization of lead–target complexes, identification of the structure of target biomolecules, and elucidation of reaction mechanism, for example.[25] Single-crystal X-ray diffraction still represents the elected tool to obtain high-quality structural information of proteins, although very few crystal structures have been solved to a high resolution for metallodrug–protein adducts, primarily attributed to the intrinsic difficulty in obtaining good quality crystals for metallodrug–protein adducts.[106] Other techniques, such as Mössbauer spectroscopy, X-ray absorption spectrometry (XAS), electron paramagnetic resonance (EPR), and neutron scattering provide additional and complementary information for the detailed structural characterizations of the metal sites.[2,107]

9.3.1 Platinated-DNA Adducts

Upon hydrolysis, cisplatin can react with nucleophile-containing molecules, preferably N- or S-donor ligands, and, less commonly, the O-ligand.[108] *In vivo* N-donor ligands, *i.e.* the N7 atoms of guanine and adenine, are the primary binding sites for platinum complexes in double-stranded DNA. An in-depth understanding of the reaction pattern of cisplatin with DNA is therefore crucial for understanding the mechanism of the anticancer drug. Cisplatinated DNA can result in mono- and bifunctional adducts and DNA–protein cross-links. The bifunctional adducts can take the form of intra- or interstrand cross-links, causing major local distortions, such as bending and unwinding of DNA structures. Other than the differences in the carrier ligand, the platinated DNA adducts formed by cisplatin, carboplatin, and oxaliplain are nearly identical in terms of the site and type of adducts formed, *i.e.* 60–65% intrastrand GG, 25–30% intrastrand AG, 5–10% intrastrand GxG, and 1–3% interstrand.[109] The importance of the knowledge of the DNA distortion induced by cisplatin was highlighted by the finding that a number of cellular proteins containing high mobility group (HMG)-domain recognized specifically the adducts at d(GpG) and d(ApG) sites.[110] The HMG domain is composed of approximately 80 amino acids, and contains some well-conserved hydrophobic amino acids and three α-helical regions forming the shape of an L with the angle between its arm being $\sim 80°$.[111]

The first crystal structure of a cisplatinated double-stranded duplex d(CCTCG*G*TCTCC)·d(GGAGACCAGAGG) revealed the platinum to be displaced from the planes of the guanine rings by around 1 Å (Figure 9.5A),[112,113] resulting in a strained environment. The DNA has a wide and shallow minor groove, an important recognition element for protein binding.

The helix bends by $\sim 50°$ towards the major groove, and the dihedral angle between the guanine bases is 30°. One of the ammine ligands bound to platinum is hydrogen bonded to a phosphate oxygen atom. The base pairs at the platination site are propeller twisted, but retain their hydrogen bonds. The corresponding solution structure was determined by 2D NMR spectroscopy with restrained molecular dynamics refinement (Figure 9.5B).[114] The major difference between the NMR and the X-ray structure is the magnitudes of the overall helix bend angle, 78° and 39/55°, respectively, which was proposed to be due to the crystal packing constraints.[77] The base pairing at the platination site is also more distorted in the NMR structure. The platinum is displaced by 0.8 Å from the planes of the guanine bases. The DNA has a flat, wide minor groove, and, in general, the global helixes are very similar to those in the X-ray structure. The oxaliplatin derivative of the same duplex (Figure 9.5C) was determined to have a virtually identical X-ray structure with the cisplatinated form,[112,115] which could not explain the differential activities of these two platinum compounds. Therefore, it is possible that the solution structures of cisplatin- and oxaliplatin-GG are more relevant for understanding the biological differences between the platinated adducts (Figure 9.5D and E).[114,116–118] When comparing the solution structures from the same double-stranded duplex d(CCTCAG*G*CCTCC) (GGAGGCCTGAGG) with either cisplatin or oxaliplatin,[116,117] several significant conformational differences were observed between the cisplatin- and oxaliplatin-GG adducts in terms of buckle for the 5′ G6•C19 base pair, opening for the 3′ G7•C18 base pair, twist at the A5G6•T20C19 base pair step, slide, twist, and roll at the G6G7•C19C18 base pair step, slide at the G7C8•C18G17 base pair step, G6G7 dihedral angle, and overall bend angle. One or more of these conformational differences may be important for the differential recognition of cisplatinated and oxaliplatinated adducts by some mismatch repair proteins, damage recognition proteins, DNA binding proteins or DNA polymerases.

Although interstrand DNA cross-links formed by cisplatin represents only a small fraction (1–3%) of the total lesions, they could be important for the cytotoxicity of the drug.[109] A 10-mer double-stranded oligonucleotide with the sequence of d(CCTCG*CTCTC)d(GAGAG*CGAGG) containing a single interstrand cisplatin cross-link has been analyzed by 1D and 2D NMR spectroscopy,[119] and later by X-ray diffraction (Figure 9.5F).[120] NOESY spectra and chemical shifts indicated that the interstrand cross-linkage of guanines G5 and G15 induced extrahelicity of the complementary bases C6 and C16. The solution structure showed that the stacking of the cross-linked guanines with the surrounding bases induced a bend of 49° to the minor groove and a local unwinding of 76°. Similarly, the X-ray structure shows reasonable agreement for bend angle (47°) and local unwinding (70°) with the solution structure, whereas the phosphodiester backbone conformation is substantially different at the level of intrastrand cross-links (9.2 Å versus 15.6 Å for the solution and X-ray models respectively). In both structures, the *cis*-diammineplatinum(II) moiety is placed in the minor groove of the DNA, which is quite different from the intrastrand DNA cross-links. Their disparate structural features suggest

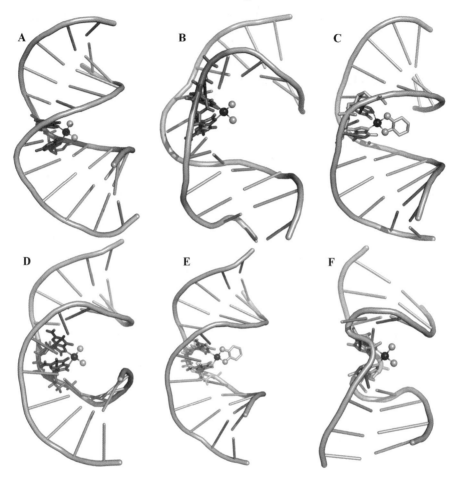

Figure 9.5 X-ray or NMR structures of classical cisplatinated or oxaliplatinated DNA adducts. The X-ray crystal[112,113] (PDB: 1AIO; A) and NMR solution[114] (PDB: 1A84; B) structures of d(CCTCTG*G*TCTCC)·d(GGA-GACCAGAGG) containing a *cis*-GG adduct, where G* denotes the location of platinated nucleotides; The X-ray crystal structure of oxaliplatinated d(CCTCTG*G*TCTCC)·d(GGAGACCAGAGG) adduct (PDB: 1IHH; C);[115] The NMR solution structures of cisplatinated (PDB: 2NPW; D)[116] or oxaliplatinated (PDB: 1PG9; E)[117] d(CCTCAG*G*CCTCC)·d(GGAGGCCTGAGG) adducts; The X-ray crystal structure of interstrand cross-link of cisplatin with d(CCTCG*CTCTC)d(GAGAG*CGAGG) (PDB: 1A2E; F). A, © 1996 American Chemical Society. B, © 1998 American Chemical Society. C, © 2001 American Chemical Society. D, © 2007 American Chemical Society. E, © 2004 Elsevier Ltd. F, © 1999 Oxford University Press

different roles for these various adducts in mediating the antitumor properties of platinum compounds.

2D [^{1}H,^{15}N] HMQC/HSQC is a better alternative in the study of binding kinetics and identification of intermediate species than conventional ^{195}Pt NMR because of the high sensitivity and information obtainable.[25] The pK_a values of mono- and diaquated cisplatin were determined by this technique via measuring ^{1}H and ^{15}N chemical shift as a function of pH,[121] and the reaction mechanism for cisplatin and the single-stranded oligonucleotide d(ACATGGTACA) and the duplex containing the complementary strand were investigated.[30] In both cases the pertinent species, dichloro, chloro-aqua, monofunctional chloro adducts, and bifunctional DNA adducts, could easily be observed. The only missing species was the aquated monofunctional Pt-DNA adduct with short lifetime, which is rapidly converted into its bifunctional adduct.[122]

However, it is difficult to make unambiguous assignments of the NMR signals from the monofunctional cisplatin adducts due to their rapid conversion to cross-linked adducts to be detectable by 2D NMR experiments. One-dimensional ^{1}H NMR is quick enough for the intermediate identification but is too crowded, especially for oligonucleotide sequences containing GG steps as both may be equally reactive to cisplain.[25] In AG and GA sequences, it is easier to assign the monofunctional adducts since the adenines and guanines are well separated in the NMR spectra, and therefore Pt-G and Pt-A monofunctional adducts are likely to be well separated. (PtCl(dien))$^+$ (dien = diethylenetriamine) can only form monofunctional adducts with DNA and thus a model for the monofunctional adduct formation of cisplatin. The kinetics of the reaction between (PtCl(dien))$^+$ and three self-complementary duplexes d(TATGGTACCATA) (I), d(TATGGATCCATA) (II) and d(TATGGCCATA) (III) were studied and it was found that (PtCl(dien))$^+$ had the following kinetic preference ($k5'/k3'$) for the two guanines, 5'-G and 3'-G, in each duplex: 8.6 (I), 1.2 (II), while in III only the 5'-G adduct was detected, suggesting that the base sequence on the 3'-side of the GG step significantly influences the binding preference of (PtCl(dien))$^+$.[123] It was noted that the reactive species of (PtCl(dien))$^+$ is (Pt(dien))$^{2+}$, in contrast to the mono-positive charged choloroaqua, the presumed reactive species of cisplatin. However, the results may be used as a guideline to predict where cisplatin preferably forms the monofunctional adducts: if a 5'-GGX step has a C or T as its 3'-neighbour, the 5'-G is preferred (especially in the case of 5'-GGC), but if the 5'-GG step has an A as its 3'-neighbour, the two Gs will have almost equal affinities. Another study established that cisplatin had a slight preference for the internal GG step (a ratio of 1.6 for internal/terminal GG cross-link), excluding the influence of steric effects in the reaction between duplex DNA with cisplatin.[124]

9.3.2 Metallodrug–Protein Interactions

Since the discovery of platinum containing compounds in the treatment of a variety of cancers, remarkable progress has been made in understanding the cellular and molecular mechanisms of cytotoxicity.[13] For the past three decades, most of the research interests were focused on the interactions of platinum anticancer drugs with DNA. Only recently, the interactions of platinum and non-platinum anticancer metallodrugs with proteins have received increasing attention,[125] especially in the regards of the *in vitro* interactions between metal-based drugs and the two major serum proteins, albumin and transferrin. In particular, Sadler and co-workers reported a pioneering NMR investigation of the reaction of cisplatin with serum albumin and the resulting adducts.[126] Subsequently, Khalaila *et al.* investigated and modeled the binding of cisplatin to transferrin.[127] It is speculated that albumin and transferrin are crucially involved in the transport of a variety of metal ions and metallodrugs through the bloodstream ensuring their transfer from absorption to utilization sites.[128–130] The *in vitro* interactions of the cysteine-rich intracellular protein Zn_7-metallothionein with cisplatin and transplatin[131] and the histidine-rich proteins Hpn[132,133] and HspA[134] with a bismuth antiulcer compound were investigated, respectively. These kinds of interactions may play a crucial role in the metabolism of various metallodrugs. Notably, drug binding to plasma proteins has a strong influence on their biodistribution, biotransformation, and pharmacokinetics, and therefore merits further characterizations.

The development of non-DNA-binding platinum(II/IV) anticancer agents is a further support for the involvement of metallodrug–protein adducts in the cytotoxicity.[135] The lack of DNA binding was demonstrated by the absence of a detectable platinum signal by atomic absorption spectroscopy using isolated DNA from human ovarian cancer cells treated with a platinum(II)-pyrophosphato complex (*trans*-1,2-cyclohexanediamine)(dihydrogen pyrophosphato)-platinum(II), (pyrodach-2) and from NMR experiments using a variety of nucleotides including double-stranded calf-thymus DNA, a synthetic 25-mer oligonucleotide, a dinucleotide (dGpG), and nucleotide monophosphates (5'-dGMP and 5'-dAMP),[135] results all representing a clear paradigm shift now expanding the DNA-based molecular targets for platinum anticancer drugs but also in strategic development of more effective and less toxic anticancer drugs.

Some specificities in metallodrug–biomolecules can be expected depending upon the inherent nature and surface exposure of different amino acids and their affinities for a metal center; for example, platinum(II) and some other soft metals having a high affinity for cysteine-rich metallothioneins. Most NMR studies relevant to platinum–protein interactions have been focused on amino acids containing S-ligands, *i.e.* Met and Cys. Therefore it was quite unexpected that it is possible to displace a platinum(II)-coordianted thioether by guanine N7,[136,137] which was further supported by a kinetic study of the reaction between $Pt(en)Cl_2$ (en = ethylenediamine) and methionine with 2D [^{1}H,^{15}N] HSQC NMR spectroscopy showing that the initially methionine S-bound ligand was readily displaced by 5'-GMP or GpG.[138] Reedijk and co-workers studied the reaction of $(Pt(dien)(H_2O))^{2+}$ with two methionine- and histidine-containing peptides, using HPLC and NMR spectroscopy.[139,140] For the two peptides His-Gly-Met and Ac-His-(Ala)$_3$-Met-NHPh investigated, a relatively rapid formation of the kinetically favorable methionine S-bound complex was observed, followed by slow intermolecular migration of the $(Pt(dien))^{2+}$ fragment to the Nε of the histidine side chain over a period of 500 h. The reversibility of Pt–S bonds in proteins was further studied with a model system allowing only intermolecular competition.[141] The adducts of $(Pt(dien)Cl)^+$ with GSH and *S*-methyl-glutathione (GSMe) were selected as representing the respective Pt-cysteine and Pt-methionine adducts in a protein. Rate constants were derived based on the ^1H chemical shifts as a function of time. It was concluded that only the sulfur atom in a Pt–sulfur adduct of the thioether type can be substituted by N7 of guanine and the Pt-cysteine model adduct appear to escape substitution in the presence of guanines.[141]

A detailed NMR and molecular dynamic analysis of a DNA duplex d(CCTCG*CTCTC)·(GAGAGCGAGG) with *trans*-EE showed that monofunctional *trans*-EE platination induced a bending of the helix axis towards the minor groove by 45°,[142] a bending angle comparable to the bifunctional adducts of cisplatin although in the latter case the bending is directed to the major grove.[112] Competitive reaction of ^{15}N-labeled *trans*-EE with GSH and GMP was followed by 2D [^1H,^{15}N] HSQC spectroscopy.[143] *Trans*-EE was found to react faster with GSH than with GMP, with half-lives ($t_{1/2}$) of 19 and 65 min, respectively. However, in a mixture of GSH and GMP, the most dominant species was identified to be a bifunctional GMP-*trans*-EE adduct. Although it is generally considered unlikely for the formation of bifunctional GMP (N-donor) adduct in the presence of GSH (S-donor), it has also been observed in a similar reaction involving *cis*-$PtCl_2(NH_3)$(picoline).[144]

Calderone *et al.* reported an X-ray crystal structure of bovine erythrocyte superoxide dismutase (beSOD) with platinum bound to the Nε of His19, two chloride and one loosely bound water molecule (Figure 9.6A).[145] Crystals of cisplatin-treated beSOD were obtained after incubation of the protein with a ten-fold molar equivalent of cisplatin for 2 weeks at 4°C. X-ray diffraction data were collected at low temperature and the structure solved with 1.8 Å resolutions. Binding was found to be highly selective. This raises the possibility that cisplatination of this enzyme follows an unusual pathway where ammonia ligands are released from platinum, which have been supported by the theoretical studies by Deubel and co-workers[146–148] showing that loss of the ammonia is indeed feasible with a specific macromolecular microenvironment, such as the less polarizable environment in the case of the platinum binding site here, where the thermodynamic *trans* influence is greatly reduced and the kinetic *trans* effect is enhanced in the platinum center, and may represent a pathway of cisplatin inactivation as both ammonia groups are normally required for its biological activity.[16]

At variance, a more classical chemical environment was found for the platinum(II) center in the cisplatin–hen egg white lysozyme (HEWL) species (Figure 9.6B),[149] where platinum was found to bind to the Nε of His15 of HEWL, two ammonia molecules in the cisplatin and another one possibly to be loosely bound water molecule. No more significant modifications of the electron density map of the protein surface were observed, ruling out the presence of additional binding site, for example, the two methionine residues (Met-12

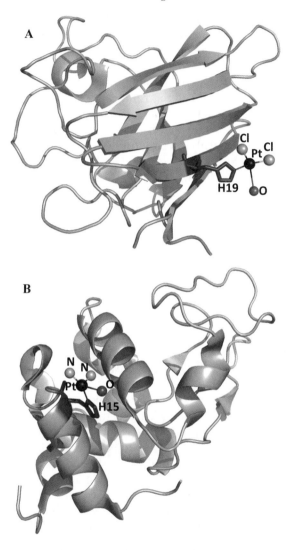

Figure 9.6 X-ray crystal structures of platinated beSOD (PDB: 2AEO; A) and hen egg white lysozyme (PDB: 2I6Z; B) with the distorted square-planar platinum(II)-binding sites highlighted. Platinum is bound to histidine-19, two chloride and one water molecule (Pt–O distance, 3.52 Å), and histidine-15, two ammonia and one water molecule (Pt–O distance, 3.91 Å) in platinated beSOD and hen egg white lysozyme, respectively. A, © 2006 Wiley-VCH Verlag GmbH & Co. KGaA, Weinheim. B, © 2007 The Royal Society of Chemistry.

and Met-105) which commonly represent preferred anchoring sites for soft platinum(II) compounds. In both cases, platination takes place predominantly to a single site located on the protein surface. Competitive binding of platinum(II) drugs to either nitrogen or sulfur donors is a matter of hot debate,[139,147,148,150] whereas the present two platinated proteins seem to favor the histidine residues. The difference between the platinum binding site in these two studies supported the theoretical finding that specific protein microenvironment determines ligand substitution, the ammonias or the chlorides to be released.[146]

There has been a strong interest in analyzing the *in vitro* interactions with plasma proteins of novel anticancer ruthenium(III) complexes *trans*-RuCl$_4$(In)$_2$(InH) (In = indazole) (KP1019) and *trans*-RuCl$_4$(Im)$_2$(ImH) (KP418) by following the signal of ^{13}C-enriched bicarbonate,[151] which found that bicarbonate was essential for KP1019–transferrin adduct formation and for KP418 only direct binding of bicarbonate to the ruthenium(III) center was apparent under slow exchange conditions on the NMR time scale. The binding of KP1019 and KP418 with apolactoferrin, a member of the transferrin family, was studied by X-ray diffraction analysis.[152,153] The KP1019 complex binds specifically to lactoferrin via the His-253 site located in the N-lobe of lactoferrin and easily accessible for coordination through its open conformation, while the indazole ligand remains coordinated to the Ru center. Coordination of KP418 involves more histidine residues, *i.e.* His-253, His-590, His-597, and His-654, the former two of which are known as the iron-binding centers.[154] There seems to be no correlation between the basicity of the heterocycle and the binding behavior when examining the structures of KP1019 and KP418.[155] Instead, π-stacking interactions and hydrogen bonding with the appropriate stereochemical arrangement of the ligand may account for the relative binding strength. Fluorescence spectra of the KP1019–transferrin and KP418–transferrin adducts seem to support the conclusion that the indazole ring is able to close the transferrin lobe whereas the imidazole adduct is not.[155] From the structural information, it was concluded that the transferrin cycle is a suitable mechanism to transport both ruthenium(III) complexes into the cell. This was supported by the variant level of cellular uptake of KP1019 under the influence of transferrin.[156] In comparison with the uptake of individual KP1019, an uptake of the Ru complex to the human colon carcinoma cell line SW480 was reduced two-fold when transferrin was loaded with KP1019 in a ratio of 1:2, whereas a four-fold higher content of Ru was observed in the cells when transferrin was loaded with an equimolar mixture of KP1019.[156] Therefore, it seems that overloading of the protein with KP1019 changes the protein structure so dramatically that it cannot be recognized by the respective receptors.

X-ray absorption is a suitable approach to provide additional and complementary information on the metallodrug–protein interactions and thus helps clarify some debatable issues, such as the structural and electronic characterization of small protein portions and the detailed description of the local environment of specific atomic probes, such as the spatial distribution, chemical speciation of elements, and the oxidation states of metals and metalloids in a wide range of environmental and biological samples.[157] Most XAS experiments are performed at energies higher than 5 keV, the typical threshold above which the K-edge absorption of elements having an atomic number greater than 22 occur. Lower energy measurements present additional experimental difficulties and require specific setups.[158] Ascone and co-workers reported an XANES investigation of the binding environment of bovine serum albumin (BSA) to trans-RuCl$_4$(Im)(DMSF)(ImH) (Im = imidazole; DMSF = dimethylsulfoxide) (NAMI-A).[159] More than 95% of NAMI-A was found to bind to transferrin or BSA upon 24 h incubation at room temperature.[160] The relative binding affinities of NAMI-A to human serum albumin (HAS) and other related biomolecules (such as DNA) confirmed the preferential interaction of NAMI-A to proteins over nucleotides,[161] similar to the binding preference of pyrodach-2.[135] NAMI-A presents a relatively high redox potential in comparison to other ruthenium(III) anticancer complexes,[162,163] and is known to be reduced into the corresponding Ru(II) species via ascorbic acid and glutathione. However, ruthenium K- and L$_3$-edge spectra proved unambiguously that the ruthenium center remained in the oxidation state of +3 upon protein binding. Sulfur K-edge spectra also remained the same during the binding, whereas chlorine K-edge spectra indicated changes in the chlorine chemical environment following NAMI-A binding to BSA with partial chloride release,[159] which was demonstrated to be attributed to the release of two chlorides followed by the further replacement by two water molecules through spectrophotometric analysis.[164] This finding is quite similar to the previous report that lysozyme appeared to form non-covalent bonding, most likely mediated by electrostatic interactions, and slowed down the intrinsic NAMI-A degradation processes, with either intact or monohydrolyzed NAMI-A through the combined use of ESI-MS, NMR and inductively coupled plasma–optical emission spectroscopy (ICP-OES).[165] Whereas, despite the similarity to lysozyme as a highly cationic protein at physiological pH and prone to interact with anions,[166] cytochrome c was found to enhance NAMI-A degradation, accelerating the progressive detachment of the various ligands from the ruthenium center with the final result of a highly degraded ruthenium-containing species in which most of the original metal ligands have been lost.[165] This process is most likely facilitated by an initial electrostatic interaction between the negatively charged NAMI-A and cytochrome c, and then progressively replaced by coordinative binding of the ruthenium(III) center to the protein. Therefore, it seems that proteins may interact with metallodrugs and affect in different ways its intrinsic degradation–activation processes under physiological relevant conditions. This characteristic is similar to the ligand exchange processes of mono-platinated cytochrome c species in the ESI-MS spectra when platinum complexes such as cisplatin, transplatin, carboplatin, or oxaliplatin were added into the protein solution in a 3:1 molar ratio,[167] and Met-65 was identified to be the major binding site for platinum(II) iminoethers on cytochrome c.[42]

Protein conformational changes induced by metallodrug binding could be studied by pulsed-field gradient diffusion NMR spectroscopy,[168] circular dichroism (CD) or fluorescence spectra.[169–171] Covalent binding of cisplatin to the Ca^{2+}-binding protein calmodulin (CaM) was shown to induce a near complete collapse in the protein secondary structures, and resulted in a decrease of hydrodynamic radius to 21.4 ± 0.2 Å from 24.5 ± 0.4 Å for the

Ca^{2+}-CaM.[168] It was suggested that cisplatin binds to CaM in a 9:1 ratio through S atoms of the Met side chains and the binding of cisplatin to CaM diminished the ability of CaM to bind to the target peptides. Similarly, binding of KP1019, KP418 or cisplatin complexes to albumin caused comparably dramatic conformational changes, such as the loss of α-helical stability of the protein possibly accounted for by the disulfide bond cleavage,[172,173] dimer formation,[170] and the concomitant strong quenching of the tryptophan-214 fluorescence intensity,[170,172,173] suggesting that conformational changes took place around this amino acid residue.

9.3.3 Platinated DNA–Protein Interactions

Cisplatin and carboplatin share the *cis*-diammine "carrier" ligands and form the same Pt–DNA adducts *in vivo* and are generally not effective in cell lines or tumors that have developed resistance to either agent, whereas oxaliplatin is a third generation platinum complex that forms Pt–DNA adducts containing *trans*-(*R,R*)-1,2-diaminocyclohexane "carrier" ligand and is often effective in cisplatin-resistance cell lines and tumors.[174] The effectiveness of oxaliplatin in cisplatin-resistant cell lines is thought to be attributed to repair or damage-recognition processes that discriminate between cisplatin and oxaliplatin, possibly through the recognition of differential structures of platinated DNA adducts.[116,117,175] The damage recognition proteins include HMG-domain and non-HMG-domain proteins, all of which share the characteristics of binding to the pre-bent DNA with a preferred DNA bending angle in the range of 70–120° in the minor groove[176] or bending the DNA in the direction of the major groove upon binding,[77] with postulated biological functions in inhibiting nucleotide excision repair and translesion DNA synthesis.[175]

Several mismatch repair, transcriptional factor, or damage recognition proteins, such as TATA binding protein (TBP),[177] hMSH2,[178] MutS,[179] and HMG-containing HMG1,[177] have been shown to bind more tightly to cisplatinated than to oxaliplatinated DNA adducts. The ability of the mismatch repair system to prefer cisplatinated over oxaliplatinated DNA adducts is believed to account for the greater efficacy of oxaliplatin in mismatch repair deficient or acquired cisplatin resistant cell lines and tumors.[178,180,181] The crystal structure of MutS bound to DNA with a mismatch shows that the binding is located in the minor groove and results in 60° bending towards the major groove.[182] An X-ray crystal structure of the cisplatinated DNA–HMG complex has been reported (Figure 9.7A).[183] Domain A of HMG1 was identified as bound through its concave surface to the widened minor groove of a 16 bp DNA probe containing a site-specific *cis*-Pt(NH$_3$)$_2$-d(GpG) adduct. Many of the DNA helical parameters for the cisplatinated DNA–HMG complex differ from those of the solution structures of both the cisplatinated and oxaliplatinated DNA adducts, possibly reflecting the severe DNA distortion induced by HMG domain binding. DNA was strongly bent (by ~61°) towards the major groove and the protein bend is not centered at the platinated adduct,

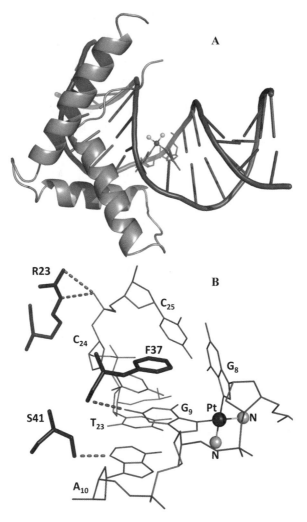

Figure 9.7 X-ray crystal structure of HMG1 domain A bound to d(CCTCTCTG*G*ACCTTCC)•d(GGAAGGTCCAGAGAGG) containing a *cis*-GG adduct (PDB: 1CKT; A). The Phe-37 residue highlighted in black intercalates into the platination site. (B) Zoomed view of protein–DNA interactions at the site of platination. The aromatic rings of Phe-37 and G_8 pack edge-to-face with a dihedral angle of 74°. The main-chain carbonyl group of Phe-37 is within hydrogen-bonding distance (3.1 Å) of the exocyclic amino group of G_9. As predicated,[187] Ser-41 and Arg-23 are involved in very tight hydrogen-bonding contacts with N3 of A_{10} and of O3 of C_{25}, respectively. A, © 2000 Macmillan Magazines Ltd. B, © 1988 American Chemical Society.

but is translocated by 2 bp to the 3' side. This unique positioning also occurs in solutions as confirmed by hydroxyl radical footprinting. The *cis*-diammineplatinum(II) fragment is anchored to the N7 atoms of two adjacent purine rings, G_8 and G_9. The Pt–N7 distances were not restrained, yet refined to the expected value of 2.0 ± 0.1 Å.[184] One of the NH_3 ligands is within 3.2 Å of the phosphate oxygen atom of G_8. The phenyl ring of Phe-37 intercalates into the DNA at a hydrophobic notch formed by the hydrophobic surfaces of the G_8–G_9 purine rings in the minor groove across from the platinated adduct (Figure 9.7B). Binding of HMG1 to the DNA is stabilized by interactions between the N-terminal helices I and II with the sugar phosphate backbone of the minor groove on the 3' side of the platinated strand. Mutation of this amino acid to alanine significantly diminished the binding affinity of this protein to the cisplatin–DNA adduct, suggesting that it is an important element for the platinated DNA–protein complex formation. The dihedral angle between guanine ring planes is 75°, larger than the observed in the X-ray and NMR structures of the cisplatin-modified duplex DNA, which makes the geometry less constrained and is similar to that encountered in the platinated dinucleotide d(pGpG) structure.[184,185] This work provides the first detailed structural information for a complex between an HMG-domain protein and cisplatinated DNA adduct and may be used in conjunction with other mechanistic work to facilitate the design of more effective anticancer agents.

The twists at the A5·G6* and G6*·G7* and the slide at the G7*·C8 base pair steps for the cisplatinated DNA–HMG complex are much closer to those of the cisplatinated adduct than the oxaliplatinated form. There is a particular biological significance for the conformational difference in the slide at the G7*·C8 base pair step, as both the affinity of the HMG domain for the cisplatinated adduct[177,186,187] and the ability of the HMG domain to discriminate between cisplatinated and oxaliplatinated DNA adducts[177] are highly dependent on the base of the 3' side of the adduct. The conformational characteristics may facilitate the recognition of the cisplatinated from oxaliplatinated adducts by the HMG domain.

9.4 Conclusions and Perspectives

The emerging fields of metallomics and metalloproteomics are dedicated to understanding the intrinsic mechanisms of biological functions of biometals with genes, proteins, metabolites, and other biomolecules within biological systems. Various nuclear analytical techniques, such as neutron activation analysis, X-ray emission spectroscopy/fluorescence, isotope dilution, and tracing techniques have been widely used for the chemical speciation analysis of metalloproteins and metallodrugs, either alone or linked with one or two of other traditional kinds of techniques; for example, separation and enrichment methods such as liquid chromatography, two-dimensional gel electrophoresis, immobilized metal affinity chromatography; and detection methods such as inductively coupled plasma mass spectrometry, matrix-assisted laser desorption/ionization mass spectrometry, and electrospray ionization mass spectrometry. In contrast, Mössbauer spectroscopy, X-ray absorption, neutron scattering, electron paramagnetic resonance, X-ray diffraction, and nuclear magnetic resonance find extensive applications for examining structure–function relationships of metalloproteins and metallodrugs. Since the development of cisplatin around 30 years ago, metallodrugs have been playing an increasingly significant role in the therapeutic and diagnostic medicine. It is essential to uncover the pharmacokinetics, metabolites and molecular mechanisms underlying the cytotoxicity of these metallodrugs to pave the way for further development of more effective and selective drugs. This chapter has highlighted various aspects of recent metallodrug research, and critically examined the potential and utilization of advanced nuclear analytical techniques, both their merits and limitations, in this area of growing importance. Despite the information gathered so far, many aspects of the mechanisms of action of the anticancer metallodrugs remain a mystery. The pathways of metallodrug resistance, both intrinsic and acquired, are not well understood at the molecular level, due to many factors involved. Alternative cellular DNA targets and proteins that bind to metallodrug–DNA adducts continue to be investigated and may prove to be important. These kinds of cellular interactions rely on further improvements in the state-of-the-art techniques.

Acknowledgements

This work was supported by the Research Grants Council of Hong Kong (HKU7512/05M, HKU7043/06P, HKU7042/07P, HKU02/06C, and HKU1/07C), the Area of Excellence Scheme of the University Grants Committee, The University of Hong Kong, National Natural Science Foundation of China (No. 20801061), and Guangdong Natural Science Foundation (No. 8451027501001233).

References

1. H. Haraguchi, *J. Anal. At. Spectrom.*, 2004, **19**, 5.
2. Y. Gao, C. Chen and Z. Chai, *J. Anal. At. Spectrom.*, 2007, **22**, 856.
3. C. Orvig and M. J. Abrams, *Chem. Rev.*, 1999, **99**, 2201.
4. B. Rosenberg, L. VanCamp, J. E. Trosko and V. H. Mansour, *Nature*, 1969, **222**, 385.
5. Z. Guo and P. J. Sadler, *Angew. Chem. Int. Ed.*, 1999, **38**, 1513.
6. R. Ge and H. Sun, *Acc. Chem. Res.*, 2007, **40**, 267.
7. H. Sun, L. Zhang and K. Y. Szeto, *Met. Ions Biol. Syst.*, 2004, **41**, 333.
8. K. H. Thompson and C. Orvig, *Science*, 2003, **300**, 936.
9. M. Ralle and S. Lutsenko, *BioMetals*, 2009, **22**, 197.
10. M. D. Hall, G. J. Foran, M. Zhang, P. J. Beale and T. W. Hambley, *J. Am. Chem. Soc.*, 2003, **125**, 7524.
11. M. D. Hall, C. T. Dillon, M. Zhang, P. Beale, Z. Cai, B. Lai, A. P. Stampfl and T. W. Hambley, *J. Biol. Inorg. Chem.*, 2003, **8**, 726.

12. C. X. Zhang and S. J. Lippard, *Curr. Opin. Chem. Biol.*, 2003, **7**, 481.
13. D. Wang and and S. J. Lippard, *Nat. Rev. Drug Discov.*, 2005, **4**, 307.
14. N. Farrell, *Met. Ions Biol. Syst.*, 2004, **42**, 251.
15. Y. Jung and S. J. Lippard, *Chem. Rev.*, 2007, **107**, 1387.
16. J. Reedijk, *Proc. Natl. Acad. Sci. U. S. A.*, 2003, **100**, 3611.
17. L. Messori, F. Abbate, G. Marcon, P. Orioli, M. Fontani, E. Mini, T. Mazzei, S. Carotti, T. N. O'Connell and P. Zanello, *J. Med. Chem.*, 2000, **43**, 3541.
18. L. Messori, G. Marcon, P. Orioli, M. Fontani, P. Zanello, A. Bergamo, G. Sava and P. Mura, *J. Inorg. Biochem.*, 2003, **95**, 37.
19. B. Hellquist, L. A. Bengtsson, B. Holmberg, B. Hedman, I. Persson and L. I. Elding, *Acta Chem. Scand.*, 1991, **45**, 449.
20. R. Ayala, E. S. Marcos, S. Diaz-Moreno, V. A. Sole and A. Munoz-Paez, *J. Phys. Chem. B*, 2001, **105**, 7588.
21. F. Jalilehvand and L. J. Laffin, *Inorg. Chem.*, 2008, **47**, 3248.
22. L. Helm, L. I. Elding and A. E. Merbach, *Inorg. Chem.*, 1985, **24**, 1719.
23. H. Sun, in *Encyclopedia of Nuclear Magnetic Resonance: Advances in NMR*, ed. D. M. Grant and R. K. Harris, John Wiley & Sons, New York, 2002, **vol. 9**, p. 413.
24. Y. Chen, Z. Guo and P. J. Sadler, in *Cisplatin: Chemistry and Biochemistry of A Leading Anticancer Drug*, ed. B. Lippert, Wiley-VCH, Zürich, 1999, p. 293.
25. J. Vinje and E. Sletten, *Anticancer Agents Med. Chem.*, 2007, **7**, 35.
26. S. J. Berners-Price and P. J. Sadler, *Coord. Chem. Rev.*, 1996, **151**, 1.
27. S. J. Berners-Price, L. Ronconi and P. J. Sadler, *Prog. Nucl. Magn. Reson. Spectrosc.*, 2006, **49**, 65.
28. A. Bax, H. Griffey and B. L. Hawkins, *J. Magn. Reson.*, 1983, **55**, 301.
29. P. Horacek and J. Drobnik, *Biochim. Biophys. Acta*, 1971, **254**, 341.
30. K. J. Barnham, P. J. Sadler, S. J. Berners-Price, T. A. Frenkeil and U. Frey, *Angew. Chem. Int. Ed.*, 1995, **34**, 1874.
31. F. Legendre, V. Bas, J. Kozelka and J. C. Chottard, *Chemistry*, 2000, **6**, 2002.
32. M. S. Davies, S. J. Berners-Price and T. W. Hambley, *Inorg. Chem.*, 2000, **39**, 5603.
33. C. R. Centerwall, J. Goodisman, D. J. Kerwood and J. C. Dabrowiak, *J. Am. Chem. Soc.*, 2005, **127**, 12768.
34. F. A. Blommaert, H. C. van Dijk-Knijnenburg, F. J. Dijt, L. den Engelse, R. A. Baan, F. Berends and A. M. Fichtinger-Schepman, *Biochemistry*, 1995, **34**, 8474.
35. U. Frey, J. D. Ranford and P. J. Sadler, *Inorg. Chem.*, 1993, **32**, 1333.
36. R. J. Knox, F. Friedlos, D. A. Lydall and J. J. Roberts, *Cancer Res.*, 1986, **46**, 1972.
37. K. J. Barnham, M. I. Djuran, P. S. Murdoch, J. D. Ranford and P. J. Sadler, *Inorg. Chem.*, 1996, **35**, 1065.

38. K. J. Barnham, U. Frey, P. D. Murdoch, J. D. Ranford and P. J. Sadler, *J. Am. Chem. Soc.*, 1994, **116**, 11175.
39. M. Coluccia and G. Natile, *Anticancer Agents Med. Chem.*, 2007, **7**, 111.
40. G. Natile and M. Coluccia, *Coord. Chem. Rev.*, 2001, **216–217**, 383.
41. M. Coluccia, A. Nassi, F. Loseto, A. Boccarelli, M. A. Mariggio, D. Giordano, F. P. Intini, P. Caputo and G. Natile, *J. Med. Chem.*, 1993, **36**, 510.
42. A. Casini, C. Gabbiani, G. Mastrobuoni, R. Z. Pellicani, F. P. Intini, F. Arnesano, G. Natile, G. Moneti, S. Francese and L. Messori, *Biochemistry*, 2007, **46**, 12220.
43. A. N. Garg, V. Singh, R. G. Weginwar and V. N. Sagdeo, *Biol. Trace Elem. Res.*, 1994, **46**, 185.
44. H. Arriola, L. Longoria, A. Quintero and D. Guzman, *Biol. Trace Elem. Res.*, 1999, **71–72**, 563.
45. E. Andrasi, M. Suhajda, I. Saray, L. Bezur, L. Ernyei and A. Reffy, *Sci. Total Environ.*, 1993, **139–140**, 399.
46. F. Lux, M. Hollstein, H. Reile, G. Bernhardt and H. Schonenberger, *Biol. Trace Elem. Res.*, 1996, **53**, 113.
47. B. Rietz, K. Heydorn and A. Krarup-Hansen, *Biol. Trace Elem. Res.*, 1994, **43–45**, 343.
48. S. Mattsson and B. J. Thomas, *Phys. Med. Biol.*, 2006, **51**, R203.
49. D. E. Salt, I. Baxter and B. Lahner, *Annu. Rev. Plant Biol.*, 2008, **59**, 709.
50. R. C. Dolman, G. B. Deacon and T. W. Hambley, *J. Inorg. Biochem.*, 2002, **88**, 260.
51. E. Wong and C. M. Giandomenico, *Chem. Rev.*, 1999, **99**, 2451.
52. L. Pendyala, W. Greco, J. W. Cowens, S. Madajewicz and P. J. Creaven, *Cancer Chemother. Pharmacol.*, 1983, **11**, 23.
53. W. P. Petros, S. G. Chaney, D. C. Smith, J. Fangmeier, M. Sakata, T. D. Brown and D. L. Trump, *Cancer Chemother. Pharmacol.*, 1994, **33**, 347.
54. R. Jenkins, *X-ray Fluorescence Spectrometry*, John Wiley, New York, 1999.
55. H. Sakurai, M. Okamoto, M. Hasegawa, T. Satoh, M. Oikawa, T. Kamiya, K. Arakawa and T. Nakano, *Cancer Sci.*, 2008, **99**, 901.
56. M. D. Hall, R. A. Alderden, M. Zhang, P. J. Beale, Z. Cai, B. Lai, A. P. Stampfl and T. W. Hambley, *J. Struct. Biol.*, 2006, **155**, 38.
57. J. K. Tunggal, D. S. M. Cowan, H. Shaikh and I. F. Tannock, *Clin. Cancer Res.*, 1999, **5**, 1583.
58. K. O. Hicks, S. J. Ohms, P. L. van Zijl, W. A. Denny, P. J. Hunter and W. R. Wilson, *Br. J. Cancer*, 1997, **76**, 894.
59. R. K. Jain, *Cancer Res.*, 1987, **47**, 3039.
60. A. J. Primeau, A. Rendon, D. Hedley, L. Lilge and I. F. Tannock, *Clin. Cancer Res.*, 2005, **11**, 8782.
61. A. I. Minchinton and I. F. Tannock, *Nat. Rev. Cancer*, 2006, **6**, 583.
62. K. G. Chen, J. C. Valencia, B. Lai, G. Zhang, J. K. Paterson, F. Rouzaud, W. Berens, S. M. Wincovitch, S. H. Garfield, R. D. Leapman, V. J.

Hearing and M. M. Gottesman, *Proc. Natl. Acad. Sci. U. S. A.*, 2006, **103**, 9903.
63. M. Shimura, A. Saito, S. Matsuyama, T. Sakuma, Y. Terui, K. Ueno, H. Yumoto, K. Yamauchi, K. Yamamura, H. Mimura, Y. Sano, M. Yabashi, K. Tamasaku, K. Nishio, Y. Nishino, K. Endo, K. Hatake, Y. Mori, Y. Ishizaka and T. Ishikawa, *Cancer Res.*, 2005, **65**, 4998.
64. K. Katano, A. Kondo, R. Safaei, A. Holzer, G. Samimi, M. Mishima, Y. M. Kuo, M. Rochdi and S. B. Howell, *Cancer Res.*, 2002, **62**, 6559.
65. J. B. Waern, H. H. Harris, B. Lai, Z. Cai, M. M. Harding and C. T. Dillon, *J. Biol. Inorg. Chem.*, 2005, **10**, 443.
66. M. M. Harding and G. Mokdsi, *Curr. Med. Chem.*, 2000, **7**, 1289.
67. S. M. Moghimi, A. C. Hunter and J. C. Murray, *FASEB J.*, 2005, **19**, 311.
68. T. Paunesku, T. Rajh, G. Wiederrecht, J. Maser, S. Vogt, N. Stojicevic, M. Protic, B. Lai, J. Oryhon, M. Thurnauer and G. Woloschak, *Nat. Mater.*, 2003, **2**, 343.
69. R. McRae, B. Lai, S. Vogt and C. J. Fahrni, *J. Struct. Biol.*, 2006, **155**, 22.
70. J. A. DiMasi, R. W. Hansen and H. G. Grabowski, *J. Health Econ.*, 2003, **22**, 151.
71. J. Caldwell, *J. Pharm. Sci.*, 1996, **2**, 117.
72. A. E. Nassar, A. M. Kamel and C. Clarimont, *Drug Discov. Today*, 2004, **9**, 1055.
73. A. E. Nassar, A. M. Kamel and C. Clarimont, *Drug Discov. Today*, 2004, **9**, 1020.
74. J. H. Beumer, J. H. Beijnen and J. H. Schellens, *Clin. Pharmacokinet.*, 2006, **45**, 33.
75. A. E. Nassar, S. M. Bjorge and D. Y. Lee, *Anal. Chem.*, 2003, **75**, 785.
76. C. L. Shaffer, M. Gunduz, T. N. O'Connell, R. S. Obach and S. Yee, *Drug Metab. Dispos.*, 2005, **33**, 1688.
77. E. R. Jamieson and S. J. Lippard, *Chem. Rev.*, 1999, **99**, 2467.
78. J. R. Yachnin, I. Wallin, R. Lewensohn, F. Sirzen and H. Ehrsson, *Cancer Lett.*, 1998, **132**, 175.
79. T. Hagemeister and M. Linscheid, *J. Mass Spectrom.*, 2002, **37**, 731.
80. M. Zeisig and L. Moller, *J. Chromatogr. B Biomed. Sci. Appl.*, 1997, **691**, 341.
81. A. R. Timerbaev, C. G. Hartinger and B. K. Keppler, *Trends in Anal. Chem.*, 2006, **25**, 868.
82. A. Sanz-Medel, M. Montes-Bayón and and M. L. Fernandez Sanchez, *Anal. Bioanal. Chem.*, 2003, **377**, 236.
83. J. Szpunar, *Analyst*, 2005, **130**, 442.
84. J. Bettmer, N. Jakubowski and A. Prange, *Anal. Bioanal. Chem.*, 2006, **386**, 7.
85. C. Prakash, C. L. Shaffer and A. Nedderman, *Mass Spectrom. Rev.*, 2007, **26**, 340.
86. E. J. Oliveira and D. G. Watson, *Biomed. Chromatogr.*, 2000, **14**, 351.
87. S. Ma, S. K. Chowdhury and K. B. Alton, *Curr. Drug Metab.*, 2006, **7**, 503.
88. E. Kantharaj, P. B. Ehmer, A. Tuytelaars, A. van Vlaslaer, C. Mackie and R. A. Gilissen, *Rapid Commun. Mass Spectrom.*, 2005, **19**, 1069.

89. C. Bancon-Montigny, P. Maxwell, L. Yang, Z. Mester and R. E. Sturgeon, *Anal. Chem.*, 2002, **74**, 5606.
90. L. Yang, Z. Mester and R. E. Sturgeon, *J. Anal. At. Spectrom.*, 2002, **17**, 944.
91. C. Siethoff, I. Feldmann, N. Jakubowski and M. Linscheid, *J. Mass. Spectrom.*, 1999, **34**, 421.
92. M. Edler, N. Jakubowski and M. Linscheid, *Anal. Bioanal. Chem.*, 2005, **381**, 205.
93. M. Edler, N. Jakubowski and M. Linscheid, *J. Mass Spectrom.*, 2006, **41**, 507.
94. S. Hann, A. Zenker, M. Galanski, T. L. Bereuter, G. Stingeder and B. K. Keppler, *Fresenius' J. Anal. Chem.*, 2001, **370**, 581.
95. D. G. Sar, M. Montes-Bayón, E. B. González and A. Sanz-Medel, *J. Anal. At. Spectrom.*, 2006, **21**, 861.
96. T. Oe, Y. Tian, P. J. O'Dwyer, D. W. Roberts, M. D. Malone, C. J. Bailey and I. A. Blair, *Anal. Chem.*, 2002, **74**, 591.
97. T. Oe, Y. Tian, P. J. O'Dwyer, D. W. Roberts, C. J. Bailey and I. A. Blair, *J. Chromatogr. B Anal. Technol. Biomed. Life Sci.*, 2003, **792**, 217.
98. B. J. Vanderford and S. A. Snyder, *Environ. Sci. Technol.*, 2006, **40**, 7312.
99. F. Cuyckens, L. I. Balcaen, K. De Wolf, B. De Samber, C. Van Looveren, R. Hurkmans and F. Vanhaecke, *Anal. Bioanal. Chem.*, 2008, **390**, 1717.
100. W. Brüchert, R. Krüger, A. Tholey, M. Montes-Bayón and J. Bettmer, *Electrophoresis*, 2008, **29**, 1451.
101. F. R. Luo, S. D. Wyrick and S. G. Chaney, *J. Biochem. Mol. Toxicol.*, 1999, **13**, 159.
102. P. Allain, O. Heudi, A. Cailleux, A. Le Bouil, F. Larra, M. Boisdron-Celle and E. Gamelin, *Drug Metab. Dispos.*, 2000, **28**, 1379.
103. S. Urien and J. P. Tillement, *Drug Interact. Intern. Rep. Sanofi Winthrop*, 1995, **LPH 0022**, 125.
104. A. Ilari and C. Savino, *Methods Mol. Biol.*, 2008, **452**, 63.
105. K. Wüthrich, *J. Biol. Chem.*, 1990, **265**, 22059.
106. A. Casini, A. Guerri, C. Gabbiani and L. Messori, *J. Inorg. Biochem.*, 2008, **102**, 995.
107. L. Que Jr., *Physical Methods in Bioinorganic Chemistry: Spectroscopy and Magnetism*, University Science Books, Sausalito, 2000.
108. B. Lippert, *Coord. Chem. Rev.*, 1999, **182**, 263.
109. J. J. Roberts and F. Friedlos, *Pharmacol. Ther.*, 1987, **34**, 215.
110. P. M. Pil and S. J. Lippard, *Science*, 1992, **256**, 234.
111. R. Grosschedl, K. Giese and J. Pagel, *Trends Genet.*, 1994, **10**, 94.
112. P. M. Takahara, C. A. Frederick and S. J. Lippard, *J. Am. Chem. Soc.*, 1996, **118**, 12309.
113. P. M. Takahara, A. C. Rosenzweig, C. A. Frederick and S. J. Lippard, *Nature*, 1995, **377**, 649.
114. A. Gelasco and S. J. Lippard, *Biochemistry*, 1998, **37**, 9230.
115. B. Spingler, D. A. Whittington and S. J. Lippard, *Inorg. Chem.*, 2001, **40**, 5596.

116. Y. Wu, D. Bhattacharyya, C. L. King, I. Baskerville-Abraham, S. H. Huh, G. Boysen, J. A. Swenberg, B. Temple, S. L. Campbell and S. G. Chaney, *Biochemistry*, 2007, **46**, 6477.
117. Y. Wu, P. Pradhan, J. Havener, G. Boysen, J. A. Swenberg, S. L. Campbell and S. G. Chaney, *J. Mol. Biol.*, 2004, **341**, 1251.
118. L. G. Marzilli, J. S. Saad, Z. Kuklenyik, K. A. Keating and Y. Xu, *J. Am. Chem. Soc.*, 2001, **123**, 2764.
119. F. Paquet, C. Perez, M. Leng, G. Lancelot and J. M. Malinge, *J. Biomol. Struct. Dyn.*, 1996, **14**, 67.
120. F. Coste, J. M. Malinge, L. Serre, W. Shepard, M. Roth, M. Leng and C. Zelwer, *Nucleic Acids Res.*, 1999, **27**, 1837.
121. S. J. Berners-Price, T. A. Frenkiel, U. Frey, J. D. Ranford and P. J. Sadler, *J. Chem. Soc., Chem. Commun.*, 1992, **1992**, 789.
122. F. Legendre, J. Kozelka and J. C. Chottard, *Inorg. Chem.*, 1998, **37**, 3964.
123. J. Vinje, J. A. Parkinson, P. J. Sadler, T. Brown and E. Sletten, *Chemistry*, 2003, **9**, 1620.
124. J. Vinje and E. Sletten, *Chemistry*, 2006, **12**, 676.
125. X. Sun, C. N. Tsang and H. Sun, *Metallomics*, 2009, **1**, 25.
126. A. I. Ivanov, J. Christodoulou, J. A. Parkinson, K. J. Barnham, A. Tucker, J. Woodrow and P. J. Sadler, *J. Biol. Chem.*, 1998, **273**, 14721.
127. I. Khalaila, C. S. Allardyce, C. S. Verma and P. J. Dyson, *ChemBioChem*, 2005, **6**, 1788.
128. Z. M. Qian, H. Li, H. Sun and K. Ho, *Pharmacol. Rev.*, 2002, **54**, 561.
129. A. R. Timerbaev, C. G. Hartinger, S. S. Aleksenko and B. K. Keppler, *Chem. Rev.*, 2006, **106**, 2224.
130. H. Li, H. Sun and Z. M. Qian, *Trends Pharmacol. Sci.*, 2002, **23**, 206.
131. M. Knipp, A. V. Karotki, S. Chesnov, G. Natile, P. J. Sadler, V. Brabec and M. Vašák, *J. Med. Chem.*, 2007, **50**, 4075.
132. R. Ge, R. M. Watt, X. Sun, J. A. Tanner, Q. Y. He, J. D. Huang and H. Sun, *Biochem. J.*, 2006, **393**, 285.
133. R. Ge, Y. Zhang, X. Sun, R. M. Watt, Q. Y. He, J. D. Huang, D. E. Wilcox and H. Sun, *J. Am. Chem. Soc.*, 2006, **128**, 11330.
134. S. Cun, H. Li, R. Ge, M. C. Lin and H. Sun, *J. Biol. Chem.*, 2008, **283**, 15142.
135. R. N. Bose, L. Maurmann, R. J. Mishur, L. Yasui, S. Gupta, W. S. Grayburn, H. Hofstetter and T. Salley, *Proc. Natl Acad. Sci. U. S. A.*, 2008, **105**, 18314.
136. K. J. Barnham, M. I. Djuran, P. D. Murdoch and P. J. Sadler, *J. Chem. Soc., Chem. Comm.*, 1994, **1994**, 721.
137. S. S. G. E. van Boom and J. Reedijk, *J. Chem. Soc., Chem. Comm.*, 1993, **1993**, 1397.
138. K. J. Barnham, Z. Guo and P. J. Sadler, *J. Chem. Soc., Dalton Trans.*, 1996, **1996**, 2867.
139. J. Reedijk, *Chem. Rev.*, 1999, **99**, 2499.
140. M. Hahn, D. Wolters, W. S. Sheldrick, F. B. Hulsbergen and J. Reedijk, *J. Biol. Inorg. Chem.*, 1999, **4**, 412.

141. S. S. G. E. van Boom, B. W. Chen, J. M. Teuben and J. Reedijk, *Inorg. Chem.*, 1999, **38**, 1450.
142. B. Andersen, N. Margiotta, M. Coluccia, G. Natile and E. Sletten, *Met. Based Drugs*, 2000, **7**, 23.
143. Y. Liu, J. Vinje, C. Pacifico, G. Natile and E. Sletten, *J. Am. Chem. Soc.*, 2002, **124**, 12854.
144. Y. Chen, Z. Guo, J. A. Parkinson and P. J. Sadler, *J. Chem. Soc., Dalton Trans.*, 1998, **1998**, 3577.
145. V. Calderone, A. Casini, S. Mangani, L. Messori and P. L. Orioli, *Angew Chem. Int. Ed.*, 2006, **45**, 1267.
146. J. K. Lau and D. V. Deubel, *Chemistry*, 2005, **11**, 2849.
147. D. V. Deubel, *J. Am. Chem. Soc.*, 2002, **124**, 5834.
148. D. V. Deubel, *J. Am. Chem. Soc.*, 2004, **126**, 5999.
149. A. Casini, G. Mastrobuoni, C. Temperini, C. Gabbiani, S. Francese, G. Moneti, C. T. Supuran, A. Scozzafava and L. Messori, *Chem. Commun.*, 2007, **2007**, 156.
150. M. Hahn, M. Kleine and W. S. Sheldrick, *J. Biol. Inorg. Chem.*, 2001, **6**, 556.
151. F. Kratz, M. Hartmann, B. Keppler and L. Messori, *J. Biol. Chem.*, 1994, **269**, 2581.
152. F. Kratz, B. K. Keppler, L. Messori, C. Smith and E. N. Baker, *Met. Based Drugs*, 1994, **1**, 169.
153. C. A. Smith, A. J. Sutherland-Smith, B. K. Keppler, F. Kratz, E. N. Baker and B. H. Keppler, *J. Biol. Inorg. Chem.*, 1996, **1**, 424.
154. B. F. Anderson, H. M. Baker, E. J. Dodson, G. E. Norris, S. V. Rumball, J. M. Waters and E. N. Baker, *Proc. Natl. Acad. Sci. U. S. A.*, 1987, **84**, 1769.
155. D. A. Powell and D. H. Hamilton, paper presented at the 227th ACS National Meeting, Anaheim, CA, 2004.
156. M. Pongratz, P. Schluga, M. A. Jakupec, V. B. Arion, C. G. Hartinger, G. Allmaier and B. K. Keppler, *J. Anal. At. Spectrom.*, 2004, **19**, 46.
157. I. Ascone, R. Fourme, S. Hasnain and K. Hodgson, *J. Synchrotron Radiat.*, 2005, **12**, 1.
158. A. Congiu-Castellano, F. Boffi, S. Della Longa, A. Giovannelli, M. Girasole, F. Natali, M. Pompa, A. Soldatov and A. Bianconi, *BioMetals*, 1997, **10**, 363.
159. I. Ascone, L. Messori, A. Casini, C. Gabbiani, A. Balerna, F. Dell'Unto and A. C. Castellano, *Inorg. Chem.*, 2008, **47**, 8629.
160. A. Bergamo, L. Messori, F. Piccioli, M. Cocchietto and G. Sava, *Invest. New Drugs*, 2003, **21**, 401.
161. M. Ravera, S. Baracco, C. Cassino, D. Colangelo, G. Bagni, G. Sava and D. Osella, *J. Inorg. Biochem.*, 2004, **98**, 984.
162. M. Groessl, E. Reisner, C. G. Hartinger, R. Eichinger, O. Semenova, A. R. Timerbaev, M. A. Jakupec, V. B. Arion and B. K. Keppler, *J. Med. Chem.*, 2007, **50**, 2185.
163. E. Reisner, V. B. Arion, B. K. Keppler and A. J. L. Pombeiro, *Inorg. Chim. Acta*, 2008, **361**, 1569.

164. L. Messori, P. Orioli, D. Vullo, E. Alessio and E. Iengo, *Eur. J. Biochem.*, 2000, **267**, 1206.
165. A. Casini, G. Mastrobuoni, M. Terenghi, C. Gabbiani, E. Monzani, G. Moneti, L. Casella and L. Messori, *J. Biol. Inorg. Chem.*, 2007, **12**, 1107.
166. T. Andersson, E. Thulin and S. Forsen, *Biochemistry*, 1979, **18**, 2487.
167. A. Casini, C. Gabbiani, G. Mastrobuoni, L. Messori, G. Moneti and G. Pieraccini, *ChemMedChem*, 2006, **1**, 413.
168. A. M. Weljie, A. P. Yamniuk, H. Yoshino, Y. Izumi and H. J. Vogel, *Protein Sci.*, 2003, **12**, 228.
169. L. Trynda-Lemiesz, A. Karaczyn, B. K. Keppler and H. Kozlowski, *J. Inorg. Biochem.*, 2000, **78**, 341.
170. L. Trynda-Lemiesz, H. Kozlowski and B. K. Keppler, *J. Inorg. Biochem.*, 1999, **77**, 141.
171. L. Trynda-Lemiesz, B. K. Keppler and H. Kozlowski, *J. Inorg. Biochem.*, 1999, **73**, 123.
172. T. Yotsuyanagi, N. Ohta, T. Futo, S. Ito, D. N. Chen and K. Ikeda, *Chem. Pharm. Bull.*, 1991, **39**, 3003.
173. N. Ohta, D. Chen, S. Ito, T. Futo, T. Yotsuyanagi and K. Ikeda, *Int. J. Pharm.*, 1995, **118**, 85.
174. O. Rixe, W. Ortuzar, M. Alvarez, R. Parker, E. Reed, K. Paull and T. Fojo, *Biochem. Pharmacol.*, 1996, **52**, 1855.
175. S. G. Chaney, S. L. Campbell, E. Bassett and Y. Wu, *Crit. Rev. Oncol. Hematol.*, 2005, **53**, 3.
176. A. Travers, *Curr. Opin. Struct. Biol.*, 2000, **10**, 102.
177. M. Wei, S. M. Cohen, A. P. Silverman and S. J. Lippard, *J. Biol. Chem.*, 2001, **276**, 38774.
178. D. Fink, S. Nebel, S. Aebi, H. Zheng, B. Cenni, A. Nehme, R. D. Christen and S. B. Howell, *Cancer Res.*, 1996, **56**, 4881.
179. Z. Z. Zdraveski, J. A. Mello, C. K. Farinelli, J. M. Essigmann and M. G. Marinus, *J. Biol. Chem.*, 2002, **277**, 1255.
180. D. Fink, H. Zheng, S. Nebel, P. S. Norris, S. Aebi, T. P. Lin, A. Nehme, R. D. Christen, M. Haas, C. L. MacLeod and S. B. Howell, *Cancer Res.*, 1997, **57**, 1841.
181. S. Aebi, B. Kurdi-Haidar, R. Gordon, B. Cenni, H. Zheng, D. Fink, R. D. Christen, C. R. Boland, M. Koi, R. Fishel and S. B. Howell, *Cancer Res.*, 1996, **56**, 3087.
182. M. H. Lamers, A. Perrakis, J. H. Enzlin, H. H. Winterwerp, N. de Wind and T. K. Sixma, *Nature*, 2000, **407**, 711.
183. U. M. Ohndorf, M. A. Rould, Q. He, C. O. Pabo and S. J. Lippard, *Nature*, 1999, **399**, 708.
184. S. E. Sherman, D. Gibson, A. H. J. Wang and S. J. Lippard, *J. Am. Chem. Soc.*, 1988, **110**, 7368.
185. S. E. Sherman, D. Gibson, A. H. Wang and S. J. Lippard, *Science*, 1985, **230**, 412.
186. S. M. Cohen, Y. Mikata, Q. He and S. J. Lippard, *Biochemistry*, 2000, **39**, 11771.
187. S. U. Dunham and S. J. Lippard, *Biochemistry*, 1997, **36**, 11428.

CHAPTER 10

Application of Integrated Techniques for Micro- and Nano-imaging Towards the Study of Metallomics and Metalloproteomics in Biological Systems

LILI ZHANG[a] AND CHUNYING CHEN[a,b,]*

[a] CAS Key Laboratory for Biological Effects of Nanomaterials and Nanosafety National Center for Nanoscience and Technology and Institute of High Energy Physics, No. 11, Beiyitiao, Zhongguancun, Beijing 100190, P. R. China; [b] CAS Key Laboratory of Nuclear Analytical Techniques, Institute of High Energy Physics, Chinese Academy of Sciences, Beijing 100049, P. R. China

10.1 Introduction

Nuclear analytical techniques include different methods for elemental analysis which are based on nuclear processes or simply employ nuclear instrumentation. The common feature is the use of sophisticated, highly developed equipment which has great versatility and therefore is able to cope with the specific requirements of the field of application.[1] Metallomics is a subject receiving great attention as a new frontier in the investigation of trace elements

---

*Corresponding author: Email: chenchy@nanoctr.cn; Tel: +86-10-82545560; Fax: +86-10-62656765.

Nuclear Analytical Techniques for Metallomics and Metalloproteomics
Edited by Chunying Chen, Zhifang Chai and Yuxi Gao
© Royal Society of Chemistry 2010
Published by the Royal Society of Chemistry, www.rsc.org

in biology and is expected to develop as an interdisciplinary science complementary to genomics and proteomics. In metallomics, metalloproteins, metalloenzymes, and other metal-containing biomolecules are defined as metallomes. The main research targets of metallomics are to identify the metallomes and to elucidate their biological or physiological functions in biological systems. Of these targets, the analysis of elemental distributions is a key technology in metallomics study. The nuclear imaging technique (which here refers to elemental distributions in samples), is a nuclear analytical technique that can provide visible information on distribution patterns of metals or metalloids in various biological tissues, cell and subcellular fractions, and even at the molecular level. Thus, it has played a significant role in metallomics and metalloproteomics studies. Besides the nuclear imaging technique, mass spectrometry imaging techniques, such as secondary ion mass spectrometry (SIMS), laser ablation inductively coupled plasma mass spectrometry (LA-ICP-MS), or matrix-assisted laser desorption/ionization mass spectrometry (MALDI-MS) enable measurements to be made of the elemental and/or molecular distribution and in thin sections of tissues for biological and biomedical research.[2-4] So the emergence and development of mass imaging techniques are suitable for the metallomics study as well. More importantly, with the rapid development of various imaging techniques in recent years, micro- and even nano-distributions of metals in biological samples can be achieved, which is vital for the assessment of biological functions and interaction sites of metals. In the past, many studies that investigated elemental influences on biosystems adopted nuclear or mass spectrometry imaging techniques, which involved many fields including environmental science and life sciences.

In this chapter, we will briefly outline the basic principles of different imaging methods and then focus on some examples of practical application for the study of metallomics, metalloproteomics and the interaction between metals and organisms.

10.2 X-ray Fluorescence

X-ray fluorescence (XRF) is a well-known and widely used multi-elemental analytical technique, which is capable of providing reliable quantitative information on the abundance of major, minor, and trace constituents in various samples with rapid speed and non-destructiveness to the samples. It dose not rely upon the use of radiolabeled compounds, but requires the sample be exposed to an external localized radiation dose.[5] The basic principle of XRF is that when an inner orbital electron is ejected from an atom, an electron from a higher energy level orbit will be transmitted to the lower energy level orbit. During this transition, a photon is emitted from the atom, according to Moseley's law, *i.e.* the square root of the frequency of X-ray fluorescence is proportional to the atomic number of this element. For a particular energy of fluorescent light emitted by an element, the number of photons per unit time is related to the amount of this element in the sample. These two above-mentioned characteristics constitute the foundation of XRF for elemental qualitative and quantitative analysis. So XRF can obtain two-dimensional imaging of the elemental distribution in an inhomogeneous sample.

Early XRF had a relatively large spot size and its spatial resolution ranged in diameter from several hundred micrometers up to several millimeters, which is appropriate for the analysis of samples without high resolution. For samples with uneven surfaces or small areas, the relative macro XRF is not suitable for detecting the elemental distribution. Thus, to minimize the beam size of X-rays to a smaller area became a key problem for small area analysis in any given part of specimen. The emergence of μ-XRF solved this problem, which was mainly attributed to the development of many technologies related to X-ray focusing optics. μ-XRF could provide a highly concentrated X-ray beam in small area in the past 20 years.[6] An important advantage of μ-XRF over macro XRF is that the primary beam can be adjusted to a spot size of a few micrometers in diameter, allowing elemental mapping of small structural features. Compared to macro XRF, μ-XRF gives more detailed information about elemental compositions and distributions with a high spatial resolution at the micron scale. With the development of optics, microelectronics, and computer science, modern XRF has been used widely in many fields.

As a new light source for XRF, the synchrotron radiation (SR) light source has received much attention worldwide. SR was first observed in a synchrotron accelerator in 1947. It is an electro-magnetic wave emitted from an electron traveling at almost the speed of light, toward its running direction when its path is bent by a magnetic field. Originally, synchrotron radiation was viewed primarily as an annoyance because of the need to compensate for the electron beam energy loss by using a powerful radio-frequency accelerating system. Soon researchers found that SR light had useful features such as brightness, high directionality and variable polarization and now SR has been applied to many fields including physical, chemical, and biological experiments.[7,8] The SR source has experienced three-generation fast development stages since its discovery. Now the more brilliant light source of the third-generation of SR is based on a facility designed especially for installing as many insertion devices as possible in a dedicated storage ring. Spring-8, APS and ESRF are the three largest third-generation SR sources in the world.

Spring-8 is located in Harima Science Garden City, Hyogo, Japan.[8] It is the largest and brightest SR light source in the world with a storage ring energy of 8 GeV. It has 62 beamlines at present. The whole facility was prepared in 1987–1989 and constructed in 1991–1997. It has been put in use since 1997. BL16XU, a hard X-ray undulator beamline of Spring-8, the X-ray at sample can reach 10^{10} photons s^{-1} with a beam size 1 μm or less.

The Advanced Photon Source (APS)[9] at Argonne National Laboratory is a national synchrotron X-ray research facility funded by the U.S. Department of Energy. It is the second highest storage ring energy of 7 GeV in the world and has 68 beamlines at present. The whole facility was prepared in 1986–1988 and constructed in 1989–1994. It has been in use since 1996. The APS is a synchrotron light source that produces high-energy, high-brilliance X-ray beams.

The source is optimized to put large quantities of high-energy photons into a very small area in a very short time. With more than 40 beamlines already operational, and more under development, the APS offers an exceptionally broad range of experimental conditions at a single facility. Each beamline at the APS offers a unique combination of capabilities, but some of the main considerations are energy range and tunability, special sample environments, time structures, and beam size.

The European Synchrotron Radiation Facility (ESRF)[10] is an international institute funded by 19 countries. It is located in Grenoble of France. It operates Europe's most powerful synchrotron light source and hosts 6000 scientific user visits per year for 900 different experiments. It is the third highest storage ring of 6 GeV in the world and has 56 beamlines at present. The whole facility was prepared in 1986–1987 and constructed in 1988–1994. It has been in use since 1994. The operation with high current, a low emittance electron beam and the use of long undulators allows a number of ESRF beamlines to be run with both high spectral flux and high spectral brilliance. The maximum current can reach 200 mA and the spatial resolution can reach the 90 nm range.

At present, there are 17 first-, 23 second- and 13 third-generation SR light sources in the world, as well as other 12 third-generation SR light sources in preparation or under construction. The greatest differences between the three generations SR light sources are the different electron beam spot size or electron emittance and the number of insertion devices to be used. Taking the SSRF (Shanghai Synchrotron Radiation Facility, China) under construction as an example, when the whole project is completed, the energy of storage ring will reach 3.5 GeV, being the highest among the medium energy of SR facilities. The light brightness will be 1600 times that of the BSRF (Beijing Synchrotron Radiation Laboratory, China), and the number of insertions will be more than ten to dozens while the second-generation SR light source has several insertions only.[7] Table 10.1 lists several other third-generation SR light sources with basic parameters.[11]

Like conventional XRF, the beam size of SRXRF has also been adjusted from macro to micro level, and it is entering the nano-probe age. Meanwhile, SRXRF has the advantages of tunable energy and adjustable beam size, which benefit the detection of some heavy metals with good sensitivity in micro- or even smaller structure. At present, SRXRF has successfully mapped the elemental distribution in plant and animal tissues, even for a single cell, and is closely involved in many fields such as environment science and life science. In this section, some practical examples will be given to show the applications of SRXRF in the above-mentioned fields.

10.2.1 Environmental Science

For biology, the environment is the climate, ecosystems, and species that surround the organisms. Environmental science is an expression encompassing the wide range of scientific disciplines that need to be brought together to

Table 10.1 Comparison of international middle- and high-energy synchrotron facilities[11] Reprinted with the permission from NSRRC.

Location	Name	Energy (GeV)	Emittance (nm*rad)	C (m)	Long strait	Cells	Beam size (μm)	Status
Taiwan, Hsinchu	TLS	1.5	(18)	120	6m*6	6 TBA		In operation
France, Grenoble	ESRF	6.0	(3.7)	850	6.3m*32	32 DBA	0.09	In operation
USA, Chicago	APS	7.0	(3.0)	1104	6.7m*40	40 DBA	0.05	In operation
Japan, Himeji	SPring-8	8.0	(3.0)	1432	6.6m*44; 30m*4	48 DBA	<1	In operation
Italy, Trieste	ELETTRA	2.4	7.0	259	6m*12	12 DBA	5*10	In operation
Korea, Pohang	PLS	2.5	(10.3)	281	6m*12	12 TBA	A few microns	In operation
Switzerland, Villigen	SLS	2.4	4.8	288	11.76m*3; 7m*3; 4m*6	12 TBA	1*1	In operation
Canada, Saskatoon	CLS	2.9	(18.2)	171	5.2m*12	12 DBA	(2–4)*(2–4)	In operation
USA, Stanford	SPEAR3	3.0	(12)	234	3m*12; 4.5m*4; 7.5m*2	18 DBA		In operation
France, Orsay	SOLEIL	2.75	(3.7)	354	12m*4; 7m*12; 3.6m*8	16 DBA		In operation
UK, Oxfordshire	DIAMOND	3.0	(2.7)	562	8m*4; 5m*18	24 DBA	1.5	In operation
Australia, Melbourne	ASP	3.0	7.0	216	5.4m*12	14 DBA	0.1*0.1	In operation
China, Shanghai	SSRF	3.5	(3.9)	432	12m*4; 6.5m*16	20 DBA	<2	In operation
Jordan, Allan	SESAME	2.5	(26)	133	5m*4; 3.5m*8; 1.9m*4	12 DBA		Under construction
Spain, Barcelona	ALBA	3.0	(4.3)	269	8m*4; 4m*12	16 DBA		Under construction
Taiwan, Hsinchu	TPS	3.0	(1.6)	518	12m*6; 7m*18	24 DBA		Under construction
USA, New York	NSLS-II	3.0	2.1 (0.5)	792	9.3m*15; 6.6m*15	30 DBA		Under construction

Note: (): distributed dispersion.
DBA: double-bend achromat.
TBA: triple-bend achromat.

understand and manage the natural environment and the many interactions among physical, chemical, and biological components. Neither large organisms nor microorganisms can survive if separated from the environment. So it is important to continue with the study of environmental science. SRXRF techniques have been successfully applied to this field, and now we introduce some of them.

10.2.1.1 Plants

Moss is a good biomonitor plant for inspecting environment pollution and has broad application prospects due to its small shape, simple structure, special physiological adaptation mechanism, and the high response sensitivity to environmental factors.[12–14] Zhang et al.[15] adopted SRXRF for studying the elemental distribution in micro-areas of the leaf, stem, and the whole moss tissue. Moss samples were exposed for 1 month at Shanghai Iron and Steel Research Institute (as a pollution site) and Shanghai Institute of Applied Physics (as a control site) simultaneously. Their results showed that metallic ions were not distributed uniformly in the polluted samples; for instance, excessive Pb was adsorbed and deposited in the leaf, causing serious damage to plant growth and inhibition of the normal absorption of nutritive elements, such as K and Ca. Xie et al.[16] summarized SRXRF spectroscopy as an analysis method in a study of the polar environment in recent years and expected this approach to be intensively used in the future.

There are many kinds of plant that can accumulate heavy metals or non-metal elements such as Mn, Ni, Cu, Pb, and As. On the one hand, these plants can serve as indicators for geological minerals;[17] on the other hand, they may be applied to phytoremediation, which is the treatment of environmental problems through the use of living green plants for *in situ* risk reduction and/or removal of contaminants from contaminated soil, water, sediments, and air. Phytoremediation is an emerging and promising technology to solve environmental pollution for its low cost and versatility. In order to study the site at which an element accumulates and explore the mechanism involved, many researchers have used SRXRF techniques to image the elemental distributions in various tissues of these plants. Hokura et al.[18] adopted SRXRF to map the distribution of arsenic in an arsenic hyperaccumulator fern. Prior to micro-XRF imaging of the fern samples, an X-ray beam of conventional size, $200 \times 200\,\mu m^2$, was used to determine the elemental distribution in fern fronds of various ages and found arsenic accumulated along the edges of mature pinna (Figure 10.1). Then the pinna slices were subjected to micro-XRF analysis by using an X-ray microbeam $3.5 \times 5.5\,\mu m^2$ (Figure 10.2). X-ray absorption near structure (XANES) was then used to analyze the arsenic oxidation state in fern and found that arsenic exists as the As(III) form in pinna, and as a mixture of As(III) and As(V) in rachis, while the As(V) form was present in cultivated soil. The findings indicate that the fern takes up arsenic as As(V) from soil and that the As(V) is then partially reduced to As(III) within the plant. Finally, arsenic is

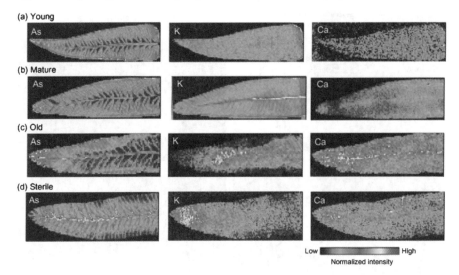

Figure 10.1 XRF imaging for As, K, and Ca of pinna at different growth stages. X-ray beam size, $200 \times 200\,\mu m^2$; scan step, $200 \times 200\,\mu m$; measurement time, $3\,s\,pixel^{-1}$. (a) Young pinna with spore, image size, 125×36 pixels; (b) mature pinna with spore, image size, 115×36 pixels; (c) old pinna, image size, 100×32 pixels; (d) sterile pinna, image size, 125×45 pixels. Elemental concentrations in each map are scaled to the maximum value for that map.[18] © 2006 The Royal Society of Chemistry.

accumulatd as As(III) form in a specific area of the pinna. Fukuda et al.[19] used the same tools to study cadmium distribution and oxidation state in *Arabidopsis halleri* ssp. *gemmifera*. The trichome of *Arabis gemmifera* is clearly seen from Figure 10.3. Isaure et al.[20] found that the trichomes (epidermal hairs) represented the main compartment of Cd accumulation in the leaves of *Arabidopsos thaliana* which were exposed Cd (Figure 10.4).

10.2.1.2 The Environment

When toxic substances enter into the environment, model organisms can be used to study their elemental distribution changes by SRXRF. Knowledge of the spatial distribution of a contaminant within an organism is important in determining the site of toxic action or in identifying processes whereby trace metals are selectively sequestered. *Caenorhabditis elegan*[21] (Figure 10.5) is a native soil invertebrate; it is relatively easy to culture and has a short life cycle; and it can be used in aqueous or soil matrixes. For such reasons it is well suited for soil toxicity testing. Jackson et al. used μ-SRXRF (beam of approximately 10 μm in diameter) to image the elemental distribution in nematodes which were exposed to Cu or Pb ions solution or their mixture solution (Figure 10.6).[22] Samber et al.[23] scanned elemental distribution in Zn exposed *Daphnia magna*[24] (Figure 10.7), which is a frequently used model organism to

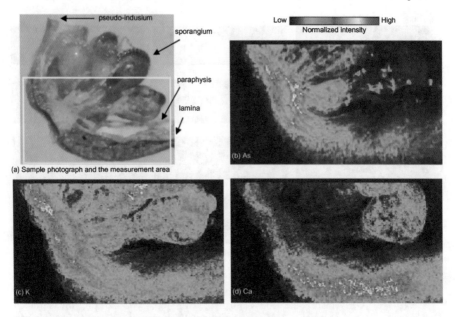

Figure 10.2 Micro XRF imaging for As, K, and Ca at the edges of mature pinna containing the lamina of pinna, pseudo-indusium, sporangium, and spore. X-ray microbeam size, $3.5 \times 5.5\,\mu m^2$; scan step, $3.5 \times 5.5\,\mu m$; measurement time, $1\,s\,pixel^{-1}$; image size, 180×125 pixels. (a) Sample photograph by optical microscopy and the measurement area. Scale bar, µm. Elemental concentrations in each map are scaled to the maximum value for that map.[18] © 2006 The Royal Society of Chemistry.

investigate the mechanisms of toxicity. The accumulation of Zn in the different tissues is clearly visible (Figure 10.8). A distinct enrichment of Zn can be observed in (A) the gill-like osmoregulatory tissue, and in the digestive system, *i.e.* (B) the gut and (C) the digestive gland.

China is universally acknowledged as a great source of rare-earth resources. The development and application of rare-earth elements and their products have invaded many areas of our lives. Whether rare earths adversely affect organisms is of concern. For example, Feng *et al.*[25] studied the results of long-term exposure to lanthanum in persistent alterations in nervous system function of Wistar rats. To fully investigate the neurotoxicological consequences of lanthanum exposure, they also used SRXRF to image brain elemental distributions and found that Ca, Fe, and Zn were significantly altered after lanthanum exposure (Figure 10.9).

10.2.1.3 Cellular Imaging

When referring to cells, the first thought is their small size, so that most must be observed under a microscope. How to image their elemental distribution with

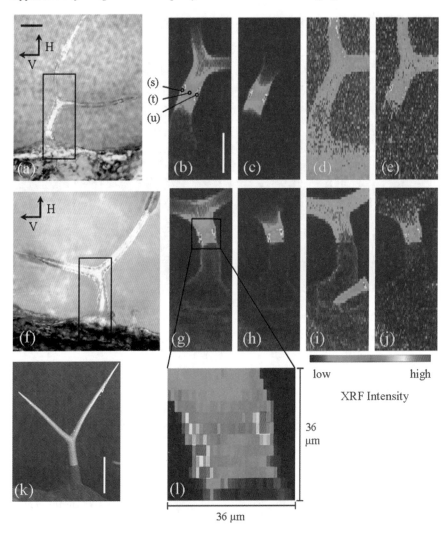

Figure 10.3 μ-XRF imaging of trichome. (a) and (f) are photographs of the measured samples showing the imaging area, (b) Cd, (c) Zn, (d) Ca, and (e) Mn of (a), (g) Cd, (h) Zn, (i) Ca, and (j) Mn of (f). For the maps, imaging area of 204 μm (H) × 81 μm (V), beam size of 3.8 μm (H) × 1.3 μm (V), step size of 3 μm (H) × 1 μm (V), step number of 68 point (H) × 81 point (V), and dwell time of 0.3 s point^{-1} were used. (k) SEM image of trichome. (l) distribution of Cd in the area shown in the box in Figure 2(g). (s)–(u) are measured points by m-XANES of Cd. The scale bar is 50 μm except for (k), for which the scale bar is 100 μm.[19] © 2008 The Royal Society of Chemistry.

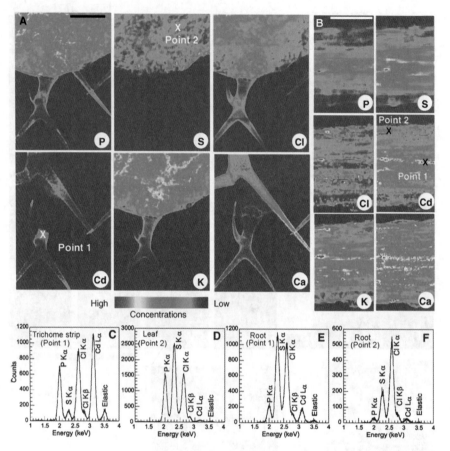

Figure 10.4 False-color μ-XRF elemental maps recorded on a leaf containing trichomes (A, scale bar = 70 μm) and on a root (B, scale bar = 25 μm), and X-ray fluorescence spectra collected on the trichome strip (C, point 1 in Figure 4A), on the leaf tissue (D, point 2 in Figure 4A), on the central vascular bundle on the root (E, point 1 in Figure 4B), and on the border of the root (F, point 2 in Figure 4B). For the maps, step size = 1 μm, dwell time = 500 ms pixel^{-1} at 3550 eV, and 100 and 250 ms pixel^{-1} at 4100 eV for leaf and root, respectively.[20] © 2006 Elsevier B.V.

high spatial resolution is a great challenge to the beam sizes usually used in, relatively, much larger samples. With the construction of third-generation SR light sources, the tiny electron beam spot size is one direction for its development, including submicro- and nano-beam size. Le Lay et al.[26] studied the elements P, S, Cl, K, and Cs in *Arabidopsis thaliana* cells which were exposed to caesium stress. Elemental maps were obtained by scanning the cells under a monochromatic beam having an energy of 5.8 keV and FWHM dimensions at the sample of 0.7 μm (H) × 0.4 μm (V). Maps show the appearance of micrometer round-shaped structures (which are projections of small intracellular volumes) indicating that the elements detected were more concentrated in

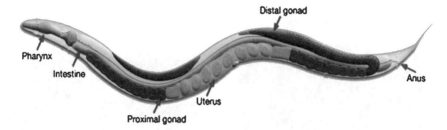

Figure 10.5 Schematic drawing of anatomical structures of an adult hermaphrodite *C. elegans*, modified from http://www.wormatlas.org/handbook/anatomyintro/anatomyintro.htm[21] © 2002–2006 Wormatlas.

numerous small volumes (grains) than in the larger part of the cells. Caesium, in particular, was co-localized with potassium preferentially in these structures in both K^+-rich and K^+-depleted media. A comparison of these X-ray fluorescence maps with photon microscopy images (Figure 10.10) strongly suggests that the granular structures revealed by X-ray fluorescence microscopy are reminiscent of structures identified as chloroplasts in photonic images.

Besides the above-mentioned studies, SRXRF as a complement to PIXE has been applied to study fine particle atmospheric pollution, characterizing combustion sources contributing to urban air pollution.[27] With the development of SR light sources and hard X-ray beam focusing techniques, SRXRF will be certainly applied to much broader research areas.

10.2.2 Life Science

Life science is the study of the phenomena, essence, characteristics, and the law of occurrence and development for life, as well as the relationship between biology and the environment. Life science is one of the most basic concerns of the natural sciences in today's world. The SRXRF technique has been broadly used in this field.

Neurodegenerative disease is a condition in which cells of the brain and spinal cord are lost. It results from deterioration of neurons or their myelin sheath which, over time, will lead to dysfunction and disabilities. Over the past 20 years, many scientists have found that dysregulation of the redox active transition metal ions is linked to neurodegenerative diseases, and landmark papers have been published.[28] The quantitative determination of essential elements (e.g. P, Cu, Fe, Zn, Mn, Co, Se, and others) in medical tissues is of growing interest in brain research and biosciences, and is relevant for studying many neurodegenerative diseases. The deficit (or the surplus) of essential elements in human tissue or in proteins is observed in neurodegenerative diseases (including Alzheimer's disease or Parkinson's disease). To clearly understand the mechanisms of these diseases and develop corresponding therapy effectively, some scientists applied the XRF technique to the study of neurodegenerative diseases, mainly including

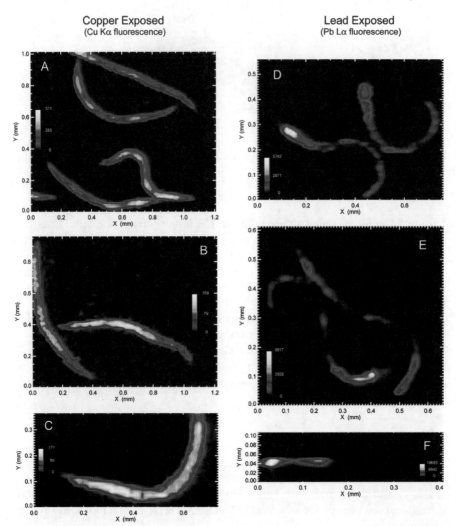

Figure 10.6 Representative SXRF maps for Cu (A–C) and Pb (D–F) exposed nematodes. Dark to light colors represent increasing fluorescence intensity proportional to an increase in the elemental concentration.[22] © 2005 American Chemical Society.

Alzheimer's disease and Parkinson's disease. We will illustrate the XRF technique applied in these two diseases.

10.2.2.1 Alzheimer's Disease

Alzheimer's disease (AD) is a progressive neurodegenerative disorder characterized by the accumulation of senile plaques and neurofibrillary tangles in

Application of Integrated Techniques for Micro- and Nano-imaging 311

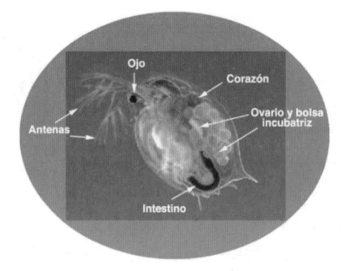

Figure 10.7 Schematic drawing of anatomical structures of an adult hermaphrodite *Daphnia magna*, downloaded from http://www.idrc.ca/openebooks/147-7/f0183-02.gif [24] (Reprinted with permission from IDRC, Canada.)

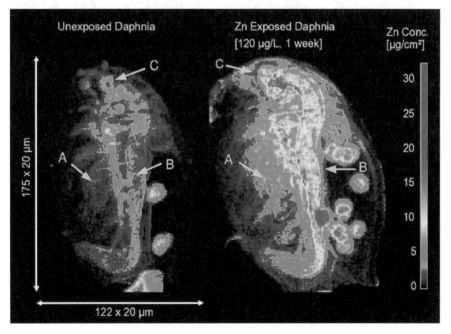

Figure 10.8 Micro-XRF dynamic scans on *Daphnia magna* samples subjected to different exposure levels of Zn, using a microbeam of approximately 15 μm (FWHM) which defines the attainable resolution during this experiment. The left image corresponds to an unexposed sample, while the sample on the right was exposed to 120 μg L^{-1} Zn during 1 week.[23] © 2007 Springer-Verlag.

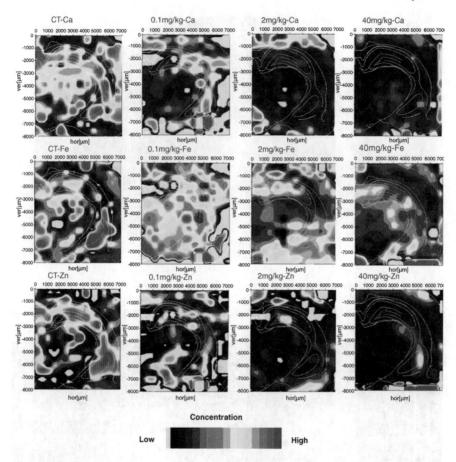

Figure 10.9 LaCl$_3$ at doses of 0, 0.1, 2, and 40 mgkg^{-1} was orally administered to rats for 6 months prior to the measurement of brain elemental distributions by SRXRF. Microscope pictures and elemental distribution of Ca, Fe, Zn for each LaCl$_3$ exposed group are shown in the figure. Half brain section area: 7000 μm × 8000 μm.[25] © 2006 Elsevier Ireland Ltd.

gray matter areas of the brain, and this disease gradually affects memory, thinking, communication, emotion, and behavior. The link between iron overloaded and AD was first discovered by Goodman in 1953.[29] To study the interaction between iron and AD, Collingwood et al.[30] reported the location and characterisation of iron compounds in human AD brain tissue sections. Iron fluorescence was mapped over a frontal-lobe tissue section from an Alzheimer's patient (Figure 10.11), and anomalous iron concentrations were identified using synchrotron X-ray absorption techniques at 5 μm spatial resolution (Figure 10.12). Concentrations of ferritin and magnetite, a magnetic iron oxide potentially indicating disrupted brain–iron metabolism, were evident. These results demonstrate a practical means of correlating iron compounds and disease pathology *in situ* and have clear implications for disease

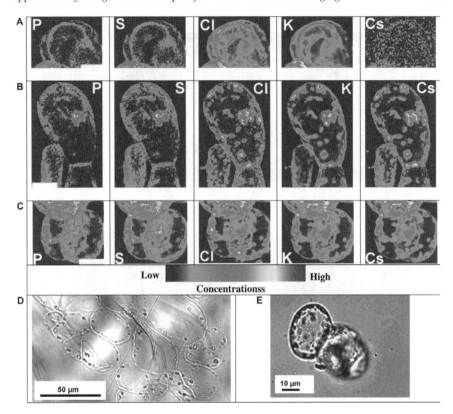

Figure 10.10 X-ray fluorescence maps of P, S, Cl, K, and Cs from isolated *Arabidopsis thaliana* cells grown in presence of: (A) 20 mM K, 0 mM Cs; (B) 20 mM K, 1 mM Cs, and (C) 0 mM K, 1 mM Cs and photonic images of control cells prepared as thin section (D) and *in vivo* (E). Relative intensities, given by the scale bar, were normalized relative to the incident beam intensity recorded on each point. The incident X-ray energy was 5.8 keV and the beam size was H = 0.7 μm × V = 0.4 μm. For (A), scale bar = 15 μm, step size = 0.5 μm, dwell time = 0.5 s point^{-1}. For (B), scale bar = 22 μm, step size = 0.7 μm, dwell time = 0.5 s point^{-1}. For (C), scale bar = 20 μm, step size = 0.5 μm, dwell time = 0.5 s point^{-1}.[26] © 2006 Elsevier Masson SAS

pathogenesis. Ide-Ektessabi *et al.* studied the Fe and Zn changes in the brain tissue of an AD patient.[31] A section of one specimen obtained from the temporal lobes of the patient was stained with ubiquitin. Ubiquitin immunoreactivity is used regularly in the identification of pathological lesions such as Lewy bodies and neurofibrillary tangles (NFTs), that are associated with neurological disorders.[32] These authors found distinct imbalances of metallic elements such as Zn and Cu as well as the Fe^{2+}/Fe^{3+} redox pair, which point to oxidative stress as a crucial factor in the development or progress of these neurodegenerative diseases (Figures 10.13 and 10.14).

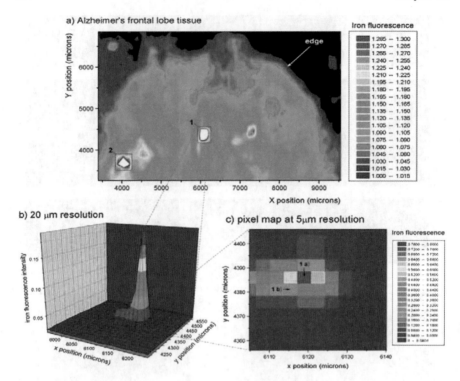

Figure 10.11 The process of locating anomalous iron concentrations by mapping iron fluorescence intensity over the tissue section, and then microfocusing in regions of interest, is illustrated. (a) Initial low-resolution (100 μm) contour map of the tissue section. (b) Medium-resolution scan of area 1, revealing that the source of the bright spot in (a) is located within a 20 μm pixel. (c) High-resolution scan within (b), locating the source of the high intensity spot to a 5 μm pixel. XANES scans of the absorption edge are performed at area 1(a), at a point slightly offset from the highest intensity, area 1(b), and at the highest intensity 5 μm pixel in area 2 which is located in an identical fashion.[30] © 2005 IOS Press and the authors.

At present, there is increasing evidence that interaction between Aβ and metal ions may play a role in this transformation. Miller et al.[33] studied the Cu and Zn co-localized with β-amyloid deposits in Alzheimer's disease. The localization of amyloid plaques was determined by staining the tissue sections with thioflavin S.[34] In the SRXRF microprobe results presented here, we find that there is a heterogeneous distribution of Zn, Cu, Fe, and Ca in the AD brain tissue, exhibiting "hot spots" of high metal content (Figure 10.15).

With the development of genetic engineering, scientists developed a series of genetically engineered mice for the disease research. APP transgenic mice were used for the pathogenesis of AD. Zhang et al.[35] used XRF to characterize Fe, Cu and Zn in organs of this kind of mouse. Liu et al.[36] used this animal model to study the metal exposure and AD pathogenesis and they first demonstrated

Application of Integrated Techniques for Micro- and Nano-imaging 315

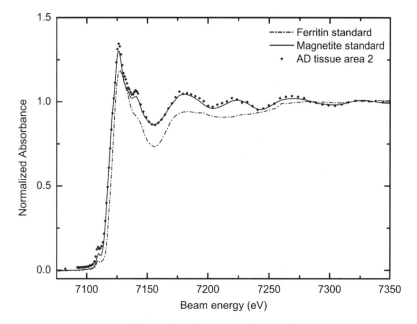

Figure 10.12 The XANES spectra from area 2 in Figure 10 is shown fitted with the XANES standard for magnetite. The excellent agreement in the traces continues far beyond the typical XANES energy region ($-20\,eV$ below to $40\,eV$ above the K edge for iron). The XANES spectrum for the physiological horse-spleen standard is included for comparison.[30] © 2005 IOS Press and the authors.

that laser capture microdissection coupled with X-ray fluorescence microscopy can be applied to determine elemental profiles in Aβ amyloid plaques (Figure 10.16).

10.2.2.2 Parkinson's Disease

Parkinson's disease belongs to a group of conditions called movement disorders. It is characterized by muscle rigidity, tremor, a slowing of physical movement and, in extreme cases, a loss of physical movement. PD is caused by the selective degeneration of neurons in the substantia nigra (SN) region of the brain. The increase in iron concentration in the SN has been reported by several research groups.[37] Ide-Ektessabi et al.[38] used SRXRF and Fe K-edge X-ray absorption near-edge structure (XANES) spectroscopy to investigate distributions and chemical states of iron in the brain tissues from monkeys that had been injected with MPTP (1-methy-l-4-phenyl-1,2,3,6-tetrahydropyridine), which can cause a PD-like syndrome. The data were measured in fluorescence mode for the biological specimens and in transmission mode for the reference samples. The results for the Fe^{2+}/Fe^{3+} ratios from the neuromelanin granules showed significant variations, which were correlated with the level of iron

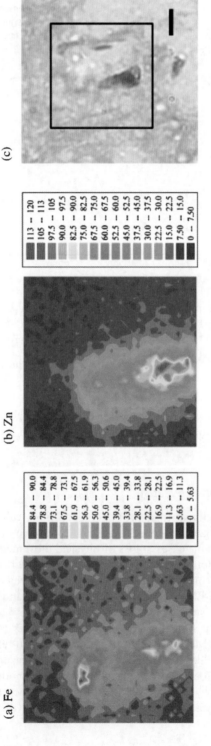

Figure 10.13 (a) and (b) XRF imaging of iron and zinc, respectively, of the temporal lobes tissue of the AD patient (ubiquitin immunoreactive). The measurement area was 40 × 40 μm, and the step for each pixel was 1 μm. The ranges of fluorescent X-ray intensity are shown in the scales (arbitrary units). (c) Optical microscopic photograph of the sample stained with hematoxylin–eosin after X-ray analysis. The scale bar is 20 μm.[31] © 2005 The Japan Society for Analytical Chemistry.

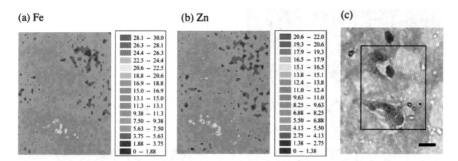

Figure 10.14 (a) and (b) XRF imaging of iron and zinc respectively of the temporal lobes tissue of the AD patient (poorly ubiquitin immunoreactive). The measurement area was 40 × 50 μm, and the step for each pixel is 1 μm. (c) Optical microscopic photograph of the sample stained with hematoxylin–eosin after X-ray analysis. The scale bar is 20 μm.[31] © 2005 The Japan Society for Analytical Chemistry.

concentration. Cells containing high level of iron had high level of Fe^{2+}. Popesu et al.[39] applied XRF to map and quantify iron, zinc, and copper in brain slices from PD patient and unaffected subjects for the purpose of carrying out systematic studies on metal pathology in neurodegenerative disease in general and PD in particular. The normal (N) brain slice was half of a coronal section of the forebrain through the genu of the internal capsule, rostral tip of dorsal thalamus, and rostral third of the hypothalamus. The matching PD slice was slightly caudal to the N brain. The midbrain was sliced through the inferior colliculus, trochlear nucleus, and superior cerebellar peduncle with the matching PD being slightly cranial. N2 was a coronal section of the forebrain through the basal ganglia at the level of the anterior commissure and rostral aspect of the hypothalamus. Maps show the abundance and location of iron, zinc, and copper were obtained in comparable brain slices from two subjects of advanced age, one normal (N) and one with PD (PD), and in a small section from a normal young brain (N2). They found that the location and amount of iron in brain regions known to be affected in PD agree with analyses using other methods (Figures 10.17 to 10.19).

Waern et al.[40] carried out their research in APS of the Argonne National Laboratory on the intracellular distribution of transition metals in V79 Chinese hamster lung cells treated with subtoxic doses of the organometallic anticancer complexes Cp_2MCl_2. Fluorescence-detected X-ray elemental distribution maps were collected at 295 K under a He atmosphere, using either a 10.0 keV or a 20.5 keV monochromatic X-ray incident beam, focused to 0.5 μm using a dual-zone plate and an order sorting aperture device. Low-resolution scans of the dimensions 30 × 30 μm, 3 μm step size and 0.5 s point^{-1}, were initially collected in order to accurately locate the cells. Once an appropriate cell had been identified, high-resolution scans were acquired with a 0.5 μm step size, 5 s point^{-1} (20 s point^{-1} for thin sections), and usually the dimensions 15 × 15 μm being dependent on cell size. Their XRF results agree with independent

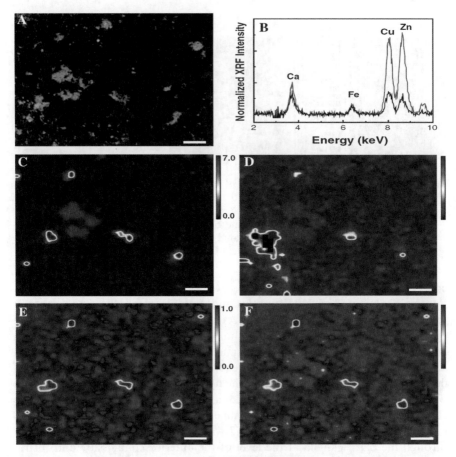

Figure 10.15 (A) Epifluorescence image of human AD tissue stained with thioflavin S. (B) SXRF microprobe spectra from a thioflavin positive area and a thioflavin-negative area. SXRF microprobe images of (C) Ca, (D) Fe, (E) Cu, and (F) Zn content in the same tissue. In the Fe image, several pixels were excluded from the analysis due to Fe contamination in the Al substrate. For all images, scale bar is 100 μm.[33] © 2005 Elsevier Inc.

chemical studies that have concluded that the biological chemistry of each of the metallocene dihalides is unique. Both studies prove that by using the SRXRF technique with a sub-micrometer beam size, the elemental distribution at a single cell level can be obtained with high spatial resolution.

At present, the SRXRF beam size at APS and ESRF can reach 50 nm and 90 nm, respectively; the constructing SSRF will have a beam size below 200 nm. The characteristics of this unique nanoprobe fulfill the requirements for mapping elemental distributions at a size compatible with the analysis of most cellular compartments. With the smaller beam size, SRXRF can be fully used for various samples from large tissues to small cells. Ortega et al.[41] determined the iron storage within dopamine neurovesicles with 90 nm spatial resolution

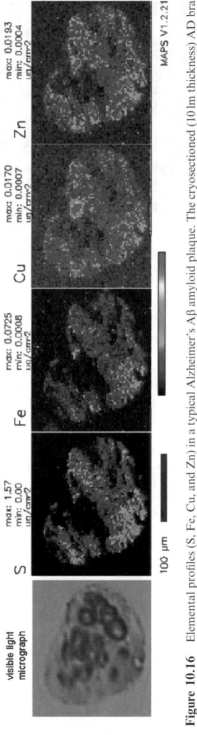

Figure 10.16 Elemental profiles (S, Fe, Cu, and Zn) in a typical Alzheimer's Aβ amyloid plaque. The cryosectioned (10 lm thickness) AD brain tissues were stained with 0.1% thioflavin-T for amyloid plaques. The amyloid plaque-bearing human brain tissues were procured by LCM (Arcturus Pixcell IIE platform) and mounted on Si3N4 membrane grids (2.0 × 2.0 mm). Guided by the optical amyloid plaque images, the samples were excited with incident synchrotron X-ray of 10 keV for elemental Kα characteristic emission lines. Elemental profiles (S, Fe, Cu, and Zn) were obtained using synchrotron scanning X-ray fluorescence microscopy (SR-XRF) at the Advanced Photon Source (APS) of the Argonne National Laboratory (ANL).[36] © 2006 Elsevier Inc.

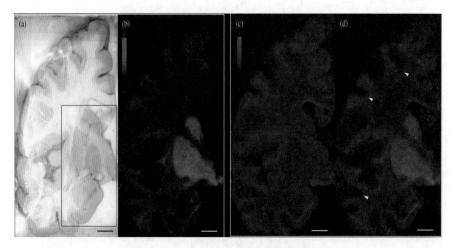

Figure 10.17 Metal levels in the brain vary widely by region. Normal (N) human forebrain, coronal section: (a) optical image; (b) iron map; (c) zinc map; (d) overlay of iron and zinc; square, area magnified in Figure 10.18 (a–c); arrow heads, subcortical white matter; scan speed 22.8 ms pixel^{-1}; scale bar = 7.5 mm; shaded scales (b and c) represent the normalized total Kα fluorescence counts, proportional to total metal present, from black (lowest) to lightest shade (highest).[39] © 2009 Institute of Physics and Engineering in Medicine.

using SRXRF in ESRF (Figures 10.20 and 10.21). This unique spatial resolution, combined with a high brightness, enables chemical element imaging in subcellular compartments. Carmona *et al.*[42] also used the same setup at ESRF for nano-image trace metals in dopaminergic single cells and neurite-like processes.

As a sample is not destroyed during SRXRF, the technique is very suitable for applications in biomedical sciences because the samples being measured can be used for histopathological examination before or after elemental analyses. Homma-Takeda *et al.*[43] attached TUNEL-stained renal sections (10 μm) to mylar film which was then subjected to SRXRF analysis. They found proximal tubular-selective apoptosis induced by inorganic mercury, which was also confirmed by *in situ* SRXRF showing that the apoptotic cells localized in the proximal tubules did contain higher levels of mercury.

10.3 Particle Induced X-ray Emission

Particle Induced X-ray Emission or Proton Induced X-ray Emission (PIXE) is an accelerator-based technique which offers sensitive, multi-elemental and fast analysis. The basic principle is similar to XRF. When samples are bombarded with particles generated by accelerators, electrons in the inner shells of an atom are ejected, and electrons in the outer shells then fill vacancies in the inner shell, with the emission of X-rays whose energy is characteristic of the parent atom.

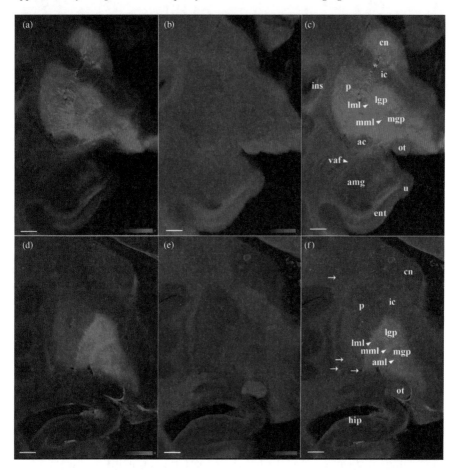

Figure 10.18 Metal content is abnormal in PD basal ganglia. Basal ganglia, coronal section normal (N) brain (a)–(c) (expanded view of Figure 10.17); PD brain (d)–(f). (a), (d) Iron maps; (b), (e) zinc map; (c) (f) overlay of iron and zinc; cn, caudate nucleus; p, putamen; lgp, lateral globus pallidus; mgp, medial globus pallidus; lml, lateral medullary lamina; mml, medial medullary lamina; aml, accessory medullary lamina; ic, internal capsule; ins, insula; ot, optic tract; ac, anterior commissure; vaf, ventral amygdalofugal fibers; amg, amygdaloid nucleus; u, uncus; ent, enthorinal cortex; hip, hippocampus; *, blood vessel; arrows, perivascular iron; scale bar = 5 mm; shaded scales (a, b, d, e) represent the normalized total Kα fluorescence counts proportional to total metal present, from black (lowest) to lightest shade (highest).[39] © 2009 Institute of Physics and Engineering in Medicine.

By measuring the energy of the characteristic X-rays emitted, the elemental composition of the sample can be determined.[44] PIXE is also a non-destructive analytical technique which can be used to image the distribution of major, minor, and trace constituents in various samples. The method is sensitive to

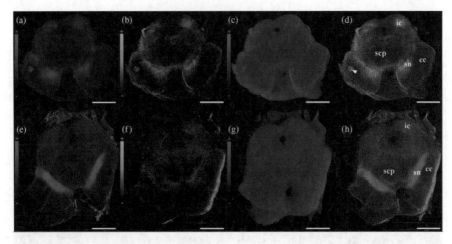

Figure 10.19 Metal distribution is abnormal in PD midbrain. Transverse section, normal (N) midbrain ((a)–(d)) and PD midbrain ((e)–(h)): (a) and (e) iron map; (b) and (f) copper map; (c) and (g) zinc map; (d) and (h) overlay of iron, copper and zinc; sn, substantia nigra; cc, crus cerebri; scp, spinocerebellar peduncle; ic, inferior colliculus; scan speed 22.8 ms pixel^{-1}; shaded scales (a, b, c, e, f and g) represent the normalized total Kα fluorescence counts, proportional to total metal present, from black (lowest) to lightest shade (highest); scale bar = 5 mm.[39] © 2009 Institute of Physics and Engineering in Medicine.

most elements ($Z \geq 12$), and its relative sensitivity reaches ppm level (parts per million). Researchers have applied this technique in environmental studies, biomedical and archaeological studies *etc*.[45] PIXE with a micro beam size is particularly important in biomedical applications for it allows information concerning localization and distribution patterns of many elements to be obtained at the same time.[46] Like XRF, when using PIXE to measure elemental distribution on uneven surfaces or small areas of a sample, or where high spatial resolution images are required, the μ-PIXE technique can be adopted. μ-PIXE is especially suitable for scanning the sample with a lateral resolution reaching down to the sub-micrometer range, localizing the elements of interest, and allowing simultaneous mapping of these elements at high sensitivities.[47,48] It can provide a deeper insight into metabolic processes. Similarly, adoption of conventional PIXE or μ-PIXE is based on the research purposes and the expected results to obtain. Here we divide the measured samples into plants, animals, and cells to briefly show the application of PIXE in environmental and biological studies.

10.3.1 Plants

Fundamental processes of plant physiology are affected or regulated by mineral nutrients.[49] Hence understanding biological roles of nutrient or toxic elements

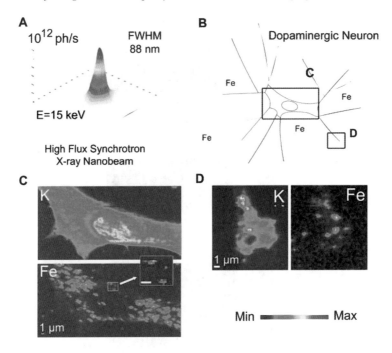

Figure 10.20 Synchrotron X-ray chemical nano-imaging reveals iron sub-cellular distribution. The synchrotron X-ray fluorescence nanoprobe endstation installed at ESRF was designed to provide a high flux hard X-ray beam of less than 90 nm size (FWHM, full width at half maximum). The intensity distribution in the focal plane is shown in (A); dopamine producing cells were exposed *in vitro* to 300 μM FeSO$_4$ during 24 h (B). Chemical element distributions, here potassium and iron, were recorded on distinct cellular areas such as cell bodies (C), neurite outgrowths, and distal ends (D). Iron was found in 200 nm structures in the cytosol, neurite outgrowths, and distal ends, but not in the nucleus. Iron-rich structures are not always resolved by the beam and clusters of larger dimension are also observed. Min–max range bar units are arbitrary. Scale bars = 1 μm.[41] © 2007 PLoS ONE.

and their mechanisms in plant metabolism is of fundamental importance in both basic and applied plant studies. So far, there are many studies on minerals accumulative or resistant mechanism of plants. For instance, Mesjasz-Przybyłowicz *et al.*[50] adopted 3.0 MeV protons with the beam size focused to approximately 3 × 3 μm to map the elemental distribution of S, K, and Zn in seed part of *Biscutella laevigata* L. collected from zinc dump near Olkusz, Upper Silesia, hoping to clarify the mechanism of plant adaptation to a hostile environment with rich heavy metals. They conclude that it is likely the endosperm that acts a barrier, actively controlling elemental uptake and preventing access of toxic levels of elements into the embryonic parts.

In fact, for this kind of studies, one can also adopt similar method to that given by Hokura *et al.*[18] First, conventional PIXE is applied to image the

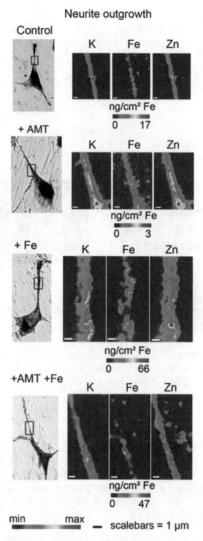

Figure 10.21 Nano-imaging of potassium, iron, and zinc in neurite outgrowths. Each series of images are representative of the entire cell population for each condition (control, 1 mM AMT and/or 300 mM FeSO$_4$). The scanned area (left images, squares) is shown on a bright field microscopy view of the freeze-dried cell. Iron is located within dopamine vesicles of 200 nm size or more in control cells with a large number of Fe-dopamine structures in Fe exposed cells. The iron concentration is close to the limit of detection in neurites of AMT cells. Min–max range bar units are arbitrary for potassium and zinc distributions. For iron distribution the maximum threshold values in micrograms per square centimeter are shown for each shaded scale. Scale bars = 1 μm.[41] © 2007 PLoS ONE.

Application of Integrated Techniques for Micro- and Nano-imaging 325

elemental distribution in the fronds to get the main overall changed parts in the plant. Then, a microbeam technique is used for mapping those given changed parts to gain higher spatial resolution images. This strategy is useful to the study of elucidating the interaction between metals or metalloids and biosystems.

10.3.2 Animals

The μ-PIXE can image the elemental distribution in animal sample with such a high spatial resolution that its overall appearances can be seen clearly. For example, Augustyniak et al.[51] used μ-PIXE to study the elemental distribution in grasshopper *Stenoscepa* sp., which is an insect species feeding on the South African Ni-hyperaccumulating plants. A proton beam of 3.0 MeV energy and 200–300 pA current was focused to $3 \times 3\,\mu m^2$ spot and raster scanned over the grasshoppers' tissues. Their μ-PIXE results showed that the highest Ni level was distributed in the gut and Malpighian tubules (Figure 10.22). From the elemental maps of K, Cl, and Ni in *Stenoscepa* sp., the whole body structure is clearly visible.

An exciting investigation is that Kertész et al.[52] used μ-PIXE to study TiO_2 penetration in the epidermis of human skin xenografts. A 2 MeV proton beam with currents of 80–100 pA was focused to $2.5 \times 2.5\,\mu m^2$ and scanned over the areas of interest ($100 \times 100\,\mu m^2$). They came to the conclusion that the human skin xenograft was a model particularly well adapted to their penetration studies. In 1989, Lindh et al.[44] published a paper "Distribution of lead in the

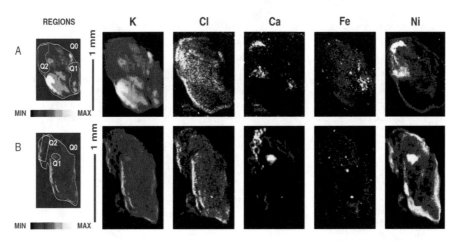

Figure 10.22 True elemental maps of *Stenoscepe* sp. Regions selective within specimens are marked on the left side: (A) brain Q0 (Q1, optic lobes; Q2, cerebral ganglia); and (B) gut Q0 (Q1, inorganic concretions supposedly taken in with food by the grasshopper; Q2, fragment with Malpighian tubules). Maps acquired with accumulated proton charge of 3.495 μC (map A) and 2.033 μC (map B).[51] © 2008 John Wiley & Sons, Ltd.

cerebellum of sucking rats following low and high dose lead exposure" in which they adopted μ-PIXE (a 3 μm beam of 2.5 MeV protons) imaging the lead distribution in the cerebellum of sucking SD rats. Before the elemental imaging, they inspected the stained sections in a light microscope and selected a suitable area for μ-PIXE analysis.

10.3.3 Cellular Imaging

μ-PIXE is a suitable imaging technique to measure the elemental distributions in tissues even cells. Rombouts et al.[53] carried out the IBC microbeam facility using tubules. A beam of 2.5 MeV protons can be focused to 1–2 μm diameter to investigate the interaction between *Schizosaccharomyces prombe* with the culture medium. From Figure 10.23, the distribution of Na in the medium around the cells showed evidence the action of the Na pump.

PIXE also has some applications in neuroscience. Scientists found that subpopulations of neurons are less vulnerable against degeneration, and one of these possesses a specialized extracellular matrix arranged as a perineuronal net (PN) which consists mainly of hyaluronic acid, glycoproteins, and sulfated proteoglycans and so displays a highly negatively charge. It is assumed that PNs are able to bind metal ions and thus reduce the oxidative effect of neurons. Reinert et al.[54] used μ-PIXE to investigate the concentration and distribution of

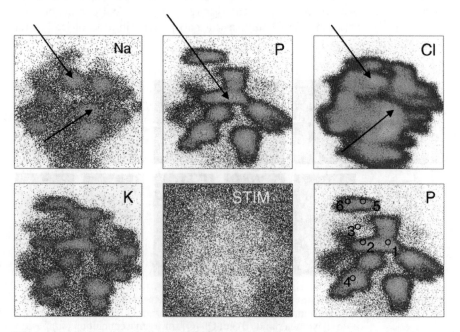

Figure 10.23 Elemental maps and STIM image of cells grown in normal medium and analysed in water and glucose. Scan size is $50 \times 50 \, \mu m^2$. Positions for point analyses are indicated in second P map.[53] © 2007 Elsevier B.V.

iron in rat brain loaded with colloidal iron with special emphasis on the PN in the extracellular matrix and found that the PN has the ability to bind large amounts of iron. Fiedler et al.[55] performed a μ-PIXE study on PN-ensheathed neurons in selected brain areas. PN-ensheathed neurons were detected by lectin-histochemical staining with Wisteria floribunda agglutinin (WFA). The staining was intensified by DAB–nickel by an established method enabling the visualization of the PNs by nuclear microscopy. Their first results of subcellular analysis (Figures 10.24 and 10.25) showed that the intracellular iron concentration of PN-ensheathed neurons tends to be slightly increased in comparison to neurons without PNs. The difference in intracellular iron concentrations could be an effect of the PNs. From their results we can see that PIXE is an extremely useful technique for scientific examinations in brain by spatially resolving iron and other elements of interest on subcellular levels. Thus it is an available imaging technique that can be explore the biosystem's response mechanism to metal or metalloid at the cell or subcellular level.

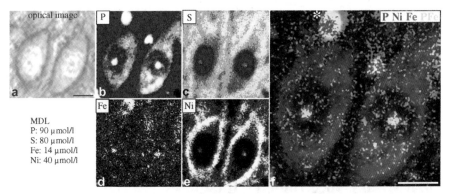

Figure 10.24 Elemental PIXE maps (b–f) of two PN-ensheathed neurons in the brainstem. (a) Optical image of the analysed neurons stained for PNs (WFADAB-Ni, black pigment) surrounding the cells. (b) The phosphorus map indicates the cytoplasm of the cells as well as the intranuclear nucleolus. (c) Sulfur nearly omits the nuclei in contrast to the extracellular matrix. (e) In the nickel map the PNs are selective recognizable due to the Ni-enhanced staining of the PNs. (f) Three elemental map of phosphorus, nickel, and iron. A co-localization of phosphorus and iron is reflected in light-coloured regions, especially the nucleolus in both neurons. Total cellular iron concentration of the right neuron (0.86 mmol L^{-1}) is contributed by the cytoplasm (0.8 mmol L^{-1}), the nucleus (1.05 mmol L^{-1}) and the nucleolus (2.08 mmol L^{-1}). The iron concentration of the surrounding PN was the same as that of the cell. The merged map also reveals an iron-rich glia cell (asterisk), probably an oligodendrocyte, because this cell type is believed to store the highest cellular iron concentrations. Collected charge: 5.5 μC; beam current: 600 pA; spatial resolution: 800 nm. Scale bar: 10 μm.[55] © 2007 Elsevier B.V.

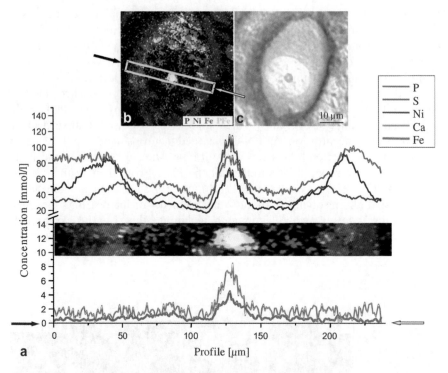

Figure 10.25 (a) Elemental profiles of P, S, Ni, Cu, and Fe along a traverse through a PN-ensheathed neuron of the brainstem using GeoPIXE II. (b) Merged PIXE elemental map containing phosphorus, nickel, and iron of the profiled neuron. (c) Optical image of the analysed neuron (WFADAB-Ni stained PN, black pigment).[55] © 2007 Elsevier B.V.

10.4 Mass Spectrometry Imaging

Mass spectrometry (MS) is an analytical technique for the determination of the elemental composition of a sample or molecule. The basic MS principle is ionizing chemical compounds to generate charged molecules or molecule fragments and measuring of their mass-to-charge ratios (m/z).[56] MS along with its hyphenated techniques is capable of high throughout, sensitivity, accuracy, and selectivity for the analysis of structure and composition of almost any product.

Today, two-dimensional mass spectrometry analysis of biological tissues by means of a technique called mass imaging, mass spectrometry imaging (MSI), or imaging mass spectrometry has been used in mapping the distribution of elements or chemical groups in samples. Matrix-assisted laser desorption ionization mass spectrometry (MALDI-MS), secondary ion mass spectrometry (SIMS), and laser ablation inductively coupled plasma mass spectrometry (LA-ICP-MS) are three commonly used MSI techniques.

10.4.1 Matrix-assisted Laser Desorption Ionization Mass Spectrometry

Of these MSI techniques, matrix-assisted laser desorption ionization mass spectrometry (MALDI-MS) is a soft ionization technique which does not produce fragments, or, at least, produces fewer ions fragments. In addition, its high sensitivity, accuracy, and spatial resolution made it suitable to measure the composition of compounds or biomacromolecules. So it can determine the distribution of peptides, proteins, or other chemicals and so on within thin slices of biological samples.[57,58] Robinson *et al.*[59] used MALDI-MS to measure the localization of water-soluble carbohydrates in wheat stems *in situ* and their method (Figure 10.26) may be useful for researchers of metallomics to carry out the study between metal or non metal and organism.

10.4.2 Secondary Ion Mass Spectrometry

Compared to MALDI-MS, secondary ion mass spectrometry (SIMS) is a less soft ionization technique which can determine the elemental, isotopic, or

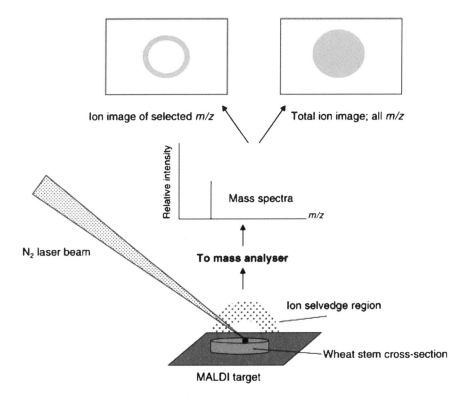

Figure 10.26 Abstract representation of the imaging matrix-assisted laser desorption ionization (MALDI) mass spectrometry technique.[59] © 2006 New Phytologist.

molecular composition of the sample surface. The briefly principle of SIMS is using a beam of "primary" ions or of neutral atoms (FAB) with a few keV energy to sputter the atoms contained in the surface of the specimen. Part of the matter is sputtered as to monoatomic or polyatomic ions. These "secondary" ions are measured with a high mass spectrometry and secondary ions of a selected mass are used to produce an enlarged image of the distribution of this ion at the sample surface.[60] SIMS is the most sensitive surface analysis techniques which can detect elements present in the parts per billion range. At present, SIMS has been applied in relation to drug delivery, the use of biomarkers, and the determination of physiological functions of elements in tissues or cells.[61–63]

10.4.3 Laser Ablation Inductively Coupled Plasma Mass Spectrometry

Laser ablation inductively coupled plasma mass spectrometry (LA-ICP-MS) is a type of mass spectrometry that is highly sensitive and capable of the determination of a range of metals and several non-metals at concentrations below one part in 10^{12}. Together with inductively coupled plasma, it is used as a method to produce ions (ionization) with a mass spectrometer for separating and detecting the ions.[64]

The whole measure process is when the sample material of interest is vaporized in a laser plasma by focused laser radiation and transported with argon into the inductively coupled plasma ion source of an ICP-MS. The positively charged ions are extracted from the inductively coupled plasma through an interface into the high vacuum of the mass spectrometer and separated in the mass-spectrometric separation system according to their mass/charge ratio and energy/charge ratio and detected by a secondary electron multiplier.[65]

The schematic of LA-ICP-MS measurement[66] is shown in Figure 10.27, which can be used for other type of samples.

There are many studies using LA-ICP-MS to image the essential and toxic elements in biological tissues in recent years. Among these studies, Becker's group has done a great deal of excellent work and published a series of articles on the application of LA-ICP-MS; the samples tested, including plant and animal samples, are introduced below.

10.4.3.1 Plants

Becker *et al.* used LA-ICP-MS (a Nd:YAG laser, wavelength: 266 nm, repetition frequency: 20 Hz, spot size: 300 μm) for quantitative imaging of toxic and essential elements in thin sections of tobacco tissues (Figure 10.28).[67] Wu *et al.*[66] studied essential element accumulation in the newly formed, fully grown and oldest leaves of a Cu-tolerant plant *Elsholtzia splendens* after Cu treatment by LA-ICP-MS. The materials of the leaves were ablated by a Nd:YAG laser

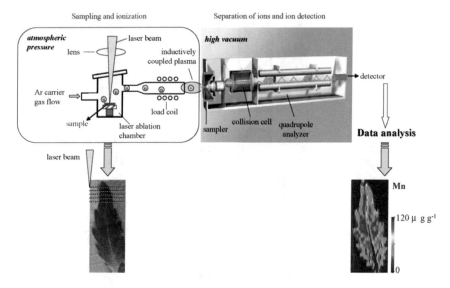

Figure 10.27 The schematic of LA-ICP-MS measurement on the leaves of *E. splendens*.[66] © 2008 Elsevier B.V.

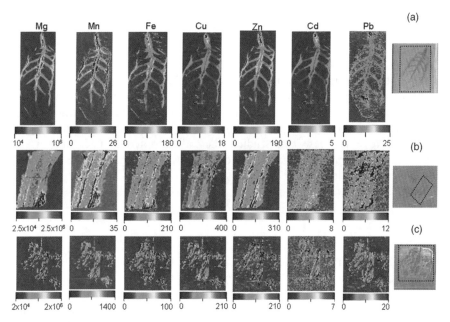

Figure 10.28 Laser ablation images of metals concentration in plant tissues. (a) leaf, (b) shoot, and (c) root. Concentrations are in µg g^{-1}, except Mg that is shown as ion intensity (cps). Pictures on the right column are photography of the glass slides with the fixed samples, whereas the area within the dashed lines was scanned.[67] © 2008 Elsevier B.V.

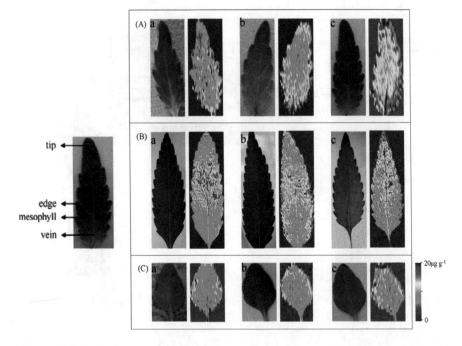

Figure 10.29 Quantified images of Cu in the newly formed (A), fully grown (B) and oldest (C) leaves of *E. splendens* before (a) and after Cu stress for 0.5 h (b) and 2 h (c).[66] © 2008 Elsevier B.V.

(wavelength: 266 nm, repetition frequency: 20 Hz, spot size: 160 μm) and then transported by argon as carrier gas into the inductively coupled plasma (ICP). From Figure 10.29 we can see Cu accumulated in the newly formed leaves increased more quickly compared with the fully grown and the oldest leaves after Cu stress. In these two studies, they all adopted a macro beam size (300 μm and 160 μm respectively), but for samples with large area this is enough to satisfy our requirements.

10.4.3.2 Animals

Much information on elemental changes can be obtained by LA-ICP-MS measurements which show close relationship to their roles in physiological or pathological processes. When using LA-ICP-MS to determine the elemental distribution in small area, a relatively small beam size can be used to gain high resolution images. Becker *et al.* adopted a 50 μm beam size to measure the elemental distribution in slug slices,[68] and human brain tissues from the hippocampus.[69] LA-ICP-MS enables studies of absorption, distribution, metabolism, and elimination (ADME) of the drugs containing trace elements such as cisplatin on cross-sections of entire small animals or, once identified, of critical organs.[70] It is very useful for the pharmacokinetics of metal-containing drugs.

The type of analysis of a sample will greatly depend on the beam size. For example, in the analysis of a very small area of the specimen, good spatial resolution with a smaller beam size is required to ensure that only the small area or feature of analytical interest is measured. As the laser power increases, the beam size increases as well.[71] The typical spatial resolution of LA-ICP-MS achieved by most commercially available laser ablation system ranges from 5 to 300 μm,[72-74] and in some cases this would be insufficient for the analysis of fine structures and regions such as in a single cell. So how to get high spatial resolution (especially for nano-imaging) images with a good signal-to-noise ratio is a challenge for LA-ICP-MS.

10.4.4 Near-field LA-ICP-MS: A Novel Elemental Analytical Technique for Nano-imaging

The emergence of near-field LA-ICP-MS as a novel elemental analytical technique provides us with a possibility which can realize the elemental imaging at the nanometer scale. Zoriy and Becker[75] reported an analytical method utilizing a near-field effect (to enhance the incident light energy on the thin tip of an Ag needle) in a LA-ICP-MS (NA-LA-ICP-MS) (Figure 10.30). They produced the thin needles with a tip diameter in the hundreds of nm range. By detecting copper isotopic standard reference material NIST SRM 976 and tungsten–molybdenum alloy NIST SRM 480 in the nm resolution range, they observed craters ranging from 200 nm to about 2 μm in diameter that were dependent on the needle used as well as on the "sample-to-tip" distance. That is, by using the NA-LA-ICP-MS the spatial resolution can be 200 nm to 2 μm, which enables nano-imaging for a single cell.

However, this is just the beginning of nano-imaging for NA-LA-ICP-MS. It requires a large amount of research work for real nano-imaging to be realized.

For the MSI techniques, they can also be applied in metallomics and metalloproteomics with XRF and PIXE. But compared to these two non-destructive techniques, MSI is a partly complete or complete devastating technique to the sample. That is, when the sample is measured over, we can not use it to do other measurements. The most advantage of XRF and PIXE over MSI is their non-destructive and non-invasive characteristics with little or no damage to the sample. So XRF and PIXE are more suitable to study biological samples than MSI if the samples need other further studies.

XRF, PIXE, and MSI are all two-dimensional (2D) imaging techniques, and either tissues or cells are commonly prepared as thin slices with constant thickness to obtain a precise elemental distribution. The experiments are usually carried out *in vitro* as researchers can gain little information from three-dimensional (3D) imaging. In the next section, we briefly introduce tomography techniques for 3D imaging *in vivo*.

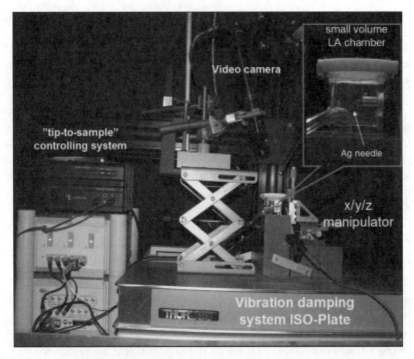

Figure 10.30 Photograph of the NF-LA-ICP-MS experimental setup.[75] © 2008 John Wiley & Sons, Ltd.

10.5 Tomography

Tomography is imaging by sections or sectioning. The method is widely used in medicine, archaeology, biology, geophysics, oceanography, materials science, astrophysics, and other fields. In most cases it is based on the mathematical procedure called tomographic reconstruction. More modern variations of tomography involve gathering projection data from multiple directions and feeding the data into a tomographic reconstruction software algorithm processed by a computer. Different types of signal acquisition can be used in similar calculation algorithms in order to create a tomographic image.[76] Actually, tomography is primarily for imaging the structures of organisms. In some tomographic inspection processes, a tracer or contrast agent is needed. So this method also involves 3D imaging of some elements in the body. Here we briefly introduce single-photon emission computed tomography (SPECT) and positron emission tomography (PET), which are of particular interest because of their unique ability to image trace amounts of important biomarkers.

SPECT is a nuclear medicine tomographic imaging technique using gamma rays. It is very similar to conventional nuclear medicine planar imaging using a gamma camera. However, it is able to provide true 3D information. This information is typically presented as cross-sectional slices through the patient, but can be freely reformatted or manipulated as required.[77] Carlson et al.[78]

used micro-SPECT/CT to trace the ^{123}I uptake in mice which were inoculated human pancreatic cancer xenografts. Intratumoral radioisotope uptake was achieved via intratumoral injection of an attenuated measles virus vector expressing the NIS gene (MV-NIS). On various days after MV-NIS injection, ^{123}I planar and micro-SPECT/CT imaging was performed (Figure 10.31). They concluded that micro-SPECT/CT can be used to accurately quantify intratumoral radioisotope uptake *in vivo* and is more reliable than planar or micro-SPECT imaging alone, because of its ability to more accurately define tumor margins and regions of interest.

PET can produce a 3D image or picture of functional processes in the body. This system detects pairs of gamma rays emitted indirectly by a

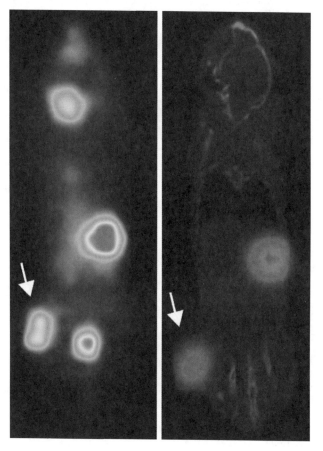

Figure 10.31 Images of mice with intratumoral ^{123}I uptake (arrows) on planar (left) and micro-SPECT/CT (right). Note the other physiologic areas of increased uptake in the thyroid gland and bladder (planar image) and stomach (planar and micro-SPECT/CT images). The background activity in the control flank was negligible, as determined by dose calibrator measurements and ROI analysis.[78] © 2006 Academy of Molecular Imaging.

positron-emitting radionuclide (tracer), which is introduced into the body on a biologically active molecule. Images of tracer concentration in 3D space within the body are then reconstructed by computer analysis.[79] For example, PET with ^{18}F-FDG, a tracer of glucose metabolism, has been highly successful for imaging a wide variety of tumors and for monitoring response to therapy.[80] L-*Methyl*-^{11}C-methionine (^{11}C-methionine) is also used to image a variety of tumors, including lymphomas and tumors of the brain, lung, and head or neck. Nuñez *et al.*[81] combined ^{18}F-FDG and ^{11}C-methionine PET scans in patients with newly progressive metastatic prostate cancer. Their findings reflected the different biological characteristics of the lesions in a heterogeneous tumor such as prostate cancer and suggested that a time-dependent metabolic cascade may occur in advanced prostate cancer, with initial uptake of ^{11}C-methionine in dormant sites followed by increased uptake of ^{18}F-FDG during progression of disease.

Table 10.2 Analytical features of the most popular techniques using accelerator-based sources[82] © 2008 Springer-Verlag.

Technique	Ion beam	Energy	Sensitivity	Spatial resolution	Selectivity
PIXE	H$^+$	1–4 MeV	<1 ppm	*ca.* 1 μm	Maximum sensitivity in atomic ranges: $10 < Z < 35$ and $75 < Z < 85$
RBS	^{4}He$^+$, H$^+$	≤2 MeV	<100 ppm	*ca.* 1 μm	$Z < 15$
NRA	H$^+$, D$^+$	0.4–3 MeV	10–100 ppm depending on element	*ca.* 1 μm	Low-Z elements
PIGE	H$^+$	MeV	0.1–1 ppm, depending on element	*ca.* 1 μm	Low-Z elements such as: Li, F, Na, Mg, and Al
SIMS	Ar$^+$, Xe$^+$, O$_2^+$, Cs$^+$	eV to keV	0.1 ppb	<0.1 μm with TOF-SIMS	Multielemental
STIM	H$^+$ (or heavy ions)	MeV	<1000 ppm	Sub-100 nm	Multielemental
μXRF	Photons (X-rays)	keV	<10 ppb	*ca.* 1 μm	Multielemental
FEL	Photons	eV to MeV	100 ppm	Diffraction limited (*ca.* 1–10 μm)	Infrared-active vibrational modes

FEL, free electron laser; NRA, nuclear reaction analysis; PIGE, particle-induced gamma-ray emission; RBS, Rutherford backscattering spectrometry; SIMS, secondary ion mass spectrometry; STIM, scanning transmission ion microscopy.

The emergence of tomography provide a more accurate localization and quantitative estimate based on the tracer or contrast agent in live animals, and its development will be more beneficial to the progress of diagnostics.

10.6 Conclusions

In fact, the elemental imaging techniques described herein are mainly focused on 2D imaging methods. At the moment, the imaging techniques applied more commonly primarily use accelerator sources, including XRF and PIXE. The great advantage of these techniques is their non-destructive characteristics to samples, which is very suitable for biological specimens. However, there is much research that can not be carried out very well due to limited experimental time and the spatial resolution provided by a synchrotron radiation station. In Table 10.2[82] we list some characteristics of these techniques. In addition, MS imaging has developed rapidly in recent years, especially LA-ICP-MS. Although samples are partly or completely destroyed during analysis, this technique has much merit, such as high sensitivity, isotope analysis, simultaneous determination for elemental concentrations, and so on. According to different research purposes and experimental conditions, researchers can choose different imaging techniques for their work. From a strict definition, tomography is not an elemental imaging technique as its targets are certain isotopes. However, we can obtain a lot of useful information about the physiological and pathological information of the animal being tested.

In conclusion, with the development of imaging techniques, high-throughput techniques of 2D or 3D features are required for metallomics metalloproteomics. The development of nuclear or non-nuclear analytical imaging techniques will greatly accelerate the study of metallomics and metalloproteomics.

Acknowledgments

The authors acknowledge the financial support by the Ministry of Science and Technology of China as the National Basic Research Programs (2006CB705603, 2010CB934004 and 2009AA03J335), the Natural Science Foundation of China (10975040), and the NSFC/RGC Joint Research Scheme (20931160430).

References

1. R. Moro and G. Gialanella, *Phys. Scr.*, 1990, **T32**, 233–236.
2. R. M. A. Heeren, L. A. McDonnell, E. Amstalden, S. L. Luxembourg, A. F. M. Alteraar and S. R. Piersma, *Appl. Surf. Sci.*, 2006, **252**, 6827–6835.
3. L. A. McDonnell, S. R. Piersma, A. F. Maarten Altelaar, T. H. Mize, S. L. Luxembourg, P. D. E. M. Verhaert, J. van Minnen and R. M. A. Heeren, *J. Mass Spectrom.*, 2005, **40**, 160–168.

4. A. Brunelle, D. Touboul and O. Laprévote, *J. Mass Spectrom.*, 2005, **40**, 985–999.
5. J. D. Robertson, E. Ferguson, M. Jay and D. J. Stalker, *Pharm. Res.*, 1992, **9**, 1410–1414.
6. K. Nakano, A. Matsuda, Y. Nodera and K. Tsuji, *X-Ray Spectrom.*, 2008, **37**, 642–645.
7. http://ssrf.sinap.ac.cn/
8. http://www.sping8.or.jp/
9. http://www.aps.anl.gov/
10. http://www.esrf.eu/
11. http://www.nsrrc.org.tw/chinese/tps_convenience.aspx
12. M. Aceto, O. Abollino, R. Conca, M. Malandrino, E. Mentashi and C. Sarzanini, *Chemosphere*, 2003, **50**, 333–342.
13. B. Wolterbeek, *Environ. Pollut.*, 2002, **120**, 11–21.
14. H. Th. Wolterbeek and T. G. Verburg, *Environ. Monit. Assess.*, 2002, **73**, 7–16.
15. Y. X. Zhang, T. Cao, A. Iida, C. J. Yang, M. Wang, Q. C. Cao, G. L. Zhang and Y. Li, *Nuc. Techniq.*, 2007, **30**, 730–734.
16. Z. Q. Xie, L. G. Sun, B. B. Cheng and L. Zhang, *Chin. J. Polar Res.*, 2004, **16**, 99–105.
17. H. L. Canon, *Science*, 1960, **132**, 591–598.
18. A. Hokura, R. Omuma, Y. Terada, N. Kitajima, T. Abe, H. Saito, S. Yoshida and I. Nakaia, *J. Anal. At. Spectrom.*, 2006, **21**, 321–328.
19. N. Fukuda, A. Hokura, N. Kitajima, Y. Terada, H. Saito, T. Abe and I. Nakai, *J. Anal. At. Spectrom.*, 2008, **23**, 1068–1075.
20. Marie-Pierre Isaure, B. Fayard, G. Sarret, S. Pairis and J. Bourguignon, *Spectrochim. Acta Part B*, 2006, **61**, 1242–1252.
21. http://www.wormatlas.org/handbook/anatomyintro/anatomyintro.htm.
22. B. P. Jackson, P. L. Williams, A. Lanzirotti and P. M. Bertsch, *Environ. Sci. Technol.*, 2005, **39**, 5620–5625.
23. B. De Samber, G. Silversmit, R. Evens, K. De Schamphelaere, C. Janssen, B. Masschaele, L. Van Hoorebeke, L. Balcaen, F. Vanhaecke, G. Falkenberg and L. Vincze, *Anal. Bioanal. Chem.*, 2008, **390**, 267–271.
24. http://www.idrc.ca/openebooks/147-7/f0183-02.gif.
25. L. X. Feng, H. Q. Xiao, X. He, Z. J. Li, F. L. Li, N. Q. Liu, Y. L. Zhao, Y. Y. Huang, Z. Y. Zhang and Z. F. Chai, *Toxicol. Lett.*, 2006, **165**, 112–120.
26. P. Le Lay, M. P. Isaure, J. E. Sarry, L. Kuhn, B. Fayard, J. L. Le Bail, O. Bastien, J. Garin, C. Roby and J. Bourguignon, *Biochimie*, 2006, **88**, 1533–1547.
27. D. D. Cohen, R. Siegele, Ed Stelcer, D. Carton, A. Stampfl, Z. Cai, P. Ilinski, W. Rodrigues, D. G. Legnini, W. Yun and B. Lai, *Nucl. Instrum. Methods Phys. Res. Sect. B*, 2002, **189**, 100–106.
28. A. I. Bush and C. C. Curtain, *Eur. Biophys. J.*, 2008, **37**, 241–245.
29. L. Goodman, *J. Nerv. Ment. Dis.*, 1953, **118**, 97–130.

30. J. F. Collingwood, A. Mikhaylova, M. Davidson, C. Batich, W. J. Streit, J. Terry and J. Dobson, *J. Alzheimers Dis.*, 2005, **7**, 262–272.
31. A. Ide-Ektessabi and M. Rabionet, *Anal. Sci.*, 2005, **21**, 885–892.
32. A. Alves-Rodrigue, L. Gregori and M. E. Fiqueiredo-Pereira, *Trends Neurosci.*, 1998, **21**, 516–520.
33. L. M. Miller, Q. Wang, T. P. Telivala, R. J. Smith, A. Lamzirotti and Judit Miklossy, *J. Struct. Biol.*, 2006, **155**, 30–37.
34. R. Guntern, C. Bouras, P. R. Hof and P. G. Vallet, *Cell. Mol. Life Sci.*, 1992, **48**, 8–10.
35. Z. Y. Zhang, N. Q. Liu, J. Zhang, H. Zhu, C. Qin, Z. Y. Zou and X. W. Tang, *X-Ray Spectrom.*, 2006, **35**, 253–256.
36. G. J. Liu, W. D. Huang, R. D. Moir, C. R. Vanderburg, B. Lai, Z. C. Peng, R. E. Tanzi, J. T. Rogers and X. D. Huang, *J. Struct. Biol.*, 2006, **155**, 45–51.
37. A. Ide-Ektessabi, S. Fujisawa and S. Yoshida, *J. Appl. Phys.*, 2002, **91**, 1613–1617.
38. A. Ide-Ektessabi, T. Kawakami and F. Watt, *Nucl. Instrum. Methods Phys. Res. Sect. B*, 2004, **213**, 590–594.
39. B. F. G. Popescu, M. J. George, U. Bergmann, A. V. Garachtchenko, M. E. Kelly, R. P. E. McCrea, K. Lüning, R. M. Devon, G. N. George, A. D. Hanson, S. M. Harder, L. D. Chapman, I. J. Pickering and H. Nichol, *Phys. Med. Biol.*, 2009, **51**, 651–653.
40. J. B. Waern, H. H. Harris, B. Lai, Z. J. Cai, M. M. Harding and C. T. Dillon, *J. Biol. Inorg. Chem.*, 2005, **10**, 443–452.
41. R. Ortega, P. Cloetens, G. Devès, A. Garmona and S. Bohic, *PLoS One*, 2007, **9**, e925.
42. A. Carmona, P. Cloetens, G. Devès, S. Bohic and R. Ortega, *J. Anal. At. Spectrom.*, 2008, **23**, 1093–1088.
43. S. Homma-Takeda, Y. Takenaka, Y. Kumagai and N. Shimojo, *Environ. Toxicol. Pharmacol.*, 1999, **7**, 179–187.
44. U. Lindh, N. G. Conradi and P. Sourander, *Acta Neuropathol.*, 1989, **79**, 149–153.
45. M. G. Budnar and J. L. I. Campbell, *X-Ray Spectrom.*, 2005, **34**, 263–264.
46. R. Cesareo, *Nuclear Analytical Techniques in Medicine*, Elsevier, Amsterdam, 1988, p. 123.
47. M. Nečemer, P. Kump, J. Ščančar, R. Jačimović, J. Simčič, P. Pelicon, M. Budnar, Z. Jeran, P. Pongrac, M. Regvar and K. Vogel-Mikuš, *Spectrochim. Acta Part B*, 2008, **63**, 1240–1247.
48. A. G. Kachenko, B. Singh, N. P. Bhatia and R. Siegele, *Nucl. Instrum. Methods Phys. Res. Sect. B*, 2008, **266**, 667–676.
49. H. Marschner, *Mineral Nutrition of Higher Plants*, Academic Press, London, 1995, p. 3.
50. J. Mesjasz-Przybyłowicz, K. Grodzińska, W. J. Przybyłowicz, B. Godzik and G. Szarek-Łukaszewska, *Nucl. Instrum. Methods Phys. Res. Sect. B*, 2001, **181**, 634–639.

51. M. Augustyniak, W. Przybyłowicz, J. Mesjasz-Przybyłowicz, M. Tarnawska, P. Migula, E. Głowacka and A. Babczyńska, *X-Ray Spectrom.*, 2008, **37**, 142–145.
52. Zs. Kertész, Z. Szikszai, E. Gontier, P. Moretto, J.-E. Surlève-Bazeille, B. Kiss, I. Juhász, J. Hunyadi and Á. Z. Kiss, *Nucl. Instrum. Methods Phys. Res. Sect. B*, 2005, **231**, 280–285.
53. P. M. M. Rombouts, I. Gomez-Morilla, G. W. Grime, R. P. Webb, L. Cuenca, R. Rodriguez, M. Browton, N. Wardell, B. Underwood, N. F. Kirkby and K. J. Kirkby, *Nucl. Instrum. Methods Phys. Res. Sect. B*, 2007, **260**, 231–235.
54. T. Reinert, M. Morawski, T. Arendt and T. Butz, *Nucl. Instrum. Methods Phys. Res. Sect. B*, 2003, **210**, 395–340.
55. A. Fiedler, T. Reinert, M. Morawski, G. Brückner, T. Arendt and T. Butz, *Nucl. Instrum. Methods Phys. Res. Sect. B*, 2007, **260**, 153–158.
56. D. O. Sparkman, *Mass Spectrometry Desk Reference*, Global View Pub, Pittsburgh, 2000.
57. R. L. Caldwell and R. M. Caprioli, *Mol. Cell Proteomics*, 2005, **4**, 394–401.
58. S. Khatib-Shahidi, M. Andersson, J. L. Herman, T. A. Gillespie and R. M. Caprioli, *Anal. Chem.*, 2006, **78**, 6448–6456.
59. S. Robinson, K. Warburton, M. Seymour, M. Clench and J. Thomas-Oates, *New Phytol.*, 2007, **173**, 438–444.
60. C. Dérue, D. Gibouin, M. Demarty, M. C. Verdus, F. Lefebvre, M. Thellier and C. Ripoll, *Microsc. Res. Technol.*, 2006, **69**, 53–63.
61. S. Chandra, D. R. Lorey and D. R. Smith, *Radiat. Res.*, 2002, **157**, 700–710.
62. J. -L. Guerquin-Kern, F. Hillion, J.-C. Madelmont, P. Labarre, J. Papon and A. Croisy, *Biomed. Eng. Online*, 2004, doi: 10.1186/1475-925X-3-10.
63. E. Gazi, J. Dwyer, N. Lockyer, P. Gardner, J. C. Vickerman, J. Miyan, C. A. Hart, M. Brown, J. H. Shanks and N. Clarke, *Faraday Discuss.*, 2004, **126**, 41–59.
64. http://en.wikipedia.org/wiki/Inductively_coupled_plasma_mass_spectrometry
65. http://www.fz-juelich.de/zch/la-icp-ms
66. B. Wu, Y. X. Chen and J. S. Becker, *Anal. Chim. Acta*, 2009, **633**, 165–172.
67. J. S. Becker, R. C. Dietrich, A. Matusch, D. Pozebon and V. L. Dressler, *Spectrochim. Acta Part B*, 2008, **63**, 1248–1252.
68. J. S. Becker, A. Matusch, C. Depboylu, J. Dobrowolska and M. V. Zoriy, *Anal. Chem.*, 2007, **79**, 6074–6080.
69. J. Dobrowolska, M. Dehnhardt, A. Matusch, M. Zoriy, N. Palomero-Gallagher, P. Koscielniak, K. Zilles and J. S. Becker, *Talanta*, 2008, **74**, 717–723.
70. J. S. Becker, M. Zoriy, J. S. Becker, J. Dobrowolska and A. Matusch, *J. Anal. At. Spectrom.*, 2007, **22**, 736–744.
71. S. Elliott, *ICP-MS Instruments at Work*, ICP-MS-14, 1997, pp. 1–4, www.varianinc.com/media/sci/apps/icpms14.pdf.
72. J. S. Becker, *Inorganic Mass Spectrometry: Principles and Applications*, Wiley, Chichester, 2007.

73. M. V. Zoriy, M. Kayser, A. Izmer, C. Pickhardt and J. S. Becker, *Int. J. Mass. Spectrom.*, 2005, **242**, 297–302.
74. J. Koch and D. Günther, *Anal. Bioanal. Chem.*, 2007, **387**, 149–153.
75. M. V. Zoriy and J. S. Becker, *Rapid Commun. Mass Spectrom.*, 2009, **23**, 23–30.
76. http://en.wikipedia.org/wiki/Tomography
77. http://en.wikipedia.org/wiki/Single_photon_emission_computed_tomography.
78. S. K. Carlson, K. L. Classic, E. M. Hadac, C. E. Bender, B. J. Kemp, V. J. Lowe, T. L. Hoskin and S. J. Russell, *Mol. Imaging Biol.*, 2006, **8**, 324–332.
79. http://en.wikipedia.org/wiki/Positron_emission_tomography.
80. V. T. DeVita, S. Hellman and S. A. Rosenberg, *Cancer: Principles and Practice of Oncology*, J.B. Lippincott, Philadephia, PA, 1993.
81. R. Nuñez, H. A. Macapinlacm, H. W. D. Yeung, T. Akhurst, S. D. Cai, I. Osman, M. Gonen, E. Riedel, H. I. Scher and S. M. Larson, *J. Nucl. Med.*, 2002, **43**, 46–55.
82. C. Petibois and M. C. Guidi, *Anal. Bioanal. Chem.*, 2008, **391**, 1599–1608.

CHAPTER 11

Nuclear-based Metallomics in Metallic Nanomaterials: Nanometallomics

YU-FENG LI,[a] LIMING WANG,[b] LILI ZHANG[b] AND CHUNYING CHEN[a,b,*]

[a] CAS Key Laboratory of Nuclear Analytical Techniques, Institute of High Energy Physics, Chinese Academy of Sciences, Beijing 100049, China; [b] CAS Key Laboratory for Biological Effects of Nanomaterials & Nanosafety, National Center for Nanoscience and Technology of China, Chinese Academy of Sciences, Beijing 100049, China

11.1 Introduction

Metallomics is the systematic study of metallomes and the interactions and functional connections of metal ions and their species with genes, proteins, metabolites, and other biomolecules within organisms and ecosystems.[1,2] The goal of metallomics is to provide a global and systematic understanding of the metal uptake, trafficking, role, and excretion in biological systems, potentially to be able to predict all of these *in silico* by using bioinformatics.

The term "metallome" was coined by Williams who referred to it as an element distribution, equilibrium concentrations of free metal ions, or as a free element content in a cellular compartment, cell, or organism.[3] The meaning of the term metallome was then proposed to be extended to the entirety of metal and metalloid species present in a cell or tissue type.[4] Recently, it was referred

---

*Corresponding author: Email: chenchy@nanoctr.cn; Tel: +86-10-82545560; Fax: +86-10-62656765.

Nuclear Analytical Techniques for Metallomics and Metalloproteomics
Edited by Chunying Chen, Zhifang Chai and Yuxi Gao
© Royal Society of Chemistry 2010
Published by the Royal Society of Chemistry, www.rsc.org

as "the entirety of metal and metalloid species present in a cell or tissue type, their identity, quantity and localization".[1]

However, the metals or metalloids (metallome) studied in metallomics are generally referred to metal compounds or naturally occurring macro-scale (bulk) ones. In this chapter, we will focus specifically on metals or metalloids and metallic materials in nano-scale, through which we proposed the term "nanometallomics".

Nanoscale materials (*i.e.* nanomaterials) are materials with morphological features smaller than a one tenth of a micrometer in at least one dimension.[5] The common nanomaterials include zero dimension (such as quantum dots), one dimension (layers, such as thin film or surface coating), two dimensions (nanowires and nanotubes), and three dimensions (such as fullerenes). An important aspect of nanotechnology is the vastly increased ratio of surface area to volume present in many nanoscale materials which makes possible new quantum mechanical effects; for example; the "quantum size effect" where the electronic properties of solids are altered with great reductions in particle size. Materials defined at nanoscale may show very different properties from those at macroscale enabling unique applications. For instance, opaque substances become transparent (copper); inert materials attain catalytic properties (platinum); stable materials turn combustible (aluminium); solids turn into liquids at room temperature (gold); insulators become conductors (silicon). Materials such as gold, which is chemically inert at normal scales, can serve as a potent chemical catalyst at nanoscale near-edge.

Owing to the fascinating properties of nanomaterials, they are widely used in many fields such as electronics, sunscreens and cosmetics, paints, water purification, food additives, medicine or clinical detection and diagnosis, decoration materials and automobile industry, catalysts, and the energy industry.[6] With the wide application of engineered nanomaterials, we have more chances to be exposed to nanomaterials directly or indirectly, including the environmental contamination from the synthesis, manufacturing, use or consumption, and release of nanomaterials and the contact with natural nanomaterials. Nanoparticles may enter organisms including humans, animals, plants, and microorganisms that exist in the environment, by active or passive pathways. Once they have entered into biological systems, knowledge of the spatial distribution and localization of minerals within tissues/cells is paramount in order to understand the processes underlying the interaction of biological systems and nanomaterials for the evaluation of exposure levels and toxicological effects. Therefore, concerns on the potential risk of nanomaterials when entering the human body directly during manufacturing processes or indirectly via the environment and food chains, are raised.[7-11] The OECD Chemicals Committee has established the Working Party on Manufactured Nanomaterials (WPMN) to address this issue and to study the practices of OECD member countries in regards to nanomaterial safety.

To address the potential risk of nanomaterials and to better understand the interactions between nanomaterials and biological systems, dedicated analytical approaches are required to detect, locate, identify, and quantify metal species, as well as to reveal their biological functions in the body.

In this chapter, we will introduce some nuclear-based analytical techniques to study the characterization and potential risk of nanomaterials, especially the metallic ones, and their interactions with biological systems. However, it should be noticed that the study of interactions of nanomaterials with biological systems is a very new and emerging field. Publications on risk evaluation and ADMET (absorption, distribution, metabolism, excretion, and toxicity) studies of these nanomaterials sometimes can not be replicated and contradictory results can also be found.

11.2 Nanometallomics and its Study Area

Like the definition of metallomics, nanometallomics is the study of a nanometallome, interactions and functional connections of metal ions and their species with genes, proteins, metabolites, and other biomolecules within organisms and ecosystems. Nanometallomics can be regarded as a branch of metallomics which focuses on the biological effects of nano-scaled metal(loid)s and metallic nanomaterials. Besides the commonly interdisciplinary research area of metallomics with an impact on geochemistry, clinical biology and pharmacology, plant and animal physiology and nutrition, nanometallomics also incorporates nanoscience and nanotechnology, which is the representative of leading new technologies in the 21st century.[12]

Nanometallomics will study (1) quantification of nano-scaled metal(loid)s and metallic nanomaterials of interest in biological systems; (2) distributions of studied nano-scaled metal(loid)s and metallic nanomaterials in biological systems; (3) the speciation of given nano-scaled metal(loid)s; (4) the structural analysis of the nanometallome; (5) the elucidation of reactions and related mechanisms of nanometallome; (6) metabolisms of nanometallome; and (7) the specific nano-scaled metal(loid)-assisted function biosciences in medicine, environment science, food science, agriculture, toxicology, and biochemistry.

Corresponding to the objects of elementomics study, different analytical techniques can be used to reach these goals. The application of advanced nuclear analytical techniques on metalloproteomics study has been reviewed by Gao et al.[13] In the following parts, nuclear analytical techniques, which can achieve some of the above goals of nanometallomics, especially analytical techniques for characterization, elemental quantification and distribution, and structural analysis of metallic nanomaterials, will be introduced.

11.3 Nuclear Analytical Techniques for Characterization of Metallic Nanomaterials

As we know, the unique function and effects of nanoparticles are associated with their surface and lattice structures by referring to coordination geometry. X-ray absorption spectroscopy (XAS) is a useful tool for exploring the electronic configuration, site symmetry and the coordination environment of the absorbing atom as introduced in Chapter 6. We can obtain useful information from XAS, including oxidation state, electronic configuration, site symmetry of the absorbing atom, and coordination geometry.

In the field of nanotechnology, inductively coupled plasma–mass spectrometry (ICP-MS) is commonly applied during the synthesis of nanoparticles.[14] Moreover, several recent approaches have expanded the field of ICP-MS application to the size characterization of nanoparticles. More information about ICP-MS can be found in Chapter 4.

Besides, neutron and X-ray scattering can also be used for the characterization of nanometallome. More information on neutron and X-ray scattering can be found in Chapter 7. The application of X-ray scattering in characterization of nanometallomes will be shown here later.

11.3.1 Size Characterization

Degueldre et al.[15] used ICP-MS for the investigation of the size distribution colloidal gold in single-particle mode. This method employs the determination of discrete ion clouds generated by the atomization of a single nanoparticle in the plasma, causing an intensity signal at the detector proportional to the size of that particle (Figure 11.1). Single-particle introduction has been derived from the nebulization of very dilute solutions, so that statistically only one particle reaches the plasma within a certain time range. The authors investigated particles with diameters of 80–250 nm and determined a detection limit

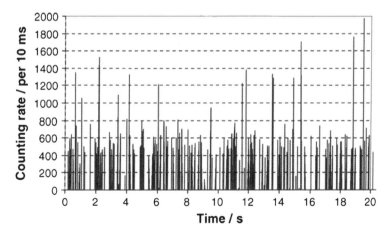

Figure 11.1 Analysis of gold colloids by ICP-MS in a single particle mode signal S_{197} for $^{197}Au^+$ recorded in a time scan for 150 nm gold colloid particle sample recorded during 20 s, counting per 10 ms. Conditions: experimental gold colloidal suspension, 10 ms detection time for $^{197}Au^+$ ion detection, $q_{col} = 2 \times 10^{-5}$ cm^3 s^{-1}, $q_{sol} = 5 \times 10^{-3}$ cm^3 s^{-1}, $q_{Ar} = 19$ cm^3 s^{-1}.[15] © 2005 Elsevier B.V.

for singly introduced particles of 25 nm. The results were confirmed with scanning electron microscopy and single-particle counting. However, because gold nanoparticles used in bioapplications are usually smaller, the authors stated that smaller nanoparticles might be investigated, if the overall ion transfer efficiency in the mass spectrometer was raised.

Helfrich et al.[16] developed a reliable method for the size characterization of Au nanoparticles based on the combination of two different separation techniques (liquid chromatography and gel electrophoresis), coupled on-line to ICP-MS. Separation by liquid chromatography shows good reproducibility with size-dependent behavior for retention (RSD < 1%). The results are in a good agreement with complementary methods like dynamic light scattering (DLS) and transmission electron microscopy (TEM).

Tiede et al.[17] developed and validated a hyphenated methodology which utilizes the extensive size separation range of hydrodynamic chromatography (here: 5–300 nm) combined with ICP-MS. The quality of the particle sizing data obtained from their study was enabled through the production of a range of gold nanoparticles (sterically stabilized to prevent aggregation in environmental matrices), which were validated for use as external size calibration standards as well as internal retention time markers, using TEM. The method has also been tested on solutions containing other commonly used nanoparticles (TiO_2, SiO_2, Al_2O_3 and Fe_2O_3). Overall, the data showed that, by using ICP-MS with collision cell technology, the methodology would be helpful in investigating the fate of a significant range of nanoparticle types.

11.3.2 Oxidation State Analysis

It is necessary to know the oxidation states of nanomaterials. For example, the oxidation state can affect numerous properties of nanomaterials including electronic properties of metal or metal oxide materials, optical and magnetic properties, electrochemical properties, chemical reaction activity, catalysis capability, and biocompatibility. Moreover, redox reactions are involved in the synthesis of nanoparticles, especially metal nanoparticles such as Au, Ag, Pt nanoparticles.[18] When chemists retrieve more information about the oxidation states of correlated elements and components in products, they may understand the mechanism and control the process much better. In addition, for environmental remediation, the oxidation state may change in the process of biomineralization especially for metal compounds being reduced into metal nanoparticles.[19]

X-ray absorption near-edge structure (XANES) can be used to analyze the oxidation states of many elements and to quantify the components of different valences for a given element in the samples. In order to perform the analysis of oxidation state, we can obtain the information by comparing the absorption edge position of the absorbing atom with that of standard samples. Moreover, we can use linear combination XANES (LC-XANES) to quantify and determine oxidation states of a given element in the samples.[19]

XANES can also be employed to study the composite and the oxidation state of the atoms of alloyed nanomaterials. Matsuo et al.[20] utilized Au L_{III} edge

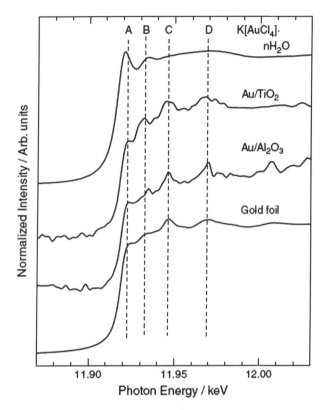

Figure 11.2 Normalized Au L$_3$ edge XANES spectra of gold particles deposited on TiO$_2$ and Al$_2$O$_3$. The reference includes gold foil and powder of potassium tetrachloroaurate(III) hydrate.[20] © 2003 John Wiley & Sons, Ltd.

XANES spectra to measure the oxidation states of Au atom absorbed on the surface of materials. As a result, they proved that Au(III) ions adsorbed both on TiO$_2$ and Al$_2$O$_3$ were reduced to Au(0) in the terms of Au/TiO$_2$ and Au/Al$_2$O$_3$ nanoparticles smaller than 1 nm. The Au(III) and Au(0) foil have different absorption edge positions as characteristics, while Au(0) foil and Au coated nanoparticles have an absorption edge position in common considered as the same covalence state (Figure 11.2).

In order to develop a novel contrast agent for *in vivo* MR images, Su *et al.*[21] synthesized Au$_3$Cu$_1$ (gold and copper) nanoshells as the first bimetallic MR contrast agents. The Cu^{3+} ions in Au$_3$Cu$_1$ nanoshells effectively pose a superparamagnetic effect because Cu atoms in Cu colloidal solutions are oxidized by the HAuCl$_4$ during the preparation. The authors took advantage of the XANES technique to study the oxidation states of Au and Cu elements in Au$_3$Cu$_1$ nanoshells (Figure 11.3). Compared with the XANES spectrum for Au and Cu foils separately, the edge energy for Au (L$_{III}$-edge) shifted slightly (-0.23 eV) and the edge energy for the Cu K edge changed strikingly ($+6.88$ eV). They explained that a small Au E_0 shift may associate with the

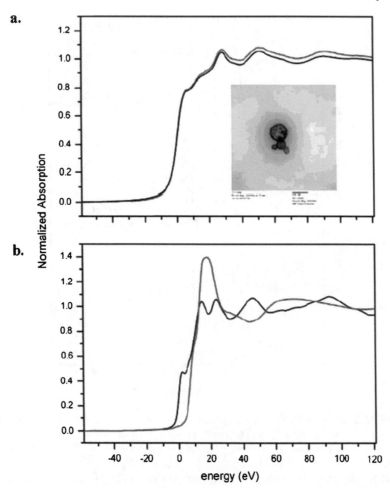

Figure 11.3 XANES spectra of Au_3Cu_1 nanoshells. All XANES were used to measure the (a) Au L_{III} edge of Au_3Cu_1 nanoshells (paler line) and Au foil (darker line, $E_0 = 11919$ eV) and the (b) Cu K edge of Au_3Cu_1 nanoshells (paler line), and Cu foil (darker line, $E_0 = 8979$ eV). The inset shows a transmission electron microscopy (TEM) image of Au_3Cu_1 hollow nanostructures.[21] © 2007 American Chemical Society.

inner chemical property of Au(0) atoms and the Cu K-edge shift correlates to a higher oxidation state (+3). For application *in vivo*, the surface of the Au_3Cu_1 nanoshell was modified with PEI and PAA to be more biocompatible.

11.3.3 Electronic Configuration and Coordination Geometry

Coordination geometry for nanomaterials depends on the coordinating atom species and numbers, bond distances and bond orientation, and oxidation state.

XAS does not rely on long-range order and is highly sensitive to the local electronic environment, coordination geometry, and bond distances of absorbed atoms in materials, which makes it a quite precise fingerprint for structure determination.[22]

XAS can serve as an efficient approach to study local geometries and electronic structures around the absorbing atoms in nanoparticles.[23,24] The size, shape, and crystal structure of nanomaterials associate with the coordination geometry of atoms, which can affect their chemical properties (chemical reactivity, catalytic activity, specific recognition) and physical properties (optical, electrical, and thermal). The size of metal oxide nanomaterials can affect 1s to nd pre-edge transition at K edges including quadrupole and dipole transitions to $(n+1)$ p states with nd sub-band. The transition is composed of several peaks and is determined by the symmetry and oxidation state of metal atoms. Surface modification and size can affect coordination geometry of atoms in nanomaterials, especially for the surface atoms.

For example, the photocatalytic capacity, the optical and electronic properties of TiO_2 nanoparticles can be changed by surface modification, as a result of the formation of a charge transfer complex between the surface modifier and nanocrystalline TiO_2. Rajh et al.[24] tried to modify ascorbic acid on the surface of three kinds of TiO_2 (a flat TiO_2 surface, 50 nm sized particles, 1.9 nm sized articles); only the smaller particles could be bound with ascorbic acid. By using FTIR spectroscopy these authors found that a five-membered ring formed at the surface Ti atoms after modification of smaller particles. The authors presumed that the coordination environment of surface Ti atoms might cause the size effect on reactivity of nanoparticles and they confirmed the assumption by XAFS techniques. According to XANES spectra, the coordination numbers of Ti sites associate with pre-edge peak intensities and peak positions. The intensities of a pre-edge peak decreased and edge peak shifted to higher energy when the coordination number increased from 4 to 6.[25] In this investigation, XANES pre-edge showed that Ti and its nearest O atoms formed a six coordinate (octahedral) in 50 nm particles and flat anatase TiO_2 surface. However, Ti-O atoms were pentacoordinate (square-pyramidal) in 1.9 nm particles, calibrating from the intensity of the middle pre-edge peak. When ascorbic acid absorbed on 1.9 nm particle surface, Ti atoms bound to O atoms and partly restored to an octahedral coordination because of the shift of peak position and reduced intensity. This proved that distortion of the Ti-O bond mainly occurred on 1.9 nm particle surfaces. Meanwhile, EXAFS results suggested that a Ti-O double bond (0.179 nm) formed in 1.9 nm nanoparticles, and Ti-O bonds (0.196 nm) were for 50 nm particles and the flat TiO_2 surface (Figure 11.4). Additionally, the small TiO_2 nanoparticles had a large curvature with 66% TiO_2 units on the surface and it led to distorted bond length. Therefore, the surface sites were more reactive and had the potential to bind with some functional groups of molecules. The active Ti atoms on the surface sites preferred to bind to oxygen atoms of ascorbic acid and it formed bidentate binding. Finally, the absorption of ascorbic acid changed the Ti-O double bond and partial surface Ti atoms would be relaxed to original anatase environment.

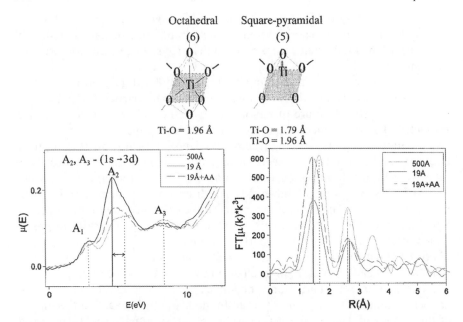

Figure 11.4 (Top) The coordination geometry of Ti atoms in flat surface and 2 nm TiO_2 particles. (Bottom left) XANES pre-edge spectra of Ti K edge ($E = 4.966$ keV) for 50 nm TiO_2 nanoparticles, 1.9 nm particles before and after modification of ascorbic acid. A1, A2, and A3 stands for pre-edge peaks. The middle peak intensity can reflect the coordination numbers of Ti atoms. (Bottom right) The bond distances of Ti and O atoms for three kinds of TiO_2 above from EXAFS results.[24] © 1999 American Chemical Society.

Meneses et al.[26] utilized in situ time-resolved XAFS techniques to detect the first stage of crystallization of nickel oxide (NiO) nanoparticles. They obtained a mixture of amorphous NiO in a Ni-salt and gelatin system below 100°C, while above 300°C heating caused metal–organic compounds to break and to expel bubbles on the surface, which initiates the nucleation process. Finally, the crystallized NiO nanoparticles started to grow. The authors used Ni K-edge XANES to analyze the biopolymer exposed to continuous heating. In the process from the Ni-salt/gelatin to the NiO amorphous phase and a NiO crystalline phase at 400°C, the initial and final post-edge peaks were different and determined the initiation of the nickel oxide crystalline phase. After 5 min at this temperature, the Ni-O interaction increased progressively according to XANES and EXAFS spectra. Based on the post-edge peak condition shift and coordination distance alteration, the first (Ni-O) shell and second (Ni-Ni) shells formed and then Ni atoms recombined O atoms to start to build up nanoparticles. Simulated XANES spectrum could estimate the neighboring O atom numbers around the Ni atom and show the dynamic process of nanoparticle growth.

Adora et al.[27] found H_2PtCl_6 (IV) was reduced to $PtCl_4^{2-}$ (II) and Pt atoms (0) during electrochemical preparation of metallic platinum nanocrystallites with XANES. EXAFS showed that Pt-Pt bonds, not Pt-O and Pt bonds formed in nanoparticles. Hwang et al.[28] used *in situ* XAFS to explore the formation mechanism of platinum–copper (Pt-Cu) bimetallic clusters in the water-in-oil microemulsion system. XANES showed that Pt clusters were derived of Pt(0) atoms reduced from Pt^{4+} to Pt^{2+} and Cu clusters were from the reduction of Cu^{2+} to Cu(0). EXAFS spectra indicated that Cu clusters formed on the surface of Pt with lower Cu-Pt coordination and without Pt-Cu coordination in bimetallic nanoclusters in microemulsions. The authors proposed a five-step model about the formation of Pt-Cu nanoclusters. It started from $PtCl_6^{2-}$ complex ions in the micelle center and Cu^{2+} ions located at the rear side. Subsequently, $PtCl_6^{2-}$ ions were reduced to $PtCl_4^{2-}$ ions through a micelle exchange process and then were reduced to Pt atoms with the significantly changed ordination number of NPt-Cl and NPt-Pt. The continuous addition of reductant led to the reduction of Cu^{2+} ions to Cu atoms and they formed a Pt-Cu cluster on the Pt surface. Finally, Pt atoms aggregated and formed Pt nanoclusters. Other interesting investigations about XAFS used in nanomaterial characterization can be found in a paper by Kocharova et al.[29]

Itoh et al.[30] elucidated the location of deuterium atoms in nano-crystalline FeTiDx by neutron and X-ray diffraction. A remarkable rearrangement of the metal atoms due to deuterium absorption was observed in the pair distribution functions, $g(r)$, obtained by X-ray diffraction (Figure 11.5). The result indicates that a disordered grain boundary is developed through deuterium absorption. The $g(r)$ function obtained by neutron diffraction indicates the occurrence of

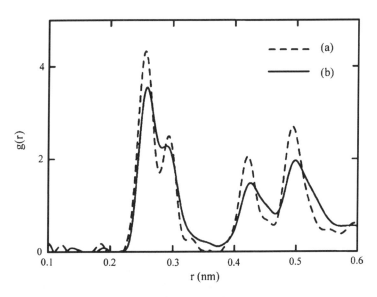

Figure 11.5 Pair distribution functions, $g(r)$, obtained by X-ray diffraction with a photon energy of 113.68 keV: (a) FeTi after 5h of milling; (b) nanocrystalline $FeTiD_{0.97}$.[30] © 2005 Elsevier B.V.

two types of deuterium sites located inside the grains and in the grain boundaries, respectively. It is found that about 50% deuterium atoms are situated in the grain boundaries and occupy tetrahedral sites consisting mainly of Ti atoms.

11.4 Quantification and Distribution of Metallic Nanomaterials

11.4.1 Neutron Activation Analysis for Quantification

Neutron activation analysis (NAA) is one of the most important nuclear techniques for nanometallome quantification, as it can simultaneously measure more than 30 elements in a sample. The detection limits of NAA range from 10^{-6} to 10^{-13} g g^{-1}.[31] In typically instrumental NAA, stable nuclides (^{A}Z, the target nucleus) in the sample undergo neutron capture reactions in a flux of (incident) neutrons. The radioactive nuclides (^{A+1}Z, the compound nucleus) produced in this activation process will, in most cases, decay through the emission of a beta particle (β^-) and gamma ray(s) with an inherent half-life. A high-resolution gamma-ray spectrometer is used to detect these "delayed" gamma rays from the artificially induced radioactivity in the sample for both qualitative and quantitative analysis.[31] One of the principal advantages of NAA is that it is nearly free of any matrix interference effects as the vast majority of samples are completely transparent to both the probe (the neutron) and the analytical signal (the gamma ray). Moreover, because NAA can most often be applied instrumentally (no need for sample digestion or dissolution), there is little, if any, opportunity for reagent or laboratory contamination.[32] Detailed information about NAA and its application in metallomics can be found in Chapter 2.

Ge et al.,[33] from our group, used NAA as a non-destructive standard method to quantify metallic impurities in carbon nanotubes (CNTs). Considerable amounts of iron, nickel, molybdenum, and chromium in the CNTs were found, which implies that these elements were dominantly used in the synthesis process. Small amounts of other impurity elements like manganese, cobalt, copper, zinc, arsenic, bromine, antimony, lanthanum, scandium, samarium, tungsten, and thorium are also found, which are presumed to have come from sources in chemical and physical manipulations used during the production process or in the precursors of the synthesis (Table 11.1). Although these commercial CNTs have been processed to reduce metal and amorphous carbon, even these as-purified samples still contain significant quantities of residual metals, which maybe contribute to the potential toxicological effects of CNTs.

NAA is widely used as a non-destructive method for the determination of elements of geological, environmental, biological samples. Unlike other traditional multielement analytical techniques, NAA normally does not require any sample pretreatment. Possessing many merits of high efficiency, precision, and accuracy, NAA is also one of the primary analytical methods to certify the

Table 11.1 The contents of impurity elements in the carbon nanotubes determined by NAA.

Nanotube	Sample							
	Cr	Mn	Fe	Co	Ni	Cu	Zn	As
SWCNT	1180 ± 42	72 ± 6	20000 ± 10	6.92 ± 0.1	959 ± 17	70 ± 6	28.1 ± 2.54	6.7 ± 0.1
MWCNT	2.14 ± 0.5	0.59 ± 0.03	64.4 ± 28.5	6.73 ± 0.08	3570 ± 26	0.53 ± 0.08	9.9 ± 1.1	1.37 ± 0.11
MWCNT-carboxyl	146 ± 9.7	48 ± 0.4	7750 ± 396	1.21 ± 0.03	954 ± 55	272 ± 10	584 ± 30	1.05 ± 0.13

Nanotube	Sample							
	Br	Mo	Sb	La	Sc	Sm	W	Th
SWCNT	6.8 ± 1.1	6030 ± 212	1.16 ± 0.05	1.1 ± 0.1	0.05 ± 0.009	0.087 ± 0.02	5.35 ± 0.7	0.107 ± 0.027
MWCNT	2.6 ± 0.06	97.8 ± 0.78	ND	4.74 ± 0.16	ND	ND	0.53 ± 0.27	0.087 ± 0.009
MWCNT-carboxyl	1.51 ± 0.27	1307 ± 3.9	ND	1.61 ± 0.04	0.012 ± 0.0006	0.18 ± 0.007	2.25 ± 0.08	ND

© 2008 American Chemical Society.
Note: the contents of Mn and Cu were determined by ICP-MS. ND, lower than detection limit.[33]
Data are expressed as mean ± standard deviation of three independent determinations ($\mu g\,g^{-1}$).

concentration of elements in standard reference materials (SRMs). Despite the distinct advantages of NAA, the low availability of the large nuclear reactor facility required for irradiation of samples is a major limitation, along with the running costs and radioactivity hazards. Therefore, due to the lack of reference materials for CNTs, in this study, we also made comparisons of the data of NAA to ICP-MS and give information of how much metals are present in the nanotubes and how to optimize the pretreatment procedures for digesting the CNTs before ICP-MS determination.[33]

Neutron activation of engineered nanoparticles is also a tool for tracing their environmental fate and uptake in biological systems. Recently, it was speculated that ultrafine particles may translocate from deposition sites in the lungs to systemic circulation. This could lead to accumulation and potentially adverse reactions in critical organs such as liver, heart, and even brain, consistent with the hypothesis that ultrafine insoluble particles may play a role in the onset of cardiovascular diseases, as growing evidence from epidemiological studies suggests. Aerosols of ultrafine iridium particles (15 and 80 nm count median diameter) radiolabeled with ^{192}Ir were generated with a spark generator (GFG 1000, Palas). The radioisotope ^{192}Ir is a beta and gamma emitter with a half-life of 74 days and gamma energies of 296, 308, 316, 468, and 588 keV (26, 29, 73, 47, and 5% efficiency, respectively). After inhalation exposure by young adult rats, excreta were collected quantitatively. At time points ranging from 6 h to 7 days, rats were sacrificed, and a complete balance of ^{192}Ir activity retained in the body and cleared by excretion was determined gamma counter. Both batches of ultrafine iridium particles proved to be insoluble (<1% in 7 days). During week 1 after inhalation, particles were predominantly cleared via airways into the gastrointestinal tract and feces. Additionally, minute particle translocation of <1% of the deposited particles into secondary organs such as liver, spleen, heart, and brain was measured after systemic uptake from the lungs. The translocated fraction of the 80-nm particles was about an order of magnitude less than that of 15-nm particles.[34]

Studies regarding the environmental impact of engineered nanoparticles (ENPs) are hampered by the lack of tools to localize and quantify ENPs in water, sediments, soils, and organisms. Neutron activation of mineral ENPs offers the possibility of labeling ENPs in a way that avoids surface modification and permits both localization and quantification within a matrix or an organism. Oughton et al.[35] demonstrates the suitability of neutron activation for Ag, Co/Co$_3$O$_4$, and CeO$_2$ nanoparticles. They studied the dietary uptake and excretion of a Co nanopowder (average particle size, 4 nm; surface area, 59 m^2 g^{-1}) in the earthworm *Eisenia fetida*. Cobalt ENPs were taken up to a high extent during 7 days of exposure (concentration ratios of 0.16–0.20 relative to the ENP concentration in horse manure) and were largely retained within the worms for a period of 8 weeks, with less than 20% of absorbed ENPs being excreted. Following dissection of the worms, ^{60}Co was detected in spermatogenic cells, cocoons, and blood using scintillation counting and autoradiography (Table 11.2).

Table 11.2 Radiochemical characteristics of some selected activation products (APs) formed during neutron activation of Ag, Ce, and Co and affecting the usefulness of the method[a]

Element	AP	Half-life	Decay process	Cross-section (σ)	β_{max} (MeV)	Branching ratio (%)	Daughter recoil (eV)	Specific activity of radioactive nuclide ($Bq\,g^{-1}$)	Irradiated activity[b] ($Bq\,AP\,mg^{-1}\,ENP$)	Isotope ratio[b] (AP/target nuclei)
Co	^{60}Co	5.27 years	beta, gamma	17	0.317	99.9	2.9	4.2×10^{13}	6.7×10^4	1.6×10^{-6}
	^{60m}Co[c]	10.46 min	beta, gamma	20	1.549	95.8	14.2	1.1×10^{19}	$1.6\times.10^5$	$3.7\times.10^{-6}$
									ND	ND
Ce	^{139}Ce	137.6 days	EC, gamma	1.1				2.5×10^{14}	1.5×10^4	4.1×10^{-10}
	^{141}Ce	35.2 days	beta, gamma	0.57	0.582	29.8	1.7	1.0×10^{15}	1.5×10^6	5.2×10^{-8}
Ag	^{110m}Ag	249.9 days	beta, gamma	4.9	0.437	70.2	5.3	$1.8\times.10^{14}$	1.1×10^5	6.0×10^{-9}
	^{110}Ag[c]	24.6 s	beta, gamma	89	0.530	98.6		1.5×10^{19}	ND	ND

Reproduced from Oughton et al.,[35] © 2008 SETAC.
[a]β_{max} = maximum energy of beta particles; EC = electron capture; ENP = engineered nanoparticles; ND = not determined.
[b]Experimentally obtained results, two separate irradiations for Co. Nanoparticles from the second activation were used for the uptake study.
[c]Properties of these APs have been included because they can contribute to chemical modifications of the irradiated nanoparticles.

The experimental procedures for the neutron activation of Fe_2O_3 nanoparticles are used in our laboratory. For the metabolic study, Fe_2O_3 nanoparticles were irradiated in a heavy-water nuclear reactor (China Institute of Atomic Energy, Beijing, China) at a neutron flux of $5\times10^{13}\,n\,cm^{-2}\,s^{-1}$ for 7 days. After 2 weeks of decay, the $^{59}Fe_2O_3$ nanoparticles were dispersed in a saline solution at a concentration of $100\,mg\,mL^{-1}$, and then sonicated for 10 min and vortexed for 5 min before intratracheal instillation. The radioactive labeled, 22-nm-sized $^{59}Fe_2O_3$ particle can be intratracheally instilled into rats to identify the particokinetic profile of inhaled Fe_2O_3 nanoparticles *in vivo*.[36]

11.4.2 ICP-MS for Quantification

ICP-MS can quantify multielements rapidly and simultaneously in one run, which is extremely sensitive, due to the efficient ionization from plasma coupled with the sensitive detection of the mass spectrometer. At its best, parts per trillion detection limits are achievable. ICP-MS can detect most elements in biological systems, but sulfur, phosphorus, and halogens are not efficiently ionized by the ICP owing to their high ionization energies.

Meng *et al.*[37] studied the copper content in mouse kidney after oral administration of nano-copper particles at the dose of $70\,mg\,kg^{-1}$ body weight. They found that massive copper enriches in renal tissue 24 h after the mice exposed to nano- and ion-copper, the concentration rise up to $13.0\pm4.1\,\mu g\,g^{-1}$ and $12.6\pm2.2\,\mu g\,g^{-1}$, respectively, which are equivalent to about three times of the control level ($4.0\pm0.8\,\mu g\,g^{-1}$). The copper content in renal tissue drops from $12.6\pm2.2\,\mu g\,g^{-1}$ to $6.5\pm1.3\,\mu g\,g^{-1}$ in the ion-copper group at 72 h; however, in the nano-copper group, it still maintains a high copper content in the kidney ($11.5\pm2.5\,\mu g\,g^{-1}$). These imply that the rate of elimination of nano-copper is very low in kidney (Figure 11.6), with only $1.5\,\mu g\,g^{-1}$ reduction within 48 h.

The percentages of Cu particles dissolved in artificial biological microenvironments are shown in Figure 11.7. The dissolution rates of nano-Cu particles in artificial cerebrospinal fluid are much higher than others and nearly reached 0.4%. The overall results show that the dissolution of copper nanoparticles is higher than their micro-sized counterparts. Considering very low dissolution rates of both micro and nano-Cu particles in nasal cavity, in fact only a small amount of Cu particles could be dissolved in nasal mucous.[38] The real-time pH detection in neutral biological fluids with micro- and nano-Cu particles, pH value remains stable. However, when nano-Cu particles were added to artificial acid juice (pH 1.75) to mimic activity in the stomach microenvironment, and temperature of the suspending solution was kept at 37°C, the pH value of the solution rapidly increase to be 5.5 within 50 min. In an *in vivo* study for oral administration, the stomachs of mice exposed to nano-copper swelled up and presented a cyan color.[38] However, mice exposed to micro-copper and ion-copper were almost the same as the control. The result suggests that nano-copper may remain in the stomach for longer; in other

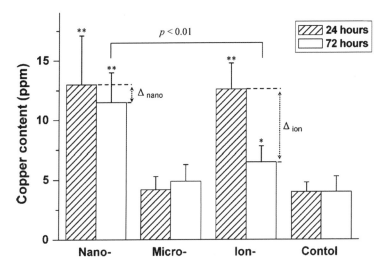

Figure 11.6 ICP-MS was used to measure the content of Cu in mouse renal tissues which are treated by nano-, micro-, and ion-copper at the dose of 70 mg kg^{-1} body weight. The asterisk indicates the copper content is significantly higher versus control (*$P<0.05$; **$P<0.01$). Both nano- and ion-copper lead to higher Cu content in renal tissue, by contrast, no difference is observed between micro-group versus control. In ion copper group, the Cu content in kidney sharply declines 72 h after the oral gavage (Δ_{ion}), however, the Cu content remains stable when the mice were treated by nano-copper even 72 h after the exposure (Δ_{nano}).[37] © 2007 Elsevier Ireland Ltd.

words, the duration of interaction with the acidic gastric juice may lead to a persistent generation of heavy metal ions *in vivo*.

11.4.3 Distribution of Metallic Nanomaterials in Biological Systems

11.4.3.1 ICP-MS Analysis

Based on nanometallome quantification in different tissues, ICP-MS can also be applied to study the body distribution of nanometallome.

To evaluate the toxicity of TiO$_2$ nanoparticles, the acute toxicity of nano-sized TiO$_2$ particles (25 and 80 nm) on adult mice was investigated compared with fine TiO$_2$ particles (155 nm).[39] Due to the low toxicity, a fixed large dose of 5 g kg^{-1} body weight of TiO$_2$ suspension was administrated by a single oral gavage according to the OECD procedure. The contents of titanium in each tissue of female mice 2 weeks post-exposure to different sized TiO$_2$ particles was determined by ICP-MS. Biodistribution experiments showed that TiO$_2$ was mainly retained in the liver, spleen, kidneys, and lung tissues (Figure 11.8),

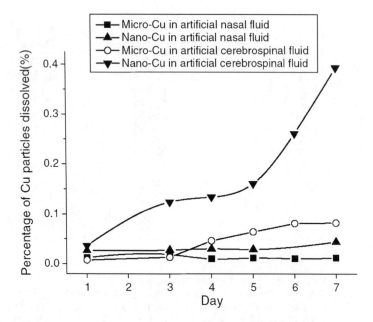

Figure 11.7 Dissolution rate of micro- and nano-Cu particles in artificial micro-environmental fluids.[38] © 2009 American Scientific Publishers.

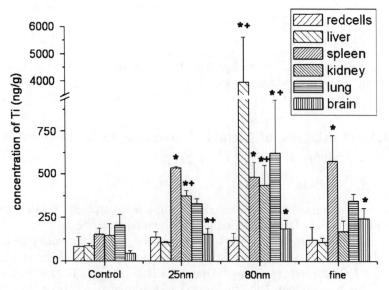

Figure 11.8 The contents of titanium in each tissue of female mice 2 weeks post-exposure to different sized TiO_2 particles by a single oral administration. *Represents significant difference from the control group (Dunnett's, $P<0.05$), and +represents significant difference from the fine group (Student's, $P<0.05$).[39] © 2006 Elsevier Ireland Ltd.

which indicated that TiO_2 particles could be transported to other tissues and organs after uptake by the gastrointestinal tract.

To study the time-dependent translocation and potential impairment on the central nervous system of TiO_2 nanoparticles, female mice were intranasally instilled with two types of well-characterized TiO_2 nanoparticles (*i.e.* 80 nm rutile and 155 nm anatase; purity >99%) every other day.[40] The Ti content in the olfactory bulb, cerebral cortex, hippocampus, and cerebellum was determined by ICP-MS. The concentrations of most brain parts were significantly higher than any of the controls for all post-exposure time periods. In the olfactory bulb, Ti contents increase gradually with time. In the hippocampus, Ti contents show a significant increase after exposure for 2 days, and remain for 10 and 20 days of exposure. However, after 30 days, it reaches the highest points in the hippocampus if compared to other parts of the brain. Ti contents in the cerebral cortex show an increase from the time point of 10 days, and then remain at a similar level until 30 days. In the cerebellum, Ti contents show a persistent increase with prolonged exposure time. In the case of Ti levels in sub-brain regions measured at 30 days of exposure, the ranking of TiO_2 deposition is hippocampus > olfactory bulb > cerebellum > cerebral cortex (Figure 11.9).

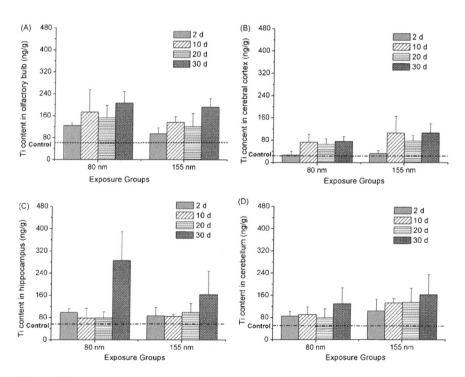

Figure 11.9 Ti content in the olfactory bulb (A), cerebral cortex (B), hippocampus (C) and cerebellum (D) of mice ($n=6$) intranasally instilled 80 and 155 nm TiO_2 particles on the time points of 2, 10, 20, and 30 days.[40] © 2008 Elsevier Ireland Ltd.

It was indicated that the instilled TiO_2 directly entered the brain through the olfactory bulb in the whole exposure period, and was deposited especially in the hippocampus region.

Liu et al.[38] evaluated the overall toxicity of nasally instilled nanoscale copper particles (23.5 nm) in mice. The results of ICP-MS analysis for copper content are shown in Figures 11.10 and 11.11. After instillation of copper nanoparticles for 1 week, in two low-dose groups, there are no statistically significant differences between experimental groups and the control group. However, in the H-Nano group, the contents of copper are significantly higher in liver (13.07 μg g^{-1}), kidneys (5.69 μg g^{-1}), olfactory bulb (4.37 μg g^{-1}) and blood (1.98 μg g^{-1}) than the control, which are in agreement with the damage in the liver, kidneys, and olfactory bulb clearly seen in pathological examinations. In other tissues and organs, no statistically significant differences are observed between the H-Nano group and the control group, which are consistent with the results of pathological observation and coefficients of specific tissues. The copper contents of the liver, kidneys, and the olfactory bulb increase significantly in the group of 40 mg kg^{-1} compared to the control group, which is in agreement with the histological changes. Therefore, the data indicate that inhaled copper particles at very high

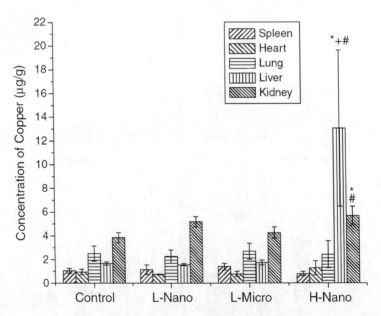

Figure 11.10 The contents of copper in each tissue of mice 1 week postexposure to different sized copper particles by nasal instilled administration of 1 or 40 mg kg^{-1} body weight. *Represents significant difference from the control group (Dunnett's $P<0.05$), #represents significant difference from the L-Micro group (Student's $P<0.05$), +represents significant difference from the L-Nano group (Student's $P<0.05$).[38] © 2009 American Scientific Publishers.

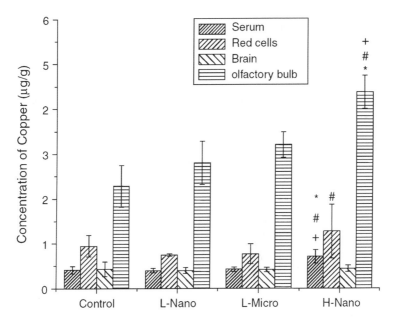

Figure 11.11 The contents of copper in each tissue of mice 1 week postexposure to different sized copper particles by nasal instilled administration at of 1 or 40 mg kg^{-1} body weight. *Represents significant difference from the control group (Dunnett's $P<0.05$), #represents significant difference from the L-Micro group (Student's $P<0.05$). +represents significant difference from the L-Nano group (Student's $P<0.05$).[38] © 2009 American Scientific Publishers.

dosage can translocate to other organs and tissues and further induce certain lesions.

Wang et al.[41] studied the acute oral toxicological impact of nano- and submicro-scaled zinc oxide powder on healthy adult mice. ZnO powder of 20 nm and 120 nm at doses of 1, 2, 3, 4, and 5 g kg^{-1} body weight were administrated. The accumulation of Zn in the organic tissues and serum of the mice in the 5 g kg^{-1} body weight dose groups are shown in Figure 11.12. After exposure to 20-nm ZnO at 5 g kg^{-1}, compared with the control, significant increases of Zn contents were found in the kidney, pancreas, and bone ($P<0.05$), and slight increases were found in the liver and heart of the ZnO-treated mice. In contrast to the 120-nm ZnO-treated mice, the Zn content in the liver, kidney, and pancreas of the 20-nm ZnO group mice was a little higher, suggesting that more Zn may be excreted from the mice that received the 20-nm ZnO than from those that received 120-nm ZnO. Among the organs observed, the highest Zn content was found in the bone. The 120-nm ZnO mice retained significantly higher Zn in the bone than did the 20-nm ZnO mice. The authors concluded that the liver, spleen, heart, pancreas, and bone are the target organs when mice are orally exposed to 20- and 120-nm ZnO.

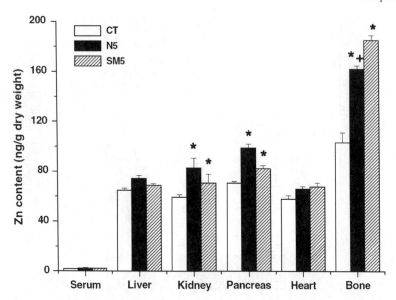

Figure 11.12 Content of Zn in serum and organic tissues of mice exposed to zinc oxide powder at the dose of 5 g kg^{-1} body weight at 14 days post-oral administration. CT: Control; N5: 5 g kg^{-1} body weight 20 nm ZnO group; SM5: 5 g kg^{-1} body weight 120 nm ZnO group.[41] © 2008 Springer Science + Business Media B. V.

11.4.3.2 X-ray Fluorescence Analysis

When a primary X-ray with sufficient energy strikes a sample, the inner electrons may be ejected. The excited atoms can emit characteristic radiation during the subsequent process of de-excitation. The emitted X-rays carry information about the elemental composition of the specimen in the irradiated region which is called X-ray fluorescence analysis (XRF). XRF is also a multi-elemental technique and more detailed information about XRF can be found in Chapter 3.

To investigate the bioaccumulation of engineered copper nanoparticles in *Caenorhabditis elegans*, which was used as a "model" organism, a scanning technique of microbeam (3×5 μm^2) synchrotron radiation X-ray fluorescence (μ-SRXRF) was used.[42] The mapping results of the whole organism indicate that the exposure to copper nanoparticles can result in an obvious elevation of Cu and K levels, and a change of bio-distribution of Cu in nematodes (Figure 11.13). Accumulation of Cu occurs in the head and at a location one-third of the way up the body from the tail compared to the unexposed control. In contrast, a higher amount of Cu was detected in other portions of the worm's body, especially in its excretory cells and intestine when exposed to Cu^{2+}. The results compared well with total Cu levels in nematodes measured by ICP-MS. The non-destructive and multielemental μ-SRXRF

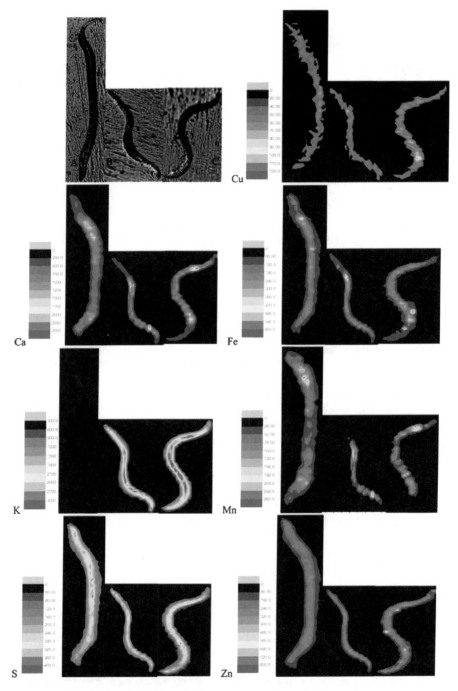

Figure 11.13 Elemental distribution in the nematodes (*Caenorhabditis elegans*), which were fed by *E. coli* OP50 + PBS (A), *E. coli* OP50 + Cu nanoparticles (B), and *E. coli* OP50 + Cu^{2+}.(C), respectively.[42] © 2008 The Royal Society of Chemistry.

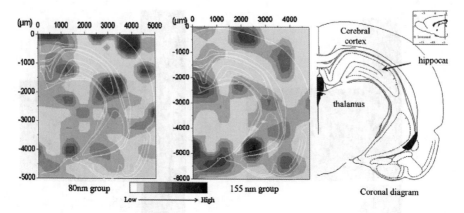

Figure 11.14 SRXRF mapping of Ti-element distribution in the brain sections 30 days after intranasal instillation of the different-sized TiO_2 particles. In the control mice, the Ti contents are lower than the detection limit of SRXRF and the mapping is not available.[43] © 2008 Elsevier Ireland Ltd.

provides an important tool for mapping the elemental distribution in the whole body of a single tiny nematode at lower levels.

Nanoscale titanium dioxide (TiO_2) is massively produced and widely used in the living environment, which may lead to a potential risk to human health. The central nervous system (CNS) is the potential susceptible target of inhaled nanoparticles, but related studies have been limited so far. Wang et al.[43] reported the accumulation and toxicity results in vivo of two crystalline phases of TiO_2 nanoparticles (80 nm rutile and 155 nm anatase; purity >99%). The female mice were intranasally instilled with 500 μg of TiO_2 nanoparticles suspension every other day for 30 days. Synchrotron radiation X-ray fluorescence analysis (SRXRF) was used to determine titanium distribution in the murine brain.

Figure 11.14 shows the distribution of Ti in mouse brain sections at 30 days after intranasal instillation of the different-sized TiO_2 particles. It was found that titanium mainly accumulated in the cerebral cortex, thalamus, and hippocampus, especially in the CA1 and CA3 regions of hippocampus after exposure to the different-sized TiO_2 particles. The significantly increased Ti content in the hippocampus results in the obviously irregular arrangement and loss of neurons in that region. Intranasal instillation of either rutile or anatase TiO_2 nanoparticles produced sustained accumulation in brain tissues, especially depositing in the hippocampus during the whole exposure process, which indicates that the TiO_2 nanoparticles can enter the brain via the olfactory bulb.[40,43]

SRXRF was also used to monitor the distribution of titanium in the olfactory bulb of mice as shown in Figure 11.15 and its effects on the distribution of Fe, Cu, and Zn in the olfactory bulb of mice in different experimental groups.

TiO_2 particles were taken up by the olfactory bulb via the primary olfactory neurons and accumulated in the olfactory nerve layer (ON), olfactory ventricle

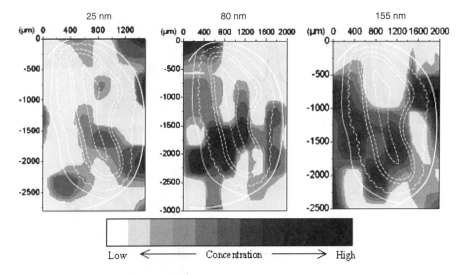

Figure 11.15 Distribution of titanium in olfactory bulb of mice in the different experimental groups analyzed by SRXRF.[44] © 2007 Akadémiai Kiadó, Budapest.

(OV), and granular cell layer of the olfactory bulb (GrO). The distribution areas of fine TiO_2 were wider than those of nano-sized TiO_2 in the olfactory bulb of the three experimental groups, which indicated that fine TiO_2 was more prone to enter the olfactory bulb through olfactory tract than nano-sized TiO_2. This is likely because nano-sized TiO_2 particles might be adsorbed in the nasal cavity and/or mucosa and only a small amount of particles were translocated into the olfactory bulb and brain through the respiratory tract and olfactory nervous system.[44]

In the 25 and 80 nm groups, the copper and zinc distributions differ significantly from that of the control group.[44] The different distribution of iron is also found in the four groups. The changes of their distributions would influence the normal metabolism in the organism. For example, the deficiency of copper could induce high cholesterol and uric-acid symptoms.

The distribution of other nanoparticles, such as nano-Fe_2O_3, was also studied using SRXRF.[45] The micro-distribution map of iron in the olfactory bulb and brain stem shows an obvious increase of Fe contents in the olfactory nerve and the trigeminus of brain stem, suggesting that Fe_2O_3 particles were possibly transported via uptake by sensory nerve endings of the olfactory nerve and trigeminus.

The μ-XRF technique has also been applied to study cell-carbon nanotube interactions.[46] Bussy and co-workers studied the distribution of unpurified and purified single-walled (SW) and multiwalled (MW) carbon nanotubes (CNT) in macrophages by monitoring the catalyst metal particle employed in most synthesis technique and finally remaining attached to or contained in nanotubes (Figures 11.16 and 11.17). The μ-XRF technique is used to study CNT localization at the single-cell level with simultaneous analysis of the biological

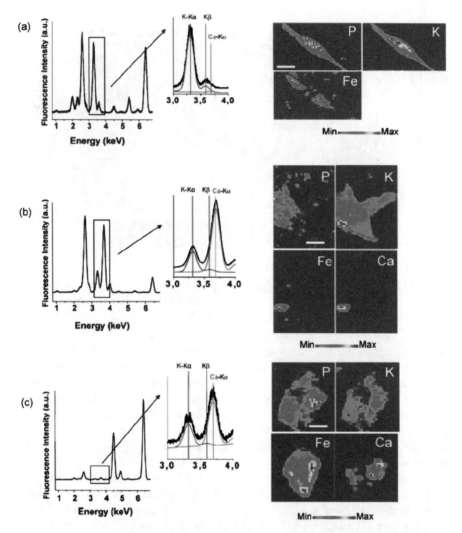

Figure 11.16 X-ray microfluorescence spectra integrated over the whole scanned area of three murine macrophages exposed for 24 h to MWCNT suspensions at concentrations of 10 (panel a) and 100 μg mL^{-1} (panels b and c). Zoomed areas around the positions of the Kα and Kβ fluorescence peaks of potassium and of the Kα peak of calcium are shown, together with fits of the potassium and calcium contributions, in blue and in purple lines, respectively. Elemental maps of phosphorus, potassium, iron, and calcium, if in detectable amounts, are drawn. The pixel size is 1 μm×1 μm (scale bar, 10 μm).[46] © 2008 American Chemical Society.

response through observation of changes in cell-elemental composition (calcium in the present study). Analysis of cell-CNT interactions by XRF is an original approach that represents a significant advance in the field of toxicology of carbon nanotubes and can provide new data to understand biological effects

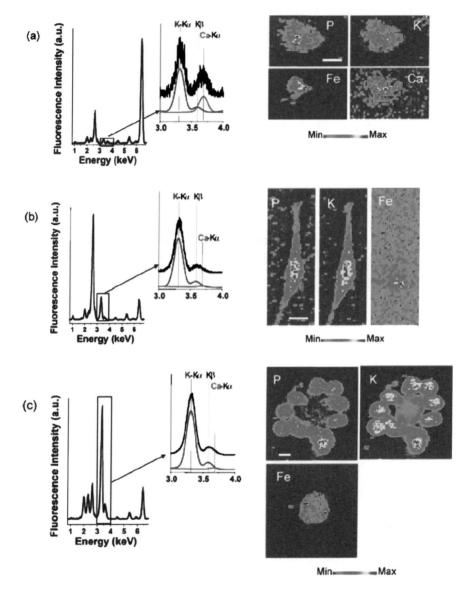

Figure 11.17 X-ray microfluorescence spectra integrated over the whole scanned area of murine macrophages exposed for 24 h to SWCNT suspensions. (a) Suspension of NP-SWCNT at $100\,\mu g\,mL^{-1}$. (b) Suspension of P-SWCNT at $100\,\mu g\,mL^{-1}$. (c) Suspension of NP SWCNT at $10\,\mu g\,mL^{-1}$. Zoomed areas around the positions of the Kα and Kβ fluorescence peaks of potassium and of the KR peak of calcium are shown, together with fits of the potassium and calcium contributions. Elemental maps of phosphorus, potassium, iron, and calcium, if in detectable amounts, are drawn for all cells. The pixel size is $1\,\mu m \times 1\,\mu m$ (scale bar, $10\,\mu m$).[46] © 2008 American Chemical Society.

of these nanomaterials as they are produced, without any artificial labeling which will affect the CNT-cell interactions.

Considering exposed cells, iron maps reveal one or several iron-rich zone(s) inside or close to the cell contours (Figures 11.16 and 11.17), whereas no such zone was observed in control cells. Iron-rich zones give the localization of the catalyst particles inside the MWCNT or attached to the SWCNT, therefore allowing CNT localization. Iron-based catalyst particles could even be detected in the case of cell exposure to P-SWCNT (Figure 11.17b), for which the amount of iron was minimized, thus highlighting the high sensitivity of the RF method. Observation of SWCNT-cell interaction is particularly interesting since we could not clearly identify these carbon materials in cells with optical microscopy or conventional TEM, as previously discussed by other investigators. In addition, in some cells exposed to SWCNT, the iron map shows colocalization of the highest Fe signal with the highest P signal (Figure 11.17b), suggesting an interaction of SWCNT with the nuclear or perinuclear region, as evidenced by Porter and co-workers. This finding could have important implications (for example, in terms of genotoxicity) and would need a dedicated investigation. XRF allows detection of a dose-response effect of the cellular Fe signal in CNT-exposed cells; a higher amount of iron with respect to phosphorus is found in cells exposed to $100\,\mathrm{mL}^{-1}$ compared to those exposed to $10\,\mathrm{mL}^{-1}$ MWCNT and for cells exposed to NP-SWCNT as compared to P-SWCNT.

11.4.3.3 Isotopic Tracing Techniques

Isotopic tracing can be used for *in vivo* or *in vitro* studies of the absorption, distribution, transportation, storage, retention, metabolism, excretion, and toxicity of nanomaterials. The isotopic tracer technique has many advantages, such as high sensitivity, good accuracy, and time saving. One or more of the atoms of the molecule of interest is substituted for an atom of the same chemical element, but of a different isotope. Moreover, isotopic tracers can easily distinguish endogenous and exogenous sources of trace elements of interest in environmental or biological samples.

Pulmonary retention and extrapulmonary redistribution of inhaled $^{59}Fe_2O_3$ nanoparticles have been considered to be important contributing factors of cardiorespiratory diseases. Radioactive, 22 nm, ferric oxide ($^{59}Fe_2O_3$) nanoparticles were intratracheally instilled into male Sprague-Dawley rats at a dose of $4\,\mathrm{mg\,rat^{-1}}$. Extrapulmonary distribution of $^{59}Fe_2O_3$ in organs and its metabolism in lung, blood, urine, and feces were measured for 50 days of exposure. Zhu and co-workers[36] found the intratracheally instilled nano-$^{59}Fe_2O_3$ could pass through the alveolar-capillary barrier into systemic circulation within 10 min and consisted of a one-compartment kinetic model. Nano-$^{59}Fe_2O_3$ in the lung was distributed to organs rich in mononuclear phagocytes, including liver, spleen, kidney, and testicle. The plasma elimination half-life of nano-$^{59}Fe_2O_3$ was 22.8 days and the lung clearance rate was $3.06\,\mathrm{\mu g\,day^{-1}}$, indicating that systemic accumulation and lung retention had occurred.

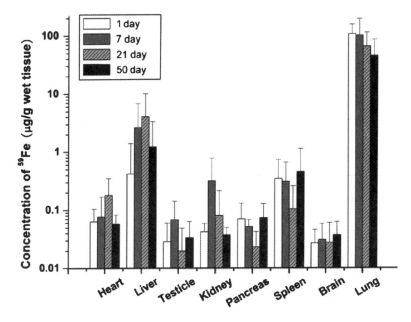

Figure 11.18 The biodistribution of ^{59}Fe in organs at days 1, 7, 21, 50 postinstillation. The values represent means ± SD ($n=8$).[36] © 2009 Society of Toxicology.

The extrapulmonary transported ^{59}Fe$_2$O$_3$ was redistributed in many organs (Figure 11.18), which indicates that ^{59}Fe$_2$O$_3$ can easily pass through a number of tissue compartments and accumulate in the extrapulmonary organs. The highest extrapulmonary ^{59}Fe levels were found in the liver, followed in decreasing order: spleen, heart, kidney, pancreas, testicle, and brain. The ^{59}Fe in the liver and heart showed time-response of accumulation from postinstilled day 1 to day 21, and then decreased at day 50. But in spleen, the ^{59}Fe showed a persistent high level up to postinstilled day 50.

Table 11.3 shows the percentage of accumulated ^{59}Fe in the organs at postinstilled day 50. The target extrapulmonary organs of ^{59}Fe were shown as organs rich in mononuclear phagocytes, such as the liver, spleen, kidney, and testicle. The organ weights were consistent at all the time points (days 1, 7, 21, and 50) of sacrificed rats except the significant decrease of lung tissue at postinstilled day 7.

Singh et al.[47] functionalized water-soluble, single-walled CNT (SWCNT) with the chelating molecule diethylenetriaminepentaacetic (DTPA) labeled with indium (^{111}In) to study the distribution of these functionalized SWCNT (f-SWCNT) after intravenous (i.v.) administration (Figure 11.19). Radioactivity tracing using gamma scintigraphy indicated that f-SWCNT are not retained in any of the reticuloendothelial system organs (liver or spleen) and are rapidly cleared from systemic blood circulation through the renal excretion route. The observed rapid blood clearance and half-life (3 h) of f-SWCNT has

Table 11.3 Biodistribution of ^{59}Fe in organs at post-instillation day 50.

Organ	^{59}Fe content (µg)	Relative content (%) of ^{59}Fe
Heart	0.067 ± 0.029	0.055 ± 0.024
Liver	15.19 ± 26.59	12.45 ± 21.81
Spleen	0.385 ± 0.590	0.320 ± 0.484
Lung	106.0 ± 95.3	86.94 ± 78.15
Kidney	0.116 ± 0.040	0.090 ± 0.033
Pancreas	0.026 ± 0.019	0.016 ± 0.016
Testicle	0.086 ± 0.076	0.070 ± 0.062
Brain	0.061 ± 0.019	0.050 ± 0.016
Total	122.0 ± 122.7	100

© 2009 Society of Toxicology.
Note. ^{59}Fe in all estimated organs were summed as 100%. Data present as relative percentage of ^{59}Fe in organs.[36]

major implications for all potential clinical uses of CNT. Moreover, urine excretion studies using both *f*-SWCNT and functionalized multiwalled CNT followed by electron microscopy analysis of urine samples revealed that both types of nanotubes were excreted as intact nanotubes.

^{99m}Tc is another radioactive tracer that has been widely used. Water-soluble functionalized multiwall carbon nanotubes (MWCNTs) were labeled with ^{99m}Tc for the first time, and the distribution and excretion of ^{99m}Tc-MWCNT-glucosamine in mice were studied.[48] It shows that MWCNTs moved easily among the compartments and tissues of the body, behaving like active molecules although their apparent mean molecular weight is tremendously large. After intraperitoneal injection, there was no severe acute toxicity responses observed in the studies because the functionalization of carbon nanotube with glucosamine improved the biocompatibility of carbon nanotube. ^{99m}Tc-MWCNT-glucosamine was quickly delivered around the whole mouse body and was excreted mainly via urine and feces (Table 11.4). It was determined that the blood circulation half-life of ^{99m}Tc-MWCNT-glucosamine is about 5.5 h. The study provides basic information for the application of water-soluble MWCNTs in biomedical and pharmaceutical sciences, such as carrier systems for therapeutic agents *in vivo*. More studies using ^{99m}Tc as the radioactive tracer can be found in the papers by Xu et al.[49] and Li et al.[50] who studied the biodistribution of ^{99m}Tc-C$_{60}$(OH)$_x$.

Wang et al.[51] labeled water-soluble hydroxylated carbon single-wall nanotubes with ^{125}I atoms, and then this radioactive tracer was used to study the distribution of hydroxylated carbon single-wall nanotubes in mice. They found that the hydroxylated carbon single-wall nanotubes moved easily among the compartments and tissues of the body, behaving as small active molecules although their apparent mean molecular weight is tremendously large. This study, for the first time, affords a quantitative analysis of carbon nanotubes accumulated in animal tissues.

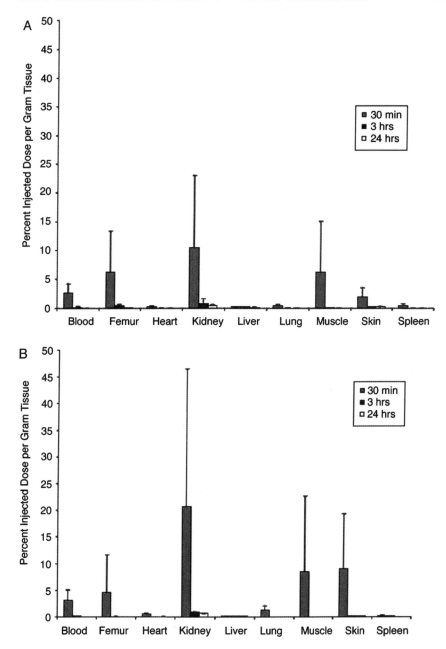

Figure 11.19 Biodistribution per collected gram of tissue of [^{111}In]DTPA SWCNT 3 (A) and [^{111}In]DTPA SWCNT 5 (B) after i.v. administration.[51] © 2006 National Academy of Sciences of the USA.

Table 11.4 Biodistribution of ^{99m}Tc-MWCNT-glucosamine in mice (% ID g^{-1}).[48]

Tissue	Time, T (h)				
	1	3	6	10	24
Blood	2.38 ± 0.242	1.82 ± 0.171	1.12 ± 0.102	0.42 ± 0.031	0.13 ± 0.009
Heart	0.86 ± 0.073	0.65 ± 0.072	0.51 ± 0.037	0.30 ± 0.021	0.29 ± 0.013
Lung	2.16 ± 0.189	1.57 ± 0.144	1.19 ± 0.097	0.61 ± 0.047	0.39 ± 0.026
Liver	2.13 ± 0.193	2.28 ± 0.232	1.84 ± 0.065	1.81 ± 0.074	0.42 ± 0.031
Spleen	1.16 ± 0.103	0.98 ± 0.093	0.77 ± 0.068	0.42 ± 0.033	0.23 ± 0.011
Kidney	1.44 ± 0.112	1.32 ± 0.121	1.05 ± 0.093	0.62 ± 0.053	0.29 ± 0.023
Stomach	9.80 ± 0.092	15.93 ± 1.680	12.91 ± 1.370	6.55 ± 0.781	2.09 ± 0.201
Intestines	1.19 ± 0.135	1.165 ± 0.027	0.98 ± 0.083	0.32 ± 0.018	0.27 ± 0.016
Coat	2.41 ± 0.251	3.49 ± 0.332	2.36 ± 0.233	2.12 ± 0.193	1.49 ± 0.132
Muscle	0.84 ± 0.071	0.48 ± 0.050	0.41 ± 0.026	0.28 ± 0.012	0.14 ± 0.013
Enterogastric area	27.43 ± 2.961	37.21 ± 3.621	53.67 ± 6.332	4.80 ± 0.436	0.81 ± 0.076
Faces	–	–	11.81 ± 1.316	30.70 ± 4.120	18.31 ± 2.012
Urine	–	–	5.43 ± 0.497	4.61 ± 0.386	4.90 ± 0.456

© 2007 Elsevier Inc.

11.5 Structural Analysis for the Bio-nano Interaction

For the determination of the metallome structure, different nuclear-base techniques can be applied, like X-ray crystallography and solution structure determination by multi-dimensional nuclear magnetic resonance (NMR). Other techniques capable of offering the data mainly include Mössbauer spectroscopy, X-ray absorption spectrometry (XAS), and electron paramagnetic resonance (EPR), and neutron scattering. However, in the structural analysis of nanometallome, XAS is the most often used techniques through literature.

One of the main sources of toxicity of metallic nanoparticles appears to be the electronic and/or ionic transfers occurring during the oxido-reduction, dissolution, and catalytic reactions either within the nanoparticles lattice or on release to culture medium. The effects of nanoscale materials on biological systems are vital for the industrial production and their safe application in daily life and biomedicine. XAS is a powerful tool to investigate bio-nano interactions and can provide structural details of biomolecules at the interface of bio-nano systems.

Iron oxide nanoparticles is a candidate material in nanomedicine for its potential application in magnetic resonance imaging, diagnosis, and treatment. Some investigations were carried out to assess the environmental and healthy effects of iron oxide nanoparticles (Fe_3O_4, γFe_2O_3, Fe_2O_3) and revealed that largely internalized nanoparticles were toxic to cells. Surface-modified magnetic nanoparticles had a decreased or discarded cytotoxicity and it suggested that the strong oxidizing surface, size, and aggregation state may affect the interaction between materials and cells.[52,53] Meso-2,3-dimercaptosuccinic acid (DMSA) coated nanomaghemites (NmDMSA) are largely taken into cells and are considered as a biocompatible material. Auffan et al.[54] found that the

DMSA coating could directly protect cells from the toxic surface of nanomaghemites. The authors used XAS to identify the physicochemical states of NmDMSA at the atomic scale to better understand the interaction of cells and nanomaterials. It was found that DMSA could form Fe-S bonds with nano-γFe_2O_3 and the interatomic distance was 0.221 nm. The strong chemical linkage was formed between surface Fe atoms in nano-γFe_2O_3 and sulfur atoms in DMSA. As a result, the Fe-S bonds changed the surface structure of nano-γFe_2O_3 and made it more dispersible in culture medium through the COO- group. Then they studied surface state of NmDMSA in medium and in endocytosis vesicles, cell components after 24 h cell culture. The EXAFS results revealed that the Fe-S coordination still existed with the same bond distance, 0.22 nm. DMSA also firmly absorbed on nanoparticle surface even after endocytosis. This study proved that NmDMSA was stable in medium and cells and could prevent a direct contact between the nanoparticle surface and cell components. Therefore, a DMSA coating on nano- γFe_2O_3 could act as a durable barrier to avoid the negative effects of the nanoparticle surface.

Auffan et al.[55] investigated the cytotoxicity of three kinds of Fe nanoparticles on *Escherichia coli*. These nanoparticles were magnetite nanoparticles (nMagnetite, $Fe_3^{II/III}O_4$), maghemite nanoparticles (nMaghemite, γFe_2O_3) and zerovalent iron nanoparticles (nZVI, Fe^0). In order to identify the correlation of nanomaterial toxicity with reactive oxygen species (ROS), the Gram-negative bacterium *E. coli* and a superoxide dismutase (SOD)-deficient mutant strain (*sodA sodB*) were used. Through the experiment, the authors found that nanomaterial toxicity for *E. coli.* and *sodA sodB* was in the order: nMaghemite < nMagnetite < nZVI. Under the same conditions, *sodA sodB* was more sensitive to nMagnetite and nZVI than *E. coli*. The reason was that a large amount of ROS might be produced in Fe^{II} and Fe^0 systems and *sodA sodB* failed to resist the stress with less viability. ROS can be produced by many approaches for the mineral surface, such as dissolution and release of metallic nanomaterial, surface catalysis of oxygen, structure defects, electron transfer (redox) reactions.[56] Fe^{II} and oxygen reaction on mineral surface is more active than $Fe^?$ ions dissolved in solution.[57] According to the Fe K-edge XANES spectrum, three compounds, including γFe_2O_3, Fe_3O_4, and $\gamma FeOOH$ had different pre-edge intensity, position of pre-edge and main edge, ramped absorption position. The pre-edge information showed that Fe^{II} in nMagnetite could be oxidated to Fe^{III} in water and in contact with *E. coli*. All the Fe^0 atoms in nZVI were highly active to oxygen atoms under the same condition and transformed into $\gamma FeOOH$ and Fe_3O_4 supported by XRD and XANES results. The process of Fe^0 and Fe^{II} oxidation resulted in ionic or electronic transfers on nanoparticle surface that might intervene the metabolism of bacterial when cell membrane and component contacted with them. The ionic or electronic transference on nZVI surface could also catalyze a series of reaction and produced Fe^{2+}, Fe^{3+}, H_2O_2, OH·and OH^-. While reduced-iron oxides could generate ROS by Fe oxidation on Fe_3O_4 surface and released Fe^{2+}. However, Fe^{III} in nMaghemite was physicochemically stable and failed to produce ROS. Consequently, the authors concluded that the toxicity of Fe nanoparticles containing ferrous and zerovalent iron resulted from the generation of reactive oxygen species or interference on electron/ion transport chains. The work suggests that surface properties of metal nanomaterials may affect their toxicity heavily by ROS generation. Therefore, suitable surface modification needs consideration in order to be more biocompatible and healthy.

Cerium toxicity has been previously studied in its Ce^{III} form,[58] and, more recently, a mechanism of cerium uptake by fibroblasts was proposed.[59] Cerium oxide is indeed used in its nanometric form as an exhaust gas catalyst, and, when the nanoparticles are diluted, they form a dispersion of individual nanoparticles. To study the impact of nanoparticles through a water path and describes the interaction between a water dispersion of nanoparticles and a model bacteria, Thill et al.[60] selected cerium oxide as the model nanoparticles, and E. coli was chosen as a widely used model organism. They found that (1) a large amount of CeO_2 NPs can be adsorbed on the E. coli outer membrane; (2) the speciation of the NPs is significantly modified after adsorption; (3) the adsorption of the NPs and their reduction are associated with a significant bacterial cytotoxicity; and (4) the toxicity of the NPs is prevented when they are put into contact with the bacteria in the presence of the growth medium.

The reduction of the nano-CeO_2 is related to the interaction with organic molecules (mainly proteins) present in the DMEM.[61] Such a reaction can occur due to the highly elevated standard potential of the chemical couple Ce^{4+}/Ce^{3+} (1.72 V) as compared to the standard potential of (in)organic molecules present in DMEM (from 0.38 V to 0.34 V). Differences were observed in the XANES spectra of nano-CeO_2 powders respectively suspended in water and in the abiotic DMEM for 24 h. A shoulder appeared on the XANES spectra of nano-CeO_2 suspended in the DMEM at the energy of the Ce^{3+} (Figure 11.20). By linear combination between the spectra of nano-CeO_2 powders suspended in water and Ce^{3+} oxalate reference compound, a contribution of 8.2% of Ce^{3+} was required to obtain a suitable fit with the XANES spectra of nano-CeO_2 suspended in the abiotic DMEM. Taking into account the size of the particles (7 nm) and the thickness of the surface layer (0.5 nm), 35–40% of the Ce atoms are localized at the surface of the nano-CeO_2. It was concluded 8.2% of the Ce^{4+} of nano-CeO_2 was therefore reduced to Ce^{3+} in the abiotic DMEM.

Environmental pollution threatens ecological systems that maintain comfortable living conditions for humans, and many studies are being undertaken to determine how to decrease the degree of pollution and recover the balance by natural processes such as biomineralization and phytoremediation.[62,63] In the past decades, many researchers found that the oxidation states of some elements in contaminants may affect toxicity, and some metallic ions and other toxic elements become less toxic when reduced to metals,[64] which suggests that biomineralization and phytoremediation are effective and natural approaches for environmental recovery.

XAS techniques have been introduced to study bioremediation and to identify the properties of nanoparticles from mineralization. XANES was applied to identify the change of key element speciation and to support the evidence that metallic ions were reduced to zero valence, provided by TEM,

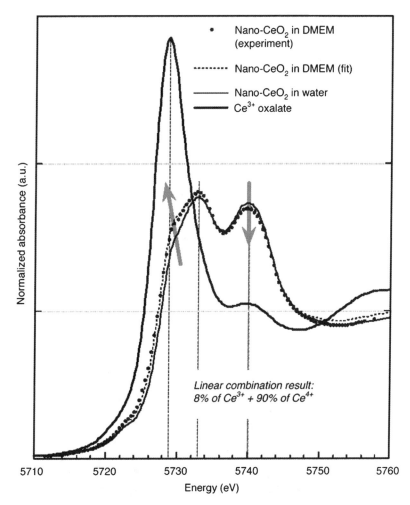

Figure 11.20 Experimental XANES spectra at the Ce L_{III} edge of pure nano-CeO_2, and nano-CeO_2 incubating in abiotic DMEM during 24 h. (·) nano-CeO_2 at $0.6 g L^{-1}$. The experimental data were fitted using linear combination of XANES spectra of nano-CeO_2 suspended in water and Ce^{3+} oxalate reference compound. The arrows highlight the decrease of the intensity of the peak at 5740 eV and the growth of the peak at 5729 eV attributing to the appearance of Ce^{3+} after incubation in the DMEM. These Ce^{3+} atoms are assumed to be on the surface of nano-CeO_2.[61] © 2009 Informa UK Ltd.

EXAFS, or XRD, while EXAFS is quite useful for identifying the local structure of the nanoparticles, including interatomic distances, and the type of coordination atom to know the composition and interatomic interaction.

Recently, some groups have used XANES and related techniques to explore the reduction process of metal cations into metal nanoparticles assisted by plants and microorganisms.[19,65–68] Such bio-reduced nanomaterials by plants and microorganisms (fungi, bacteria, and algae) include gold nanoparticles, silver nanoparticles, Au–Ag alloy, copper nanoparticles, magnetite nanoparticles, cadmium sulfide, and lead sulfide.[63,64]

Gardea-Torresdey et al.[67] reported on the formation of gold nanoparticles by living Alfalfa plants, which opened up new and exciting ways to fabricate nanoparticles. Alfalfa plants were grown in an $AuCl_4^-$ rich environment. The absorption of Au metal by the plants was confirmed by XAS (Figure 11.21) and TEM. Atomic resolution analysis confirmed the nucleation and growth of Au nanoparticles inside the plant and proved that the Au nanoparticles are in a crystalline state. Images also showed defects such as twins in the crystal structure, and in some cases icosahedral nanoparticles were found.

Lopez et al.[69] studied the binding and reduction of Au(III) by hop biomass using XAS. The XAS data confirmed the presence of Au(0) in both the native and chemically modified hop biomasses. XANES fittings show that the Au(III) was reduced to Au(0) by approximately 81%, 70%, and 83% on the native, esterified, and hydrolyzed hop biomass, respectively. In addition, the calculation of the particle radius was also in agreement with the results of transmission electron microscopy studies. The average particle could only be calculated for

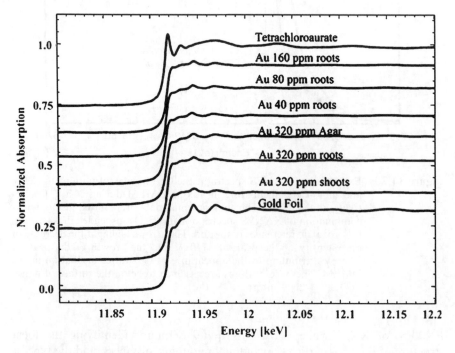

Figure 11.21 XANES of the gold alfalfa roots and shoots and gold-enriched agar samples, gold(0) foil, and the tetrachloroaurate model compound. These data show that the gold present in the alfalfa and agar samples is present as gold(0).[67] © 2002 American Chemical Society.

the native and esterified hops biomass, which showed average particle radii of 1.73 nm and 0.92 nm, respectively.

Besides the gold nanoparticles found in plants, copper nanoparticles were also found at the soil-root interface. Manceau and co-workers[68] reported that the common wetlands plants *Phragmites australis* and *Iris pseudoacorus* can transform copper into metallic nanoparticles in and near roots with evidence of assistance by endomycorrhizal fungi when grown in contaminated soil in the natural environment.

The structure of biogenic uraninite produced by *Shewanella oneidensis* strain MR-1 was also studied by XAS.[65] The biogenic uraninite formed at pH 8 with no subsequent cleaning treatment are compared to spectra for abiotic $UO_{2.00}$, $UO_{2.05}$, and $UO_{2.25}$, in Figure 11.22a and b, respectively. Qualitative comparison of the spectra showed three major findings. First, the biogenic uraninite corresponds well with stoichiometric $UO_{2.00}$. In particular, all of the FT peaks observed for $UO_{2.00}$ are present in the biogenic uraninite sample up to ca 0.8 nm, $R + \delta R$. Beyond this point no FT peaks are present in the biogenic uraninite FT. This result indicates that the overall UO_2 structure is qualitatively well preserved over a length of ~ 1 nm. Second, the ratio of the amplitude of the FT peak of the first U-U shell to the first U-O shell is reduced for biogenic uraninite in relation to $UO_{2.00}$, indicating a smaller coordination number of the U-U shell, consistent with its nanoparticulate size. Third, the spectrum for biogenic uraninite is different from those of $UO_{2.05}$ and $UO_{2.25}$. The latter two spectra span the range of

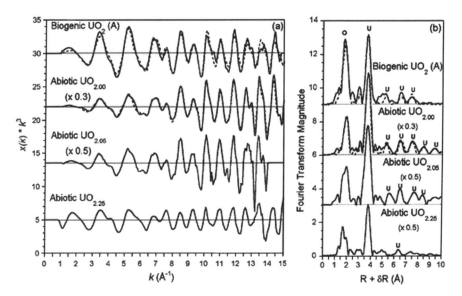

Figure 11.22 (a) EXAFS spectra (solid lines) collected at 77 K with fit to data (dashed lines) and (b) corresponding Fourier transform for biogenic uraninite formed at pH 8 with no post-formation cleaning, abiotic $UO_{2.00}$, $UO_{2.05}$ and $UO_{2.25}$. Data are scaled as indicated.[65] © 2008 American Chemical Society.

likely environmental UO_{2+x} compositions and show the distortions to local- and intermediate-range structure that are characteristic of hyperstoichiometric composition. In particular, the U-O FT peak (Figure 11.22b) is split into two subshells at around 0.225 and 0.240 nm, subtly so for $UO_{2.05}$ and strongly so for $UO_{2.25}$. In contrast, the U-O shell in biogenic uraninite is not visibly split, and thus, it can be concluded that the structure of biogenic uraninite is qualitatively more similar to stoichiometric $UO_{2.00}$ than to UO_{2+x}.

Gardea-Torresdey and co-workers[70] reported an interesting process of absorption and transformation of Ag element in living alfalfa. The plant was cultured in agar medium with a concentration of 320 ppm $AgNO_3$ for 9 days. According to the XAS technique, they found that $AgNO_3$ could be reduced to Ag(0) atoms in agar medium partly and then the roots only absorbed the Ag(0) atoms and organized them into Ag nanoparticles. Finally, the nanoparticles were transferred to shoots where Ag(0) nanoparticles converted into Ag(I). They employed XANES to study the speciation of silver in agar, roots, shoots, compared with Ag(0) foil as an internal standard and silver nitrate as a model sample. Then they confirmed that silver in roots, shoots and agar existed as Ag(0) (Figure 11.23). As an example for quantity analysis, they utilized linear combination fitting analysis of XANES and got the components of Ag(0) in agar, roots and shoots was 98.5, 94.8, and 96.7, respectively. And then they

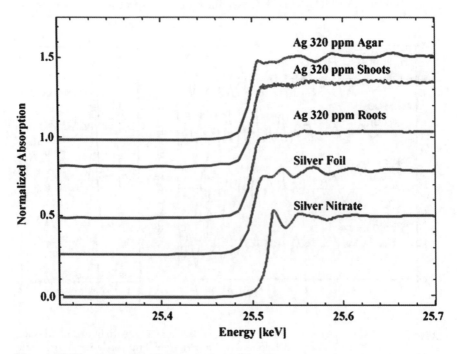

Figure 11.23 Normalized Ag K-edge XANES of five samples, including silver nitrate, silver foil, Ag in roots, Ag in shoots, Ag in agar.[70] © 2003 American Chemical Society.

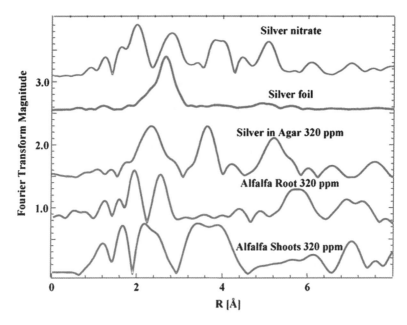

Figure 11.24 EXAFS of these samples shows the interatomic distances of Ag atoms and nearest neighboring atoms in the alfalfa and agar samples.[70] © 2003 American Chemical Society.

used EXAFS to identify the interatomic distances and the coordination number from the silver atom to the nearest neighbor atoms (Figure 11.24). As a result, the bond length of Ag-Ag in roots (0.288 nm) was similar to that in the Ag foil (0.287 nm), which suggested that the majority of Ag existed in the form of Ag(0) in roots and the reduced silver atoms organized themselves in the form of aggregation-nanoparticles supported by HRTEM images and X-ray EDS analysis.

11.6. Conclusion and Outlook

In summary, selected nuclear-based techniques for nanometallomics studies are summarized in Figure 11.25. ICP-MS, XAS, neutron and X-ray scattering can be used to characterize the nanometallome, while NAA, ICP-MS, XRF and isotopic tracing are multielemental quantification nuclear technique for nanometallome. For the distribution studies of nanometallome, besides ICP-MS, synchrotron radiation-based XRF was widely used while isotopic tracing has been used for the study of the absorption, distribution, transportation, storage, retention, metabolism, excretion and toxicity of nanometallome. Although X-ray crystallography, NMR, Mössbauer spectroscopy, electron paramagnetic resonance, and neutron scattering can be used for structural determination, XAS is the most commonly used technique for nanometallome structural

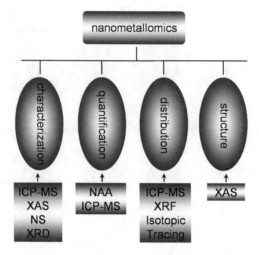

Figure 11.25 Selected nuclear techniques for nanometallomics studies.

determination so far. XAS combined with other techniques broadens our understanding of matter at the nanoscale. The techniques can provide plenty of information about elemental species, electronic configuration, and coordination structure and oxidation state. Nanomaterials pose some unique physical and chemical properties because of their surface effects and size effects due to coordination geometry and electronic configuration. Other issues need to be elucidated are how to understand the environmental and health effects of nanomaterials and what can be done to make them more controllable both for large-scale production and for health and biocompatibility aspects. XAS techniques are powerful tools to study what is happening at the nano-bio interface. As a systematic approach, XAS techniques are desired to connect microscopic materials at the atomic scale or nanometer scale to macroscopic systems including environmental and biological safety.

Nanometallomics is an emerging discipline which bridges nano-biological science with metallomics. Many techniques besides nuclear-based techniques can be applied to realize the goal of nanometallomics studies. However, since nanometallomics is such a young branch of metallomics, many techniques have not yet been applied in this field. Therefore, in the point of application of analytical techniques in nanometallomics, there is huge space for development of state-of-art techniques in nanometallomics. Further, owing to the unique properties of nanometallome, specific analytical techniques may be needed for future nanometallomics studies.

Acknowledgments

The authors acknowledge the financial support by the Ministry of Science and Technology of China as the National Basic Research Programs (2006CB705603, 2010CB934004 and 2009AA03J335), the Natural Science

Foundation of China (10975040), the NSFC/RGC Joint Research Scheme (20931160430) and the Knowledge Innovation Program of the Chinese Academy of Sciences (KJCX2-YW-M02).

References

1. S. Mounicou, J. Szpunar and R. Lobinski, *Chem. Soc. Rev.*, 2009, **38**, 1119.
2. H. Haraguchi, *J. Anal. At. Spectrom.*, 2004, **19**, 5.
3. R. J. P. Williams, *Coord. Chem. Rev. 2001*, **216–217**, *583*.
4. J. Szpunar, *Anal. Bioanal. Chem.*, 2004, **378**, 54.
5. C. Buzea, I. I. Pacheco and K. Robbie, *Biointerphases*, 2007, **2**, MR17.
6. Y. Zhao and H. S. Nalwa, eds. *Nanotoxicology-Interactions of Nanomaterials with Biological Systems,* American Scientific Publishers, California. 2006.
7. G. Brumfiel, *Nature*, 2003, **424**, 246.
8. V. L. Colvin, *Nat. Biotechnol.*, 2003, **21**, 1166.
9. R. F. Service, *Science*, 2003, **300**, 243a.
10. P. H. M. Hoet, A. Nemmar and B. Nemery, *Nat. Biotechnol.*, 2004, **22**, 19.
11. Y. Zhao, G. Xing and Z. Chai, *Nat. Nano.*, 2008, **3**, 191.
12. B. Wang, W. Feng, Y. Zhao, G. Xing, Z. Chai, H. Wang and G. Jia, *Sci. Chin. Ser. B: Chem.*, 2005, **48**, 385.
13. Y. Gao, C. Y. Chen and Z. F. Chai, *J. Anal. At. Spectrom.*, 2007, **22**, 856.
14. R. Allabashi, W. Stach, A. de la Escosura-Muñiz, L. Liste-Calleja and A. Merkoçi, *J. Nanopart. Res.*, 2008.
15. C. Degueldre, P. Y. Favarger and S. Wold, *Anal. Chim. Acta*, 2006, **555**, 263.
16. A. Helfrich, W. Brüchert and J. Bettmer, *J. Anal. At. Spectrom.*, 2006, **21**, 431.
17. K. Tiede, A. B. A. Boxall, D. Tiede, S. P. Tear, H. David and J. Lewis, *J. Anal. At. Spectrom.*, 2009, **24**, 964.
18. M.-C. Daniel and D. Astruc, *Chem. Rev.*, 2004, **104**, 293.
19. N. C. Sharma, S. V. Sahi, S. Nath, J. G. Parsons, J. L. Gardea- Torresde and T. Pal, *Environ. Sci. Technol.*, 2007, **41**, 5137.
20. S. Matsuo, T. Tsukamoto, A. Kamigaki, Y. Okaue, T. Yokoyama and H. Wakita, *X-Ray Spectrom.*, 2003, 32.
21. C.-H. Su, H.-S. Sheu, C.-Y. Lin, C.-C. Huang, Y.-W. Lo, Y.-C. Pu, J.-C. Weng, D.-B. Shieh, J.-H. Chen and C.-S. Yeh, *J. Am. Chem. Soc.*, 2007, **129**, 2139.
22. A. Prange and H. Modrow, *Rev. Environ. Sci. Biotechnol.*, 2002, **1**, 259.
23. M. Fernandez-Garcia, A. Martinez-Arias, J. C. Hanson and J. A. Rodriguez, *Chem. Rev.*, 2004, **104**, 4063.
24. T. Rajh, J. M. Nedeljkovic, L. X. Chen, O. Poluektov and M. C. Thurnauer, *J. Phys. Chem. B*, 1999, **103**, 3515.
25. F. Farges, G. E. Brown and J. J. Rehr, *Phys. Rev. B*, 1997, **56**, 1809.

26. C. T. Meneses, W. H. Flores and J. M. Sasaki, *Chem. Mater.*, 2007, **19**, 1024.
27. S. Adora, Y. Soldo-Olivier, R. Faure, R. Durand, E. Dartyge and F. Baudelet, *J. Phys. Chem. B*, 2001, **105**, 10489.
28. B.-J. Hwang, Y.-W. Tsai, L. S. Sarma, Y.-L. Tseng, D.-G. Liu and J.-F. Lee, *J. Phys. Chem. B*, 2004, **108**, 20427.
29. N. Kocharova, J. Leiro, J. Lukkari, M. Heinonen, T. Skala, F. Sutara, M. Skoda and M. Vondracek, *Langmuir*, 2008, **24**, 3235.
30. K. Itoh, H. Sasaki, H. T. Takeshita, K. Mori and T. Fukunaga, *J. Alloys Compd.*, 2005, **404–406**, 95.
31. Z. Chai, J. Sun and S. Ma, *Neutron Activation Analysis in Environmental Sciences, Biological and Geological Sciences,* Atomic Energy Press, Beijing, 1992.
32. Z. Chai and H. Zhu, eds., *Introduction to Trace Element Chemistry,* Atomic Energy Press, Beijing, 1994.
33. C. Ge, F. Lao, W. Li, Y. Li, C. Chen, Y. Qiu, X. Mao, B. Li, Z. Chai and Y. Zhao, *Anal. Chem.*, 2008, **80**, 9426.
34. W. G. Kreyling, M. Semmler, F. Erbe, P. Mayer, S. Takenaka, H. Schulz, G. Oberdörster and A. Ziesenis, *J. Toxicol. Environ. Health A*, 2002, **65**, 1513.
35. D. H. Oughton, T. Hertel-Aas, E. Pellicer, E. Mendoza and E. J. Joner, *Environ. Toxicol. Chem.*, 2008, **27**, 1883.
36. M.-T. Zhu, W.-Y. Feng, Y. Wang, B. Wang, M. Wang, H. Ouyang, Y.-L. Zhao and Z.-F. Chai, *Toxicol. Sci.*, 2009, **107**, 342.
37. H. Meng, Z. Chen, G. Xing, H. Yuan, C. Chen, F. Zhao, C. Zhang and Y. Zhao, *Toxicol. Lett.*, 2007, **175**, 102.
38. Y. Liu, Y. Gao, L. Zhang, T. Wang, J. Wang, F. Jiao, W. Li, Y. Liu, Y. Li, B. Li, Z. Chai, G. Wu and C. Chen, *J. Nanosci. Nanotechnol.*, 2009, **9**, 1.
39. J. Wang, G. Zhou, C. Chen, H. Yu, T. Wang, Y. Ma, G. Jia, Y. Gao, B. Li, J. Sun, Y. Li, F. Jiao, Y. Zhao and Z. Chai, *Toxicol. Lett.*, 2007, **168**, 176.
40. J. Wang, Y. Liu, F. Jiao, F. Lao, W. Li, Y. Gu, Y. Li, C. Ge, G. Zhou, B. Li, Y. Zhao, Z. Chai and C. Chen, *Toxicology*, 2008, **254**, 82.
41. B. Wang, W. Feng, M. Wang, T. Wang, Y. Gu, M. Zhu, H. Ouyang, J. Shi, F. Zhang, Y. Zhao, Z. Chai, H. Wang and J. Wang, *J. Nanopart. Res.*, 2008, **10**, 263.
42. Y. Gao, N. Liu, C. Chen, Y. Luo, Y.-F. Li, Z. Zhang, Y. Zhao, Y. Zhao, A. Iida and Z. Chai, *J. Anal. At. Spectrom.*, 2008, **23**, 1121.
43. J. Wang, C. Chen, Y. Liu, F. Jiao, W. Li, F. Lao, Y. Li, B. Li, C. Ge, G. Zhou, Y. Gao, Y. Zhao and Z. Chai, *Toxicol. Lett.*, 2008, **183**, 72.
44. J. Wang, C. Chen, H. Yu, J. Sun, B. Li, Y.-F. Li, Y. Gao, W. He, Y. Huang, Z. Chai, Y. Zhao, X. Deng and H. Sun, *J. Radioanal. Nucl. Chem.*, 2007, **272**, 527.
45. B. Wang, W. Feng, M. Wang, J. Shi, F. Zhang, H. Ouyang, Y. Zhao, Z. Chai, Y. Huang, Y. Xie, H. Wang and J. Wang, *Biol. Trace Elem. Res.*, 2007, **118**, 233.

46. C. Bussy, J. Cambedouzou, S. Lanone, E. Leccia, V. Heresanu, M. Pinault, M. Mayne-lhermite, N. Brun, C. Mory, M. Cotte, J. Doucet, J. Boczkowski and P. Launois, *Nano Lett.*, 2008, **8**, 2659.
47. R. Singh, D. Pantarotto, L. Lacerda, G. Pastorin, C. Klumpp, M. Prato, A. Bianco and K. Kostarelos, *Proc. Natl. Acad. Sci. U. S. A.*, 2006, **103**, 3357.
48. J. Guo, X. Zhang, Q. Li and W. Li, *Nucl. Med. Biol.*, 2007, **34**, 579.
49. J.-Y. Xu, Q.-N. Li, J.-G. Li, T.-C. Ran, S.-W. Wu, W.-M. Song, S.-L. Chen and W.-X. Li, *Carbon*, 2007, **45**, 1865.
50. Q. Li, Y. Xiu, X. Zhang, R. Liu, Q. Du, X. Shun, S. Chen and W. Li, *Nucl. Med. Biol.*, 2002, **29**, 707.
51. H. Wang, J. Wang, X. Deng, H. Sun, Z. Shi, Z. Gu, Y. Liu and Y. Zhao, *J. Nanosci. Nanotechnol.*, 2004, **4**, 1019.
52. C. C. Berry, S. Wells, S. Charles and A. S. G. Curtis, *Biomaterials*, 2003, **24**, 4551.
53. C. C. Berry, S. Wells, S. Charles, G. Aitchison and A. S. G. Curtis, *Biomaterials*, 2004, **25**, 5405.
54. M. Auffan, L. Decome, J. Rose, T. Orsiere, M. De Meo, V. Briois, C. Chaneac, L. Olivi, J.-l. Berge-lefranc, A. Botta, M. R. Wiesner and J.-Y. Bottero, *Environ. Sci. Technol.*, 2006, **40**, 4367.
55. M. Auffan, W. Achouak, J. Rose, M.-A. Roncato, C. Chaneac, D. T. Waite, A. Masion, J. C. Woicik, M. R. Wiesner and J.-Y. Bottero, *Environ. Sci. Technol.*, 2008, **42**, 6730.
56. M. A. A. Schoonen, C. A. Cohn, E. Roemer, R. Laffers, S. R. Simon and T. O'Riordan, *Rev. Mineral. Geochem.*, 2006, **64**, 179.
57. B. Wehrli, B. Sulzberger and W. Stumm, *Chem. Geol*, 1989, **78**, 167–179.
58. J. M. Sobek and D. E. Talburt, *J. Bacteriol.*, 1968, **95**, 47.
59. L. K. Limbach, Y. Li, R. N. Grass, T. J. Brunner, M. A. Hintermann, M. Muller, D. Gunther and W. J. Stark, *Environ. Sci. Technol.*, 2005, **39**, 9370.
60. A. Thill, O. Zeyons, O. Spalla, F. Chauvat, J. Rose, M. Auffan and A. M. Flank, *Environ. Sci. Technol.*, 2006, **40**, 6151.
61. M. Auffan, J. Rose, T. Orsiere, M. De Meo, A. Thill, O. Zeyons, O. Proux, A. Masion, P. Chaurand, O. Spalla, A. Botta, M. R. Wiesner and J.-Y. Bottero, *Nanotoxicology*, 2009, **3**, 161.
62. J. L. Gardea-Torresdey, J. R. Peralta-Videa, G. de la Rosa and J. G. Parsons, *Coord. Chem. Rev.*, 2005, **249**, 1797.
63. D. Mandal, M. Bolander, D. Mukhopadhyay, G. Sarkar and P. Mukherjee, *Appl. Microbiol. Biotechnol.*, 2006, **69**, 485.
64. D. E. Salt, R. C. Prince and I. J. Pickering, *Microchem. J.*, 2002, **71**, 255.
65. E. J. Schofield, H. Veeramani, J. O. Sharp, E. Suvorova, R. Bernier-Latmani, A. Mehta, J. Stahlman, S. M. Webb, D. L. Clark, S. D. Conradson, E. S. Ilton and J. R. Bargar, *Environ. Sci. Technol.*, 2008, **42**, 7898.
66. J. L. Gardea-Torresdey, K. J. Tiemann, J. G. Parsons, G. Gamez, I. Herrera and M. Jose-Yacaman, *Microchem. J.*, 2002, **71**, 193.

67. J. L. Gardea-Torresdey, J. G. Parsons, E. Gomez, J. Peralta-Videa, H. E. Troiani, P. Santiago and M. J. Yacaman, *Nano Lett.*, 2002, **2**, 397.
68. A. Manceau, K. L. Nagy, M. A. Marcus, M. Lanson, N. Geoffroy, T. Jacquet and T. Kirpichtchikova, *Environ. Sci. Technol.*, 2008, **42**, 1766.
69. M. L. López, J. G. Parsons, J. R. Peralta Videa and J. L. Gardea-Torresdey, *Microchem. J.*, 2005, **81**, 50.
70. J. L. Gardea-Torresdey, E. Gomez, J. R. Peralta-Videa, J. G. Parsons, H. Troiani and M. Jose-Yacaman, *Langmuir*, 2003, **19**, 1357.

Subject Index

absorption coefficient 165
aceruloplasminaemia 247
Advanced Photon Source 301–2
Aesculus hippocastanum L. 82
agarose gel electrophoresis 55
Agilent Technologies 9
alcohol dehydrogenase 13
alfalfa, nanoparticle formation by
    gold 376
    silver 378–9
alkaline phosphatase 13
aluminium, liver content 52
Alzheimer's disease 179
    iron dysregulation 247–8
    and iron overload 312–14
    metalloproteins 112, 113
    X-ray fluorescence 311–19
5-aminolevulinic acid synthase 246
β-amyloid 249
animals
    LA-ICP-MS 332–3
    particle-induced X-ray emission 325–6
antimony 266
    as impurity 353
    liver content 51, 52
antimony compounds, LD_{50} 5
apoferritin 152
Arabidopsis halleri ssp. *gemmifera* 305
Arabidopsis thaliana 305, 308, 313
arginase 13
argon plasma 96–7
Arp/Warp 228

arsenic 266
    accumulation in plants 304–5
    as impurity 353
    liver content 51, 52
    speciation of 80
arsenicals, LD_{50} 5
Astragalus bisulcatus 182
AT-binding cassette sub-family 246
atomic number 63
Auger electrons 64
    yield of 170
Auger emission 168
auranofin 267
Avogadro's number 139
Azotobacter vinelandii 195

barium, liver content 52
Beer's law 164
Beijing Synchrotron Radiation
    Laboratory 302
β-ray sources 104
Bijvoet pairs 225
bioinformatics 265
biomineralization 374–5
biomolecules 10–11
Biscutella laevigata L. 323
bismuth 266
bismuth citrate, colloidal 267
blasticidin-S deaminase 171
blood
    indium biodistribution 371
    technetium biodistribution 372

blood-brain barrier 248
Bohr magneton 156
Boltzman constant 131
bone, indium biodistribution 371
bone morphogenetic protein 245
bottom-up proteomics 12
Bradybaena similaris 174
Bragg diffraction 168
Bragg equation 67, 70
brain, iron biodistribution 370
Breit-Wigner formula 129
Bremsstrahlung spectrum 65, 66
bromine
   as impurity 353
   liver content 51, 52

cadmium
   liver content 51, 52
   radiolabelled 105
Caenorhabditis elegans 305, 309
   copper nanoparticle bioaccumulation 362–4
caesium, liver content 51, 52
calcium, liver content 51, 52
calmodulin 13
capillary electrophoresis 55, 98, 112
   on-line detection 88–90
carbon nanotubes 352, 353
   cellular interactions 365–8
carbonic anhydrase 13
carboplatin 266, 267
   hydrolysis of 270
carboxypeptidase 13
Caruso, Joseph A. 9
catalase 13, 147
cellular imaging
   particle-induced X-ray emission 326–8
   X-ray fluorescence 306–9
cellulose acetate electrophoresis 55
central nervous system, iron metabolism 248
cerium 53
   liver content 51
   neutron activation products 355
cerium oxide nanoparticles 374, 375
ceruloplasmin 13, 245

chemical ionization 101
chemical shift 155
chemical species 207
   definition 4
chlorine, liver content 52
chloroperoxidase 146
chromatography
   gas 98
   high-performance liquid 98, 99
   immobilized metal affinity 266
   liquid 50–3, 266
   size-exclusion 105
chromium 2, 53
   as impurity 353
   liver content 51
   micro-XAS spectrum 181–2
   subcellular distribution 54
circular dichroism, and XAS 194–6
cisplatin 266, 267
   activation of 268
   DNA adducts 279–82
   pharmacokinetics 276–7
Citrus sinensis L. Osbeck 80
Clostridium pasteurianum 177
Clostridium thermoaceticum 178
cobalt
   as impurity 353
   liver content 51, 52
   neutron activation products 355
collimators 70
collision-induced dissociation 102
collision/reaction cell 98, 113–14
Commission on Fundamental Environmental Chemistry 3
Commission on Microchemical Techniques and Trace Analysis 3
Commission on Toxicology 3
computational chemistry, and XAS 191–2
conalbumin 153
copper
   as impurity 353
   liver content 52
copper nanoparticles 356–7
   bioaccumulation 362–4
   dissolution rate 358
   formation by plants 377
   toxicity of 360–1

core holes 166
coupling constant 155
Cretan brake, arsenic transformation 178, 179–80
cryo-electron microscopy 212, 213
cytidine deaminase 229
cytochromes 146, 240
cytochrome b 245
cytochrome oxidases 13, 146
cytochrome P450 13
cytosine deaminase 229, 231

Daphnia magna 305, 311
dCMP deaminase 229
De Broglie wavelength 171
Debye-Waller factor 131, 139
Debye's solid model 131
delayed γ-ray neutron activation analysis (DGNAA) 45
density functional theory 192
deoxycytidine-5′-monophosphate (dCMP) 229
deoxythymidine-5′-triphosphate (dTTP) 229
Desulfovibrio gigas 79
detectors, X-ray fluorescence 68–9
Dicranopteris dichotoma 57, 59
differential centrifugation 48–50
Dirac functions 134
DMSA, protective effect of 373–4
DMT1 transporter 245
DNA 15
  cisplatinated 279–82
DNA polymerase 13
Doppler effect 130
Doppler velocity 132
DOTA 121
doxorubicin 273
DTPA 121
dynamic reaction cell 113

Ectothiorhodospira halophila 155
electric field gradient 135
electric monopole interaction 132–4
electric quadrupole interaction 134–6
electric quadrupole splitting 136, 142

electron impact mass spectrometry, iron metabolism 256–7
electron ionization 101
electron nuclear double resonance (ENDOR) 157
electron paramagnetic resonance 155–7, 266, 279
electron spin resonance, and XAS 199–201
electron transport 146
electrophoresis 53, 55–9
  agarose gel 55
  capillary 55, 88–90, 98, 112
  cellulose acetate 55
  polyacrylamide gel 55, 56, 186
  SDS-PAGE 85–8
  slab-gel 62
  and SRXRF 83–5
electrospray ionization (ESI) 95, 101–3
electrospray ionization-mass spectrometry (ESI-MS) 100–3, 266
electrostatic shift 132, 134
element-coded affinity tag (ECAT) 121–2
elementomics 15, 16
Elsholtzia splendens 330–2
energy-dispersive X-ray fluorescence (EDX) 21, 26–7, 71–4, 78
  applications 77–90
    electrophoresis and sample preparation 85–8
    elemental analysis 79–85
    on-line detection for separation techniques 88–90
  data processing 72–3
  element measuring 71–2
  geometrical setup 66
  sample preparation 71
  sensitivity, limit of detection and precision 73–4
  spectrum 65
Escherichia coli 229
  toxicity of iron nanoparticles 373
Eularian angles 218, 220
European Synchrotron Radiation Facility 75, 186, 302
European Virtual Institute of Speciation Analysis (EVISA) 9

EXAFSPAK 171, 270
experimental analytical precision 74
extended X-ray absorption fine structure
　(EXAFS) 24, 164, 167, 170–1, 268, 269
　fingerprints and structural information
　　177–9

faeces, technetium biodistribution 372
Fenton reaction 239
Fermi contact field 136
ferredoxins 148
ferric uptake regulator 191
ferrireductases 245
ferritin 13, 152, 240, 241–2, 245
　degradation 243
　mitochondrial 246
ferrochelatase 245, 246
ferroportin 244, 245, 248
ferroxidases 245
fluorescence 164–8
　X-ray *see* X-ray fluorescence
fluorescence yield 170
formate dehydrogenase 190
Fourier transformation 170, 214, 216–17
　phase 215
　reverse 217
frataxin 246
free electron laser 336
Fresnel zone plate 68
Friedel pairs 227
Friedreich's ataxia 246, 247
fullerenes 343

g-factor 156
Gaeumannomyces graminis var.
　tritici 180–1
gallium 266
γ rays
　nucleus resonance absorption 129
　sources 66, 104
gas chromatography 98
gas chromatography-mass spectrometry,
　iron metabolism 256–7
gel electrophoresis
　agarose 55
　polyacrylamide *see* polyacrylamide
　　gel electrophoresis

genome 16
genomics 16
glutaredoxin 5 246
glycomics 15
gold 266
　liver content 52
gold nanoparticles
　bioreduced 376
　formation by plants 376
　size characterization 345–6
gold/copper nanoshells 347–8
grazing-exit X-ray fluorescence 77, 78
growth differentiation factor 15
　(GDF15) 245
guanine deaminase 229

h-bar constant 129
Hamiltonians 132, 135–6
*Handbook of Elemental Speciation:
　Techniques and Methodology* 5
heart
　indium biodistribution 371
　iron biodistribution 370
　technetium biodistribution 372
Heisenberg uncertainty principle 128
heme oxygenases 245, 248
heme proteins, Mössbauer spectroscopy
　144–7
hemocyanin 200
hemoglobin 145
hemojuvelin 245
hepcidin 244, 245
hephaestin 245
hetero-atom tagged proteomics 16
hetero-elements, as elemental tags 121–2
high-performance liquid
　chromatography (HPLC) 98
　separation techniques 99
high-potential iron-protein 148
HKL2000 228
holo-transferrin 240
hop biomass, gold nanoparticle
　formation by 376–7
Horae program 171
horseradish peroxidase 146
6-hydroxydopamine 249
HysTag reagent 118

Subject Index

Iberis intermedia 182
IC_{50} 4
IFEFFIT program 171
immobilized metal affinity chromatography 266
indium
　biodistribution 371
　liver 52
　radiolabelled 369–70, 371
inductively coupled plasma mass spectrometry (ICP-MS) 20, 26–7, 47, 95–9, 103, 266
　characteristics of 96
　ICP as high-temperature ionization source 96–97
　iron metabolism 256–7
　isotope dilution 108
　isotopic tracer techniques 103–7
　laser ablation *see* laser ablation inductively coupled plasma mass spectrometry
　mass analyzers for 97–98
　nanometallomics 345
　　particle distribution 357–62
　　particle quantification 356–7
　schematic 96
　techniques 98–99
inductively coupled plasma–optical emission spectroscopy (ICP-OES) 287
Instrumental Methods in Metal Ion Speciation 5
instrumental neutron activation analysis (INAA) 46
International Union of Pure and Applied Chemistry *see* IUPAC
intestine, technetium biodistribution 372
iodine
　liver content 52
　radiolabelled 370
ion exchange 99
ion-trap mass analyzer 102
ionome 16
ionomics 16
iridium nanoparticles 354
Iris pseudoacorus, copper nanoparticle formation 377

iron
　biodistribution 370
　　liver 51, 52
　cellular uptake 240–1, 242–3
　chemistry 239
　homeostasis 243–5
　　impaired 247–9
　as impurity 353
　labile iron pool 242–3
　metabolism 244
　　central nervous system 248
　　mitochondrial 245–7
　　nuclear analytical techniques 250–60
　　　combined pre-separation 259–60
　　　Mössbauer spectroscopy 257–9
　　　neutron activation analysis 255–6
　　　particle-induced X-ray emission 251–5
　　　radioactive and enriched stable isotopes 256–7
　　　synchronous radiation 250–1
　physiology 240
　radiolabelled 368–70
　storage 241–3
iron overload, and Alzheimer's disease 312–14
iron oxide nanoparticles
　biodistribution 365, 369
　cytotoxicity 372–4
iron regulatory protein 243, 245
iron-omics 239–64
iron-sulfur proteins, Mössbauer spectroscopy 147–52
isobaric tagging for relative and absolute quantitation *see* iTRAQ
isoelectrofocusing 87
isomer shift 132, 140–1
isomorphous replacement 220–1, 223, 224
isotope dilution 107–15, 266, 277
　with ICP-MS 108
　species-specific method 108–10
　species-unspecific method 110–15
isotope ratio 107
isotope-coded affinity tagging (ICAT) 112, 117–19
　cleavable 118
isotopic analysis 19

isotopic tagging 115–22
isotopic tracer techniques
and ICP-MS 103–7
nanometallomics 368–72
iTRAQ 119–20
IUPAC 3–4
"Metal-focussed -omics: guidelines for terminology and critical evaluation of analytical approaches" 9

Journal of Analytical and Atomic Spectrometry 7, 8

kidney
  indium biodistribution 371
  iron biodistribution 370
  technetium biodistribution 372
kinetic energy discrimination 98
Kirkpatrick-Baez mirror 68
Kohonen neural network 83
KP1019 267
Kramers-Kronig transformation 227
Kronecker symbol 135

lactoferrin 153
Landé factor 136, 156
lanthanum 53
  as impurity 353
  liver content 51
large-angle X-ray scattering 268
laser ablation inductively coupled plasma mass spectrometry (LA-ICP-MS) 22, 99, 300, 330–3
  animals 332–3
  near-field 333–4
  plants 330–2
lattice vibration energy 131
LC_{50} 4
LD_{50} 4, 5
lead, liver content 52
lead compounds, LD_{50} 5
Lecythis ollaria 59
Leptospira interrogans 204
Lewy bodies 247, 249, 313
LIGPLOT 232, 233
linear combination XANES, nanometallomics 346

lipidomics 15
liquid chromatography 50–3, 266
liver
  elemental content 51, 52
  indium biodistribution 371
  iron biodistribution 370
  technetium biodistribution 372
lung
  indium biodistribution 371
  iron biodistribution 370
  technetium biodistribution 372

magnesium, liver content 52
magnetic dipole interaction 136
magnetic dipole splitting 142–3
magnetic hyperfine interaction 136–7
manganese
  as impurity 353
  liver content 52
mass balance studies 276
mass spectrometry 328–34
  electron impact 256–7
  electrospray ionization 100–3, 266
  gas chromatography 256–7
  inductively coupled plasma *see* inductively coupled plasma mass spectrometry
  laser ablation inductively coupled plasma 22, 99, 300
  matrix-assisted laser desorption ionization 329
  secondary ion 21, 22–3, 26–7, 300, 329–30, 336
  thermal ionization 256–7
matrix absorption 72
matrix-assisted laser desorption ionization (MALDI) 95, 97, 103, 266, 300
  mass spectrometry 329
  time of flight (MALDI-TOF) 101
Maxwell-Boltzmann distribution 156
mercury 2–3
  EXAFS analysis 178–9
  liver content 52
  radioisotope tracer 105–6
  XANES speciation 176–7
mercury compounds, LD_{50} 5

Subject Index

metabolic labeling 120–1
metabolites 10–11
metabolome 15, 16
metabolomics 16
metal-coded affinity tag (MeCAT) 121–2
metallodrug-protein interactions 282–8
metallodrugs 265–98
   biomolecule interactions 278–90
      metallodrug-protein interactions 282–8
      platinated DNA-protein interactions 288–90
      platinated-DNA adducts 279–82
   cellular distribution and metabolism 267–78
   cellular localization 271–6
   pharmacokinetics 276–8
   as prodrugs 268
metalloenzymes 12, 13
metallogenomics 15
metalloglycomics 15
metallolipidomics 15
metallomes 7–8, 16, 300, 342–3
   definition 9–11
metallometabollomics 15
Metallomics (journal) 7, 8
metallomics 7–11, 16
   analytical techniques 35
   definition 9–11
   experimental approaches 19
   history and development 7–9
   nuclear-based 265–98
   relationships with other disciplines 15, 16, 17
   research topics 18
Metallomics Center of the Americas 9
metalloproteins 11–12
   Alzheimer's disease 112, 113
   electrophoresis 89
metalloproteome 16
metalloproteomics 11–14, 15, 16
   studies on 12–14
metallothioneins 13, 112
methyl-coenzyme M reductase 198
micro-X-ray absorption spectroscopy 179–85
micro-X-ray fluorescence 75, 180, 301, 306, 307

micro-XANES 184
microglia 248–9
micronebulizers 99
Miller indexes 214
miniaturized Mössbauer spectrometer (MIMOS) 139
mitochondrial iron metabolism 245–7
mitoferrin 246
molecular neutron activation analysis (MoNAA) 48
molybdenum 12
   as impurity 353
   isotope dilution analysis 110–11
   liver content 51, 52
molybdocene dichloride 275
Moseley's law 63, 300
Mosflm 228
Mössbauer effect 20, 128–32
Mössbauer isotopes 152
Mössbauer, Rudolf L. 128, 130
Mössbauer spectrometer 137–40
   miniaturized 139
Mössbauer spectroscopy 24, 26–7, 128–62, 266, 279
   applications 143–54
      elemental speciation 153–4
      metalloprotein studies 144–53
   electric monopole interaction 132–4
   electric quadrupole interaction 134–6
   equipment 137–40
   iron metabolism 257–9
   limitations 159
   magnetic hyperfine interaction 136–7
   sample preparation 144
Mössbauer spectrum 132, 140–3
   electric quadrupole splitting 142
   isomer shift 140–2
   magnetic dipole splitting 142–3
mosses 304
mRNA 15
multi-channel analyzer 69
multi-wavelength anomalous dispersion (MAD) 225–8
   data collection 227–8
   data processing 228
   wavelength selection 226–7

multielemental quantification 20–1
muscle
  indium biodistribution 371
  technetium biodistribution 372
myelodysplastic syndrome 246
myocrisin 267
myoglobin 145, 240

NAMI A 267
nano-imaging 333–4
nanomaghemites 373–4
nanomaterials 343
  bioreduced 375–6
  see also nanometallomics
nanomedicine 275–6
nanometallomics 342–84
  nuclear analytical techniques 344–52
    electronic configuration and
      coordination geometry 348–52
    oxidation state analysis 346–8
    size characterization 345–6
  study area 344
  see also nanoparticles
nanoparticles 343
  biointeractions 372–9
  cerium oxide 374, 375
  copper 356–7, 358, 360–1, 362–4, 377
  distribution 357–72
    ICP-MS 357–62
  engineered 354
  fate of 354
  formation by plants and
    microorganisms 375–9
  gold 245–6, 376
  iridium 354
  iron oxide 365, 369, 372–4
  nickel oxide 350
  platinum 350–1
  quantification 352–7
    ICP-MS 356–7
    neutron activation analysis 352–6
  silver 376, 378–9
  titanium oxide 349–50, 357–60, 364–5
  zinc 361–2
nanotubes 343
  carbon 352, 353, 365–8
nanowires 343

near edge X-ray absorption fine
  structure (NEXAFS) 166
neurodegenerative diseases 247–8
  iron in 248–9
neurofibrillary tangles 247, 313
neuromelanin 257
neutron activation analysis (NAA)
  20–1, 26–7, 44–61, 266
  activation and analysis 45–6
  activation products 355
  delayed γ-ray neutron activation
    analysis (DGNAA) 45
  differential centrifugation 48–50
  γ spectroscopy system for 45–6
  instrumental 46
  iron metabolism 255–6
  liquid chromatography 50–3
  metallomics and metalloproteomics
    48–59
  nanometallomics 352–6
  prompt γ-ray neutron activation
    analysis (PGNAA) 45
  quantification 46
  radiochemical 46
  sensitivities 47
  strengths and limitations 47–8
neutron capture reaction 44–5
neutron scattering 266, 279
  and XAS 192–4
nickel, as impurity 353
nickel oxide nanoparticles 350
nicotinoyloxy succinimide (Nic-
  NHS) 118
nitric oxide 249
nitrocyanin 192
nitrogenase 13, 199
nMaghemite 373
nMagnetite 373
nuclear analytical techniques 15–20,
  299–300
  applications 20–9
    metallome/metalloproteome
      distribution 21–3
    multielemental quantification 20–1
    structural analysis 23–9
  effect of electronic/molecular
    structure 19–20

Subject Index

electrophoresis 53, 55–9
iron metabolism 250–60
   combined pre-separation 259–60
   Mössbauer spectroscopy 257–9
   neutron activation analysis 255–6
   particle-induced X-ray emission 251–5
   radioactive and enriched stable isotopes 256–7
   synchronous radiation 250–1
isotopic analysis 19
penetration of nuclear radiation 20
nuclear hyperfine interactions 132
nuclear magnetic resonance (NMR) 23, 24–5, 26–7, 212, 213, 266
   hydrolysis of platinum compounds 270
   and XAS 196–7
nuclear reaction analysis 336
nuclear resonance spectroscopy 154–5
nuclear spin number 154
nuclear Zeeman effect 136
nucleus resonance absorption of γ rays 129

octaethylporphyrin 155
open reading frames 203
organotin compounds, LD_{50} 5
ovotransferrin 153
oxaliplatin 266, 267
oxidation state analysis 346–8

pancreas, iron biodistribution 370
Panulirus interruptus 200
Parkinson's disease
   iron dysregulation 247–8, 251
   metal content of basal ganglia 321
   metal content of midbrain 322
   X-ray fluorescence 315–20
particle-induced γ-ray emission 336
particle-induced X-ray emission (PIXE) 21, 26–7, 67, 78, 320–8
   animals 325–6
   cellular imaging 326–8
   elemental analysis in protein fractions 79
   iron metabolism 251–5
   plants 322–5
   with proton microprobe 22

Patterson function 218, 221
Patterson maps 218, 219, 220
Peptococcus aerogenes 177
perineuronal net 326–8
peroxidases 146–7
Perutz, Max 145
pharmacokinetics 276–8
phase problem 217
Phaseolus vulgaris, ferritin 110
phosphoproteome 17
Phragmites australis, copper nanoparticle formation 377
phytoremediation 304, 374–5
PIN-diode 68
Plank's constant 129, 154
plants
   heavy metal accumulation 304
   LA-ICP-MS 330–2
   nanoparticle formation 375–9
   particle-induced X-ray emission 322–5
   X-ray fluorescence 304–5
   see also individual plants
plastocyanin 13, 196
platinated DNA
   adducts 279–82
   protein interactions 288–90
platinum compounds 266
   cellular localization 271–6
   hydrolysis of 267–71
platinum iminoethers 271
platinum nanocrystallites 350–1
platinum/copper nanoclusters 351
polyacrylamide gel electrophoresis (PAGE) 55, 56, 186
   two-dimensional 266
positron emission tomography (PET) 334, 335–6
potassium, liver content 51, 52
pre-separation procedures 259–60
prompt γ-ray neutron activation analysis (PGNAA) 45
proteins 10–11
   isotope dilution analysis 107–15
protein crystallography 26–7, 163–4, 212–38, 266
   SmdCD 229–35

structure determination 214–25
  multi-wavelength anomalous
    dispersion 225–8
  and XAS 188–91
Protein Data Bank 11–12, 212, 213
protein extraction 82
protein quantification 115–22
  chemical labeling 116–20
  hetero-elements as elemental tags
    121–2
  metabolic labeling 120–1
PROTEIN-AQUA 109
proteome 16
proteomics 16
  bottom-up 12
  hetero-atom tagged 16
  shotgun 12
  top-down 14
protoporphyrin IX 245
pulse pile-up 69
Pyrococcus furiosus 203

QconCAT 121
quadropole analyzer 102
quantum dots 343
quantum size effect 343

radiochemical neutron activation
    analysis (RNAA) 46
radioisotopes 104
  cadmium 105
  indium 369–70, 371
  iodine 370
  iron 368–70
  stable 104–5
  technetium 370
Raman spectroscopy, and XAS 197–9
rare earth elements 57
Rayleigh limit 101
reactive oxygen species 239, 249
  and nanoparticle toxicity 373
recoil energy 131
recoil-free fraction 131
RESOLVE 228
resonance Raman spectroscopy 157–8
reversed-phase partition 99

Rhodobacter sphaeroides 174
Rhodospirillum rubrum 79
riboflavin biosynthesis protein 171
Rieske protein 148
ringed sideroblasts 246–7
RNA 15
RNA polymerase 13
rubidium, liver content 51, 52
rubredoxin 148, 194
rusticyanin 109, 192, 193
Rutherford backscattering
    spectrometry 336

Saccharomyces cerevisiae 148
samarium 53
  as impurity 353
  liver content 51
Scalait 228
scandium
  as impurity 353
  liver content 51, 52
scanning electron microscopy 22
scanning transmission ion microscopy
    336
Schizosaccharomyces prombe 326
SDS-PAGE electrophoresis 85–8
secondary ion mass spectrometry (SIMS)
    21, 22–3, 26–7, 300, 329–30, 336
selenium 12
  liver content 51, 52
  micro-XANES spectrum 184–5
  micro-XAS spectrum 182–4
  radioisotope tracer 106–7
  XANES speciation 174, 176
selenium compounds, LD_{50} 5
selenoamino acids, isotope dilution
    analysis 109
selenocysteine 12
selenomethionine 109
selenoprotein P 13
selenoproteome 17
senile plaques 247
separation techniques, and XAS 186
Sesulfovibrio gigas 259
shape resonances 167
SHELXD 228

Subject Index

Shewanella oneidensis, biogenic uraninite formation by 377–8
shotgun proteomics 12
sideroblastic anemia 247
SILAC 120–1
silicon drift detector 68
silver, neutron activation products 355
silver nanoparticles
  bioreduced 376
  formation by plants 378–9
single-channel analyzer 69
single-crystal neutron diffraction spectroscopy (SCND) 24
single-photon emission computed tomography (SPECT) 334–5
size exclusion 99
size-exclusion chromatography 105
skin
  indium biodistribution 371
  technetium biodistribution 372
slab-gel electrophoresis 62
small-angle neutron scattering (SANS) 24
small-angle X-ray scattering (SAXS) 24
SmdCD
  protein crystallography 229–35
  structure 232
sodium, liver content 51, 52
SOLVE 228
speciation 4
  research papers on 6, 7, 30–4
speciation analysis 2–7
  definition 4
  Mössbauer spectroscopy 153–4
  qualitative 154
  quantitative 153
    XANES 172–7
speciation neutron activation analysis (SNAA) 48
spleen
  indium biodistribution 371
  iron biodistribution 370
  technetium biodistribution 372
Spring-8 301
stable isotope labeling by amino acids in cell *see* SILAC
stable isotopes 104–5
  iron metabolism analysis 256–7

standard reference materials 354
steap proteins 245
stomach, technetium biodistribution 372
Streptococcus mutans dCD *see* Sm-dCD
structure determination
  EXAFS 177–9
  multi-wavelength anomalous dispersion 225–8
  protein crystallography 214–25
sulfur
  isotope dilution analysis 112–15
  stable isotopes 112–13
superoxide dismutase 13, 110
synchrotron radiation 66–7, 301
  facilities 303
  iron metabolism 250–1
  tunable wavelengths 223
synchrotron radiation X-ray fluorescence (SRXRF) 21–2, 26–7, 66–7
  and electrophoresis 83–5
  elemental analysis in protein fractions 81–2
  iron metabolism 250–1
  with microbeam 21, 26–7
synchrotron radiation-induced X-ray emission (SRIXE) 272
Synechochoccus spp. 79

TATA binding protein 288
Taylor cone 102
technetium
  biodistribution 372
  radiolabelled 370, 372
testicle, iron biodistribution 370
tetramesitylporphyrin 155
Tf receptors 245
thermal ionization mass spectrometry (TIMS), iron metabolism 256–7
thermal vaporization 101
thin films 343
Thiocapsa roseopersicina 79, 259
Thlaspi spp. 174–5
Thomson scattering 225
thorium
  as impurity 353
  liver content 52

time-of-flight analysis 97–8
tin 2
titanium oxide nanoparticles 349–50
　bioaccumulation 364–5
　toxicity of 357–60
tomography 334–7
top-down proteomics 14
total electron yield 170
total reflection X-ray fluorescence 75–7, 78
　elemental analysis in protein fractions 80–1
trace elements 2–7
　fingerprinting 21
transferrin 13, 152–3, 240, 245
　isoforms 110
transmission electron microscopy 22
trichrome, micro-XRF imaging 306, 307, 308
tRNA-specific adenosine deaminases 171
tungsten 12
　as impurity 353

ubiquitin immunoreactivity 313
uraninite, biogenic 377–8
urease 13
urine, technetium biodistribution 372
USTCXAFS program 171

vanadium, liver content 52

Walker diagram 141, 142
wavelength-dispersive X-ray fluorescence (WDX) 21, 70–1, 78
WinXAS program 171
Working Party on Manufactured Nanomaterials 343

X-ray absorption near edge structure (XANES) 90, 164, 166, 167
　fingerprint studies and quantitative speciation 172–7
　linear combination 346
　nanometallomics 346–8

X-ray absorption spectroscopy (XAS) 24, 26–7, 163–211, 266, 279
　applications 172–85
　combined techniques 185–201, 202
　　circular dichroism 194–6
　　computational chemistry 191–2
　　electron spin resonance 199–201
　　neutron scattering 192–4
　　nuclear magnetic resonance 196–7
　　protein crystallography 188–91
　　Raman spectroscopy 197–9
　　separation 186
　　X-ray fluorescence 186–8
　data analysis 170–2
　fluorescence 164–8
　high-throughput (HT-XAS) 28
　measurement 168–70
　metallodrug-protein interactions 286–7
　micro-XAS 179–85
　nanometallomics 344–5
　samples and sample preparation 168
X-ray crystallography 24
X-ray diffraction see protein crystallography
X-ray emission spectroscopy 266
X-ray fluorescence 62–94, 163, 166, 266, 300–20
　detectors 68–9
　electronics 69
　energy-dispersive see energy-dispersive X-ray fluorescence
　enhancement by multiple excitation 72–3
　environmental science 302–9
　　cellular imaging 306–9
　　environment 305–6
　　plants 304–5
　facilities 64–71
　grazing-exit 77, 78
　life science 309–20
　　Alzheimer's disease 311–19
　　Parkinson's disease 315–20
　matrix absorption 72
　micro-XRF 75
　nanometallomics 362–8
　optics 67–8
　physics of 63–4

primary radiation source 65–67
synchrotron radiation *see*
   synchrotron radiation X-ray
   fluorescence
total reflection 75–7, 78
wavelength-dispersive *see* wavelength-
   dispersive X-ray fluorescence
and XAS 186–8
X-ray fluorescence microtomography
   (XRFM) 273

XDS 228
XFIT program 171

ZD0473 278
Zeeman effect 154, 156
zero-phonon process 131
zinc 12
   as impurity 353
   liver content 51, 52
zinc nanoparticles 361–2